Viscoelasticity: Mathematical Modeling, Numerical Simulations, and Experimental Work

Viscoelasticity: Mathematical Modeling, Numerical Simulations, and Experimental Work

Editors

Luís L. Ferrás
Alexandre M. Afonso

Basel • Beijing • Wuhan • Barcelona • Belgrade • Novi Sad • Cluj • Manchester

Editors
Luís L. Ferrás
University of Porto
Porto
Portugal

Alexandre M. Afonso
University of Porto
Porto
Portugal

Editorial Office
MDPI
St. Alban-Anlage 66
4052 Basel, Switzerland

This is a reprint of articles from the Special Issue published online in the open access journal *Applied Sciences* (ISSN 2076-3417) (available at: https://www.mdpi.com/journal/applsci/special_issues/Viscoelasticity_Mathematical_Modeling_Numerical_Simulations_and_Experimental_Work).

For citation purposes, cite each article independently as indicated on the article page online and as indicated below:

Lastname, A.A.; Lastname, B.B. Article Title. *Journal Name* **Year**, *Volume Number*, Page Range.

ISBN 978-3-0365-9750-8 (Hbk)
ISBN 978-3-0365-9751-5 (PDF)
doi.org/10.3390/books978-3-0365-9751-5

© 2023 by the authors. Articles in this book are Open Access and distributed under the Creative Commons Attribution (CC BY) license. The book as a whole is distributed by MDPI under the terms and conditions of the Creative Commons Attribution-NonCommercial-NoDerivs (CC BY-NC-ND) license.

Contents

About the Editors vii

Luís L. Ferrás and Alexandre M. Afonso
Viscoelasticity: Mathematical Modelling, Numerical Simulations, and Experimental Work
Reprinted from: *Appl. Sci.* **2023**, *13*, 1022, doi:10.3390/app13021022 1

Simon Ingelsten, Andreas Mark, Roland Kádár and Fredrik Edelvik
A Backwards- Tracking Lagrangian-Eulerian Method for Viscoelastic Two-Fluid Flows
Reprinted from: *Appl. Sci.* **2021**, *11*, 439, doi:10.3390/app11010439 3

Juliana Bertoco, Manoel S. B. de Araújo, Rosalía T. Leiva, Hugo A. C. Sánchez and Antonio Castelo
Numerical Simulation of KBKZ Integral Constitutive Equations in Hierarchical Grids
Reprinted from: *Appl. Sci.* **2021**, *11*, 4875, doi:10.3390/app11114875 29

Jie Liu, Martin Oberlack and Yongqi Wang
Analytical Investigation of Viscoelastic Stagnation-Point Flows with Regard to Their Singularity
Reprinted from: *Appl. Sci.* **2021**, *11*, 6931, doi:10.3390/app11156931 45

Laison Junio da Silva Furlan, Matheus Tozo de Araujo, Analice Costacurta Brandi, Daniel Onofre de Almeida Cruz and Leandro Franco de Souza
Different Formulations to Solve Giesekus Model for Flow between Two Parallel Plates
Reprinted from: *Appl. Sci.* **2021**, *11*, 10115, doi:10.3390/app112110115 63

Juliana Bertoco, Rosalía T. Leiva, Luís L. Ferrás, Alexandre M. Afonso and Antonio Castelo
Development Length of Fluids Modelled by the gPTT Constitutive Differential Equation
Reprinted from: *Appl. Sci.* **2021**, *11*, 10352, doi:10.3390/app112110352 87

Michael McDermott, Pedro Resende, Thibaut Charpentier, Mark Wilson, Alexandre Afonso, David Harbottle and Gregory de Boer
A FENE-P $k - \varepsilon$ Viscoelastic Turbulence Model Valid up to High Drag Reduction without Friction Velocity Dependence
Reprinted from: *Appl. Sci.* **2020**, *10*, 8140, doi:10.3390/app10228140 105

Brian Wojcik, Jason LaRuez, Michael Cromer and Larry A. Villasmil Urdaneta
The Role of Elasticity in the Vortex Formation in Polymeric Flow around a Sharp Bend
Reprinted from: *Appl. Sci.* **2021**, *11*, 6588, doi:10.3390/app11146588 127

Miles Skinner and Pierre Mertiny
Effects of Viscoelasticity on the Stress Evolution over the Lifetime of Filament-Wound Composite Flywheel Rotors for Energy Storage
Reprinted from: *Appl. Sci.* **2021**, *11*, 9544, doi:10.3390/app11209544 147

Angadi Basettappa Vishalakshi, Thippaiah Maranna, Ulavathi Shettar Mahabaleshwar and David Laroze
An Effect of MHD on Non-Newtonian Fluid Flow over a Porous Stretching/Shrinking Sheet with Heat Transfer
Reprinted from: *Appl. Sci.* **2022**, *12*, 4937, doi:10.3390/app12104937 165

Abdul Aabid, Sher Afghan Khan and Muneer Baig
A Critical Review of Supersonic Flow Control for High-Speed Applications
Reprinted from: *Appl. Sci.* **2021**, *11*, 6899, doi:10.3390/app11156899 179

Thippeswamy Anusha, Rudraiah Mahesh, Ulavathi Shettar Mahabaleshwar and David Laroze
An MHD Marangoni Boundary Layer Flow and Heat Transfer with Mass Transpiration and Radiation: An Analytical Study
Reprinted from: *Appl. Sci.* **2022**, *12*, 7527, doi:10.3390/app12157527 . . . . **205**

About the Editors

Luís L. Ferrás

Luís L. Ferrás is an Assistant Professor at the Department of Mechanical Engineering, Faculty of Engineering, University of Porto (FEUP), and a researcher at the Centre for Mathematics, University of Minho, Portugal. He received his Ph.D. in Science and Engineering of Polymers and Composites from the University of Minho in 2012, a Ph.D. in Mathematics from the University of Chester in 2019, and was a visiting researcher at MIT in 2016. His current research interests are numerical analysis, applied mathematics, partial and fractional differential equations, mathematical modeling, computational mechanics, computational fluid dynamics, complex viscoelastic flows, rheology, anomalous diffusion, and machine learning.

Alexandre M. Afonso

Alexandre M. Afonso graduated in Chemical Engineering from Faculty of Engineering of the University of Porto (FEUP) in 2000, with a final-year Research Project at the Universidad Politecnica de Catalunya graded with an Honor Grade (10/10). In 2005, Afonso completed an MSc in Heat and Fluid mechanics, and in 2010 he completed a PhD degree in Biological and Chemical Engineering at FEUP. Currently, Afonso is an Assistant Professor at the Department of Mechanical Engineering at FEUP.

Editorial

Viscoelasticity: Mathematical Modelling, Numerical Simulations, and Experimental Work

Luís L. Ferrás [1,2,*] and Alexandre M. Afonso [3]

1. Department of Mechanical Engineering (Section of Mathematics), FEUP, University of Porto, 4200-465 Porto, Portugal
2. Center for Mathematics (CMAT), University of Minho, 4710-057 Braga, Portugal
3. CEFT-Transport Phenomena Research Center, Department of Mechanical Engineering, Faculty of Engineering, University of Porto, 4200-465 Porto, Portugal; aafonso@fe.up.pt
* Correspondence: lferras@fe.up.pt

Citation: Ferrás, L.L.; Afonso, A.M. Viscoelasticity: Mathematical Modelling, Numerical Simulations, and Experimental Work. *Appl. Sci.* 2023, *13*, 1022. https://doi.org/10.3390/app13021022

Received: 27 December 2022
Revised: 6 January 2023
Accepted: 9 January 2023
Published: 12 January 2023

Copyright: © 2023 by the authors. Licensee MDPI, Basel, Switzerland. This article is an open access article distributed under the terms and conditions of the Creative Commons Attribution (CC BY) license (https://creativecommons.org/licenses/by/4.0/).

Viscoelastic materials are abundant in nature and present in our daily lives. Examples include paints, blood, polymers, biomaterials or food products. It is thus important to study and understand the viscoelastic behaviour of these different materials.

In this Special Issue, a total of eleven contributions (ten research papers and one review paper) from different areas of viscoelasticity (mathematical modelling, numerical simulations) are presented.

McDermott et al. [1] proposed an improved viscoelastic turbulence model in a fully developed drag reducing channel flow, where turbulent eddies are modelled by a k-ε representation, together with polymeric solutions described by the finitely extensible nonlinear elastic Peterlin (FENE-P) constitutive model. The performance of the model was evaluated using a variety of direct numerical simulation data described by different combinations of rheological parameters and was able to predict all regimes of drag reduction (low, intermediate, and high) with good accuracy. Ingelsten et al. [2] developed a new Lagrangian–Eulerian method for the simulation of viscoelastic free surface flows. The approach was developed from a method in which the constitutive equation for viscoelastic stresses was solved at Lagrangian nodes connected by flow and interpolated onto a Eulerian grid using radial basis functions. In the new method, a backwards-tracking methodology was used to allow fixed locations for the Lagrangian nodes to be chosen a priori. The proposed method was also extended to the simulation of viscoelastic free surface flows with the volume of fluid method. Bertoco et al. [3] presented the HiGTree–HiGFlow solver for numerical simulations of the KBKZ integral constitutive equation. The numerical method used finite differences and tree-based grids, which leads to greater accuracy in local mesh refinement. Wojcik et al. [4] performed fluid dynamic simulations using the FENE-P model and an incompressible Newtonian fluid to understand the role of elasticity in the formation of vortices in a narrow channel with a 90° curvature. The analysis bridged the flow behaviour of a purely elastic fluid and that of a Newtonian fluid. Their predictions were in good agreement with previous experimental and numerical works. Liu et al. [5] investigated singularities in the stress field of the flow of a viscoelastic fluid at the stagnation point for various viscoelastic constitutive models. Exact analytical solutions of two-dimensional steady wall-free stagnation point flows for the generic Oldroyd 8-constant model were obtained for the stress field using different material parameter relationships. Compatibility with the conservation of momentum was considered for all solutions.

Aabid et al. [6] studied and summarised the active control of high-speed aerodynamic flows. Vishalakshi et al. [7] studied 3D MHD fluid flows under the influence of a magnetic field with an inclined angle. Their results have been used in many real-world applications, e.g., automotive cooling systems, microelectronics, heat exchangers, etc. Anusha et al. [8] studied the two-dimensional magnetohydrodynamic problem for a steady incompressible flow over a porous medium. They concluded that the porosity and radiation parameters enhance the temperature distribution, while the suction/injection parameter suppresses the temperature distribution. Skinner et al. [9] developed a computational algorithm based on an accepted analytical model to investigate the viscoelastic behaviour of carbon fibre-reinforced polymer composite flywheel rotors with an aluminium hub mounted by press-fit. The simulations showed that over time the viscoelastic effects are likely to reduce the peak stresses in the composite rim. However, viscoelasticity also affects the stresses in the hub and at the hub–rim interface, leading to rotor failure. It was also found that the charge/discharge cycles of the flywheel energy accumulator can lead to significant fatigue loads.

Furlan et al. [10] derived different formulations to obtain a solution for Giesekus' constitutive model for a flow between two parallel plates. Bertoco et al. [11] presented a numerical study of the development length (the length from channel entry required for the velocity to reach 99% of its fully developed value) of a pressure-driven viscoelastic fluid flow (between parallel plates) modelled by the generalised constitutive Phan–Thien–Tanner equation (gPTT). They concluded that at low values of the Weissenberg number (Wi), the highest value of the development length was achieved for $\alpha = \beta = 0.5$; at high values of Wi, the highest value of the development length was achieved for $\alpha = \beta = 1.5$.

Although submissions for this Special Issue have now closed, research into the field of viscoelasticity continues to address various challenges we face today: medicine (e.g., drug delivery, foods that consider their rheology, and complex blood flow), development of new and smart materials (e.g., paints, biomaterials, and clothing), new industrial developments.

Funding: This research received no external funding.

Acknowledgments: Thanks to all the authors and peer reviewers for their valuable contributions to this Special Issue 'Viscoelasticity: Mathematical Modelling, Numerical Simulations, and Experimental Work'. We would also like to express our gratitude to all the staff and people involved in this Special Issue.

Conflicts of Interest: The authors declare no conflict of interest.

References

1. McDermott, M.; Resende, P.; Charpentier, T.; Wilson, M.; Afonso, A.; Harbottle, D.; de Boer, G. A FENE-P k–ε Viscoelastic Turbulence Model Valid up to High Drag Reduction without Friction Velocity Dependence. *Appl. Sci.* **2020**, *10*, 8140. [CrossRef]
2. Ingelsten, S.; Mark, A.; Kádár, R.; Edelvik, F. A Backwards-Tracking Lagrangian-Eulerian Method for Viscoelastic Two-Fluid Flows. *Appl. Sci.* **2021**, *11*, 439. [CrossRef]
3. Bertoco, J.; de Araújo, M.; Leiva, R.; Sánchez, H.; Castelo, A. Numerical Simulation of KBKZ Integral Constitutive Equations in Hierarchical Grids. *Appl. Sci.* **2021**, *11*, 4875. [CrossRef]
4. Wojcik, B.; LaRuez, J.; Cromer, M.; Villasmil Urdaneta, L. The Role of Elasticity in the Vortex Formation in Polymeric Flow around a Sharp Bend. *Appl. Sci.* **2021**, *11*, 6588. [CrossRef]
5. Liu, J.; Oberlack, M.; Wang, Y. Analytical Investigation of Viscoelastic Stagnation-Point Flows with Regard to Their Singularity. *Appl. Sci.* **2021**, *11*, 6931. [CrossRef]
6. Aabid, A.; Khan, S.; Baig, M. A Critical Review of Supersonic Flow Control for High-Speed Applications. *Appl. Sci.* **2021**, *11*, 6899. [CrossRef]
7. Vishalakshi, A.; Maranna, T.; Mahabaleshwar, U.; Laroze, D. An Effect of MHD on Non-Newtonian Fluid Flow over a Porous Stretching/Shrinking Sheet with Heat Transfer. *Appl. Sci.* **2022**, *12*, 4937. [CrossRef]
8. Anusha, T.; Mahesh, R.; Mahabaleshwar, U.; Laroze, D. An MHD Marangoni Boundary Layer Flow and Heat Transfer with Mass Transpiration and Radiation: An Analytical Study. *Appl. Sci.* **2022**, *12*, 7527. [CrossRef]
9. Skinner, M.; Mertiny, P. Effects of Viscoelasticity on the Stress Evolution over the Lifetime of Filament-Wound Composite Flywheel Rotors for Energy Storage. *Appl. Sci.* **2021**, *11*, 9544. [CrossRef]
10. da Silva Furlan, L.; de Araujo, M.; Brandi, A.; de Almeida Cruz, D.; de Souza, L. Different Formulations to Solve the Giesekus Model for Flow between Two Parallel Plates. *Appl. Sci.* **2021**, *11*, 10115. [CrossRef]
11. Bertoco, J.; Leiva, R.; Ferrás, L.; Afonso, A.; Castelo, A. Development Length of Fluids Modelled by the gPTT Constitutive Differential Equation. *Appl. Sci.* **2021**, *11*, 10352. [CrossRef]

Disclaimer/Publisher's Note: The statements, opinions and data contained in all publications are solely those of the individual author(s) and contributor(s) and not of MDPI and/or the editor(s). MDPI and/or the editor(s) disclaim responsibility for any injury to people or property resulting from any ideas, methods, instructions or products referred to in the content.

Article

A Backwards-Tracking Lagrangian-Eulerian Method for Viscoelastic Two-Fluid Flows

Simon Ingelsten [1,2,*], Andreas Mark [1], Roland Kádár [2] and Fredrik Edelvik [1]

1. Fraunhofer-Chalmers Research Centre for Industrial Mathematics, 412 88 Gothenburg, Sweden; andreas.mark@fcc.chalmers.se (A.M.); fredrik.edelvik@fcc.chalmers.se (F.E.)
2. Division of Engineering Materials, Department of Industrial and Materials Science, Chalmers University of Technology, 412 96 Gothenburg, Sweden; roland.kadar@chalmers.se
* Correspondence: simon.ingelsten@fcc.chalmers.se; Tel.: +46-31-772-42-78

Abstract: A new Lagrangian–Eulerian method for the simulation of viscoelastic free surface flow is proposed. The approach is developed from a method in which the constitutive equation for viscoelastic stress is solved at Lagrangian nodes, which are convected by the flow, and interpolated to the Eulerian grid with radial basis functions. In the new method, a backwards-tracking methodology is employed, allowing for fixed locations for the Lagrangian nodes to be chosen a priori. The proposed method is also extended to the simulation of viscoelastic free surface flow with the volume of fluid method. No unstructured interpolation or node redistribution is required with the new approach. Furthermore, the total amount of Lagrangian nodes is significantly reduced when compared to the original Lagrangian–Eulerian method. Consequently, the method is more computationally efficient and robust. No additional stabilization technique, such as both-sides diffusion or reformulation of the constitutive equation, is necessary. A validation is performed with the analytic solution for transient and steady planar Poiseuille flow, with excellent results. Furthermore, the proposed method agrees well with numerical data from the literature for the viscoelastic die swell flow of an Oldroyd-B model. The capabilities to simulate viscoelastic free surface flow are also demonstrated through the simulation of a jet buckling case.

Keywords: viscoelastic flow; computational fluid dynamics; volume of fluid; immersed boundary methods

1. Introduction

Viscoelastic free surface flows are of significant importance for many industrial processes. This includes polymer extrusion, additive manufacturing, seam sealing, and adhesive joining. In such processes, the viscoelastic properties of the flow, such as the flow history, can have major influence on the quality of the final product. Furthermore, production time and raw materials may be subject to a large cost. Therefore, extensive manual effort and physical testing may be necessary to optimize the process in terms of product quality, material consumption, and production cycle time.

Numerical tools can be helpful in reducing the manual preparation time, as they offer the possibility to replace a significant part of the physical testing with computer simulation. Furthermore, numerical simulations may enable testing early in the design phase. For many complex industrial applications, a large demand for suitable numerical simulation tools therefore exists. Furthermore, for such tools to be useful, high demands are put on accuracy, robustness, and computational efficiency.

A common approach for simulating viscoelastic fluid flow is to discretize the governing equations in the Eulerian frame of reference with the finite volume method (FVM) [1–3] or the finite element method (FEM) [4,5]. The Eulerian frame of reference is suitable for diffusion-dominated problems, including viscous fluid flow as well as heat and mass transfer. On the other hand, viscoelastic constitutive equations are typically hyperbolic and they involve no physical diffusion term. Thus, in this sense, the Lagrangian frame of reference is a more natural description for the constitutive equation. Therefore, Lagrangian or semi-Lagrangian methods are appealing as alternative approaches to the Eulerian frame.

Rasmussen and Hassager [6] developed a Lagrangian finite element method in order to calculate the flow of an upper convected Maxwell (UCM) fluid. The entire flow history was stored and re-meshing was required, due to deformation. Harlen et al. [7] proposed a split Lagrangian–Eulerian method for viscoelastic Stokes flow. The constitutive equation was solved at the nodes of a co-deforming mesh, while the momentum and continuity equations were solved with an Eulerian finite element method. Re-meshing due to mesh distortion was also used in their method. Halin et al. proposed the Lagrangian particle method (LPM) [8]. In their method, the momentum and continuity equations were solved with the Eulerian finite element method. The constitutive equation was solved along the trajectories of massless Lagrangian particles. The stresses in the particles were then fitted to local polynomial expressions in order to enable the evaluation of the corresponding finite element integrals in the momentum. Hence, a minimum of three particles per two-dimensional element was required for the polynomial approximations to be feasible. Later on, the adaptive Lagrangian particle method (ALPM) [9] was proposed, utilizing adaptive addition and the deletion of particles. However, a fairly large number of particles was required for the simulations to produce stable transient results. Furthermore, Wapperom et al. [10] proposed the backward-tracking Lagrangian particle method (BLPM), in which backwards integration of the velocity field allowed for the choice of fixed particle locations a priori, increasing the efficiency of their particle tracking approach.

The simulation of viscoelastic free surface flow has been approached with different methods. For example, Crochet and Keunings [11] simulated the viscoelastic die swell effect of an Oldroyd-B fluid with a mixed finite element method as early as 1982. More recent similar examples of finite element methods on non-stationary grids can be found in Balemans et al. [12] and Spanjaards et al. [13]. The marker-and-cell (MAC) approach is a method that has been used quite extensively for simulation free surface flows [14–20], in which the governing equations discretized with finite differences and the free surface is tracked while using marker particles. The method has been successfully used for the simulation, for example, of the die swell and viscoelastic jet buckling with a variety of constitutive models. The simulations are typically performed on staggered, uniform grids, which has certain limitations for complex simulation geometries and in terms of computational efficiency when high resolution is required. Furthermore, the existence of a gas phase surrounding the viscoelastic fluid is not taken into account, which may be required for some cases. The front-tracking method is another similar method [21].

The volume of fluid (VOF) method is a popular method for the simulation of Newtonian as well as viscoelastic free surface flow with finite volume discretization. VOF is a diffuse–interface method, in which the presence of two or more fluid phases is represented by their corresponding volume fractions, advected with geometric or algebraic schemes. A sharp fluid interface may then be reconstructed from the solution. A single set of momentum, continuity equations are then solved for the whole domain, and the fluid properties are locally averaged with the fluid volume fraction. Furthermore, the transport of the volume fraction may be solved with the same spatial grid as the momentum and continuity equation, which makes the method suitable for finite volume simulation of free surface flow. Some examples follow.

Habla et al. [22] developed a VOF-solver for viscoelastic free surface flow that is based on the open source solver OpenFOAM [23]. The simulations were stabilized by reformulating the viscoelastic constitutive equations with the log-conformation representation (LCR) and with both-sides diffusion (BSD) in the momentum equation. Comminal et al. [24] simulated the viscoelastic die swell effect for Oldroyd-B and Giesekus fluids with the VOF method, combined with two different schemes for the convection of the fluid volume fraction, a geometric scheme, as well as an algebraic scheme that directly solved the transport equation for the volume fraction. Furthermore, their methods were compared to the VOF solver that is available in RheoTool [25], which is an open source toolbox for viscoelastic flow that is based on OpenFOAM. The constitutive equation was also reformulated with LCR in their study. Bonito et al. [26] reported a related method, using a mathematical formulation that is similar to the volume of fluid method, while solving the governing equations with a piecewise linear finite element method, contrary to finite volume discretization.

In previous work, a Lagrangian–Eulerian framework for viscoelastic flow has been proposed [27,28]. In the method, the constitutive equation is solved at Lagrangian nodes, which are convected by the flow. The momentum and continuity equations are discretized with the finite volume method on an Eulerian octree grid and then solved with SIMPLEC iterations [29]. Boundary conditions that are imposed by objects in the domain are treated using implicit immersed boundary methods [30,31]. The contribution to the momentum equation from the viscoelastic stress is established through the unstructured interpolation of the stress tensors from the Lagrangian nodes to the cell centers of the Eulerian grid while using the radial basis function.

The ability to efficiently simulate industrial applications, such as adhesive extrusion, parts assembly, and additive manufacturing, is a main motivator to develop the new method. For this purpose, it is necessary that the numerical method is accurate, computationally efficient, and robust. The ability to model viscoelastic free surface flow in complex geometry with moving objects is also necessary. The proposed method was shown to produce results in good agreement with available analytic and numerical data from the literature [27]. It has also been shown that the Lagrangian formulation of the constitutive equation allows for the efficient parallelization of the viscoelastic stress calculation, which makes the method highly suitable for GPU-acceleration [28]. Furthermore, the use of an immersed boundary method in combination with the automatically generated grid enables simulation on arbitrary geometry with minimal pre-processing.

In the previous studies [27,28], the proposed Lagrangian–Eulerian method has been validated for single-phase flows with uniform grids. The natural next steps include enabling the simulation of free surface flow as well as simulation on refined grids. Therefore, in the current work, the proposed method is further developed in these two aspects. Firstly, the framework is extended for simulation of viscoelastic free surface flows with the volume of fluid method. Secondly, a backwards-tracking procedure is introduced to the solution of the constitutive equation, partly inspired by the backward-tracking Lagrangian particle method by Wapperom et al. [10]. The result is a more robust method, due to the structured relation between the Lagrangian nodes and the Eulerian grid. No stabilization method, i.e., neither both-sides diffusion nor reformulation of the constitutive equation, is required for the studied flows. Furthermore, the computational cost as compared to the original Lagrangian–Eulerian method is reduced, since no unstructured interpolation or node redistribution is needed and because the total number of Lagrangian nodes decreased. In the context of multiphase flows, the viscoelastic constitutive equation is only solved inside the viscoelastic phase, which further increases the computational performance.

The rest of the paper is structured, as follows. In the next section, the governing equations are presented, followed by a detailed description of the proposed numerical method. In the results section, the new method demonstrated and validated with analytic solutions and numerical data from the literature. Finally, conclusions are drawn and future work is outlined.

2. Governing Equations

Viscoelastic fluid flow is described by the incompressible momentum and continuity equations

$$\rho\left(\frac{\partial \mathbf{u}}{\partial t} + \mathbf{u} \cdot \nabla \mathbf{u}\right) = -\nabla p + \nabla \cdot (2\mu \mathbf{S} + \boldsymbol{\tau}) + \mathbf{f}, \tag{1}$$

$$\nabla \cdot \mathbf{u} = 0, \tag{2}$$

where ρ is the density, $\mathbf{u}$ velocity, μ the solvent viscosity, $\mathbf{S} = \frac{1}{2}(\nabla \mathbf{u} + (\nabla \mathbf{u})^T)$ the strain rate tensor, $\boldsymbol{\tau}$ the viscoelastic stress tensor, and $\mathbf{f}$ a body force. The viscoelastic stress is governed by a constitutive equation of the general form

$$\lambda \stackrel{\nabla}{\boldsymbol{\tau}} + F(\boldsymbol{\tau})\boldsymbol{\tau} = 2\eta \mathbf{S}, \tag{3}$$

where λ is the relaxation time, η polymeric viscosity, and F a scalar-valued function. The operator $(\stackrel{\nabla}{\bullet})$ denotes the upper-convected derivative of $\boldsymbol{\tau}$, which expands to

$$\stackrel{\nabla}{\boldsymbol{\tau}} = \frac{D\boldsymbol{\tau}}{Dt} - (\nabla \mathbf{u})^T \cdot \boldsymbol{\tau} - \boldsymbol{\tau} \cdot \nabla \mathbf{u}, \tag{4}$$

where

$$\frac{D\boldsymbol{\tau}}{Dt} = \frac{\partial \boldsymbol{\tau}}{\partial t} + \mathbf{u} \cdot \nabla \boldsymbol{\tau}, \tag{5}$$

is the Lagrangian, or material, time derivative of $\boldsymbol{\tau}$. The Lagrangian time derivative describes the rate of change of $\boldsymbol{\tau}$ in an infinitesimal material element, which moves with the flow. Convected time derivatives, such as $(\stackrel{\nabla}{\bullet})$, appear in viscoelastic constitutive equations and they ensure frame-invariance for the transport of tensorial properties. It is remarked that, while the form of (3) is not applicable for all available differential constitutive equations, it covers the constitutive models discussed in this work.

In general, the viscoelastic stress tensor may be modeled as a sum of the contribution of N modes as

$$\boldsymbol{\tau} = \sum_{k=1}^{N} \boldsymbol{\tau}_k, \tag{6}$$

where $\boldsymbol{\tau}_k$ is the stress tensor that corresponds to the kth mode. Each stress tensor $\boldsymbol{\tau}_k$ is then governed by a constitutive equation on the form of (3) with an individual set of model parameters.

3. Numerical Method

The momentum Equation (1) and the continuity Equation (2) are discretized on a collocated Eulerian grid with the finite volume method. The pressure–velocity coupling is accomplished by solving the momentum and continuity equations with SIMPLEC iterations [32]. An octree grid is used, in which the cells can be refined in desired locations by recursively dividing them into smaller cells to a given refinement level. The refinements may be static or adaptively updated, typically around moving objects or fluid interfaces. Figure 1 displays an example grid in two dimensions with one refinement level, showing the cell centers and the grid nodes. The grid, including the adaptive refinements, is automatically generated by the solver.

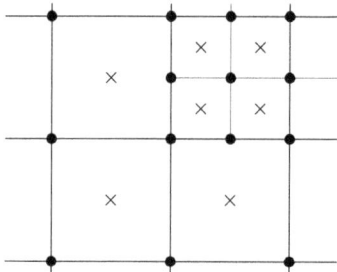

Figure 1. Two-dimensional grid with one refinement level, showing cell centers ($\times$) and grid nodes ($\bullet$).

Boundary conditions from objects in the domain are imposed while using the mirroring immersed boundary method [30,31]. The velocity field is implicitly mirrored across the immersed boundary surface, such that the imposed boundary condition is satisfied for the converged solution. Surface triangulations represent the geometrical objects.

The viscoelastic constitutive equation is solved in the Lagrangian frame of reference in material elements that are represented as Lagrangian nodes. From (3) and (4), it follows that the stress in a Lagrangian node is described by the ordinary differential equation (ODE)

$$\frac{D\boldsymbol{\tau}}{Dt} = \frac{2\eta}{\lambda}\mathbf{S} - \frac{1}{\lambda}F(\boldsymbol{\tau})\boldsymbol{\tau} + (\nabla\mathbf{u})^T \cdot \boldsymbol{\tau} + \boldsymbol{\tau} \cdot \nabla\mathbf{u}. \tag{7}$$

Furthermore, the trajectory of the node must be known when solving (7), as it involves the local velocity gradient $\nabla\mathbf{u}$. The trajectory of the node is given by the ODE

$$\frac{D\mathbf{x}}{Dt} = \mathbf{u}, \tag{8}$$

where $\mathbf{x}$ is the position of the node.

Given an initial stress state in a set of Lagrangian nodes, a straightforward approach to obtain the stress state at a future time is to simultaneously solve (7) and (8) forwards in time. Indeed, this method has been successfully employed in previous work [27,28]. Lagrangian nodes were then initialized with a given distribution density at the start of the simulation. The distribution was maintained in each time step by adding or deleting nodes, if necessary. After solving the constitutive equation at the Lagrangian nodes, the viscoelastic stress was interpolated to the Eulerian cell centers while using radial basis functions (RBF).

In this work, the method is further developed by introducing the concept of backwards-tracking. The key idea is to choose the locations of the Lagrangian nodes a priori, at which the viscoelastic stresses are then stored, and then track them backwards in time in order to obtain their initial location in each time step. The constitutive equation may then be solved forwards in time such that the stresses are obtained at for the next time step at the chosen locations. Different choices for the locations are conceivable. In the current work, the locations of the Lagrangian nodes are chosen to be at the Eulerian grid nodes.

Consider the calculation of the viscoelastic stress for the nth simulation step, which corresponds to the time interval $I_n = [t_n, t_{n+1}]$. At this point, it is assumed that momentum equation has been solved, such that the velocity field is known for $t \in I_n$.

The Lagrangian constitutive equation is solved along the trajectories of Lagrangian nodes. For a Lagrangian node, the trajectory $\mathbf{x}(t)$ can be expressed in terms of the position at the end of the time step, as

$$\mathbf{x}(t) = \mathbf{x}(t_{n+1}) - \int_t^{t_{n+1}} \mathbf{u}(t', \mathbf{x}(t'))dt', \quad t \in I_n, \tag{9}$$

where $\mathbf{x}(t_{n+1})$ is the predefined final location of the Lagrangian node which. The trajectory $\mathbf{x}(t)$ for $t \in I_n$ is first calculated by solving (8) backwards in time. Subsequently, when the trajectory is known, the viscoelastic stress $\boldsymbol{\tau}(t_n, \mathbf{x}(t_n))$ is interpolated from the viscoelastic stresses from the solution of the previous simulation step. The Lagrangian constitutive Equation (7) is then solved forwards in time and the viscoelastic stress is thus obtained at the final location $\mathbf{x}(t_n)$, i.e., at the predefined location.

Generally, the choice of locations for the Lagrangian nodes is somewhat arbitrary. However, by choosing the locations of the Eulerian grid nodes, the connectivity and structure of the octree grid may be utilized when the stress at time t_n is interpolated to the position of a Lagrangian node $\mathbf{x}(t_n)$. The stress at the node at time t_n is obtained through bilinear or trilinear interpolation, respectively, for two-dimensional (2D) and three-dimensiona (3D). Furthermore, the viscoelastic stress contribution to the discretized momentum equation may be calculated directly from node stresses. Therefore, no unstructured interpolation of viscoelastic stresses is necessary. The calculation of this contribution is described in detail further on in this section.

Figure 2 shows a schematic description of the steps that are involved in the algorithm. In summary, the performed steps are:

(a) Calculate the Lagrangian node trajectory by solving (8) backwards in time, starting at the predefined location $\mathbf{x}(t_{n+1})$.
(b) Interpolate the stress $\boldsymbol{\tau}(t_n, \mathbf{x}(t_n))$ to the Lagrangian node from the known node values at time t_n.
(c) Solve (7) forwards in time along the trajectory $\mathbf{x}(t)$, $t \in I_n$.

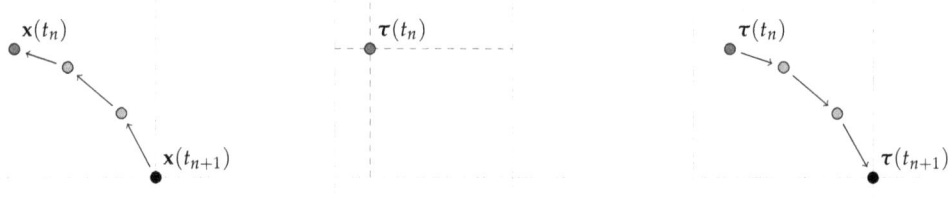

(a) Backwards-tracking   (b) Interpolation   (c) Forwards-solving

Figure 2. Schematic description of the steps that are involved in the backwards-tracking algorithm.

The ordinary differential Equations (7) and (8) may be solved with any appropriate choice of ODE solution algorithm. In this work, the fourth order Runge–Kutta method, which is commonly referred to as the RK4 method [33], is used. Furthermore, N_{sub} equally sized substeps of length Δt_{sub} are used, being defined, such that $N_{\text{sub}} \Delta t_{\text{sub}} = \Delta t = t_{n+1} - t_n$. In the current work, $N_{\text{sub}} = 3$ is used.

When solving (7) and (8), local quantities that are stored on the Eulerian grid are required along the trajectory of the Lagrangian nodes. More specifically, the velocity $\mathbf{u}$ is needed for the backwards-tracking and the velocity gradient $\nabla \mathbf{u}$ is needed to solve the constitutive equation. The velocities are stored at the Eulerian cell centers. When the local velocity is required, the cell centers containing the Lagrangian node are identified and the velocity is interpolated to its location while using bilinear or trilinear interpolation, respectively, for 2D and 3D. When the velocity gradient is required, it is calculated from the bilinear or trilinear interpolation formula using the same interpolation basis.

For the interpolation of a field ϕ to a location within a box spanned by the corners (x_i, y_j, z_k), $i, j, k \in \{0, 1\}$, the trilinear interpolation formula in order to calculate the interpolant $\hat{\phi}$ reads

$$\hat{\phi} = \sum_{i=0}^{1} \sum_{j=0}^{1} \sum_{k=0}^{1} c_x^i c_y^j c_z^k \phi_{ijk}, \tag{10}$$

where ϕ_{ijk} is the value of ϕ at the corner (x_i, y_j, z_k) and with the coefficients

$$c_l^i = \begin{cases} 1 - \tilde{x}_l / \Delta x_l, & i = 0 \\ \tilde{x}_l / \Delta x_l, & i = 1 \end{cases}, \quad l = x, y, z, \tag{11}$$

where $\tilde{x}_l$ is the lth coordinate of the interpolation position relative to the lower corner of the box and Δx_l is the size of the box that coordinate direction. The gradient of $\hat{\phi}$ follows from (10) and (11), as

$$\frac{\partial \hat{\phi}}{\partial x} = \sum_{i=0}^{1} \sum_{j=0}^{1} \sum_{k=0}^{1} \left(\frac{\partial c_x^i}{\partial x} \right) c_y^j c_z^k \phi_{ijk}, \tag{12}$$

$$\frac{\partial \hat{\phi}}{\partial y} = \sum_{i=0}^{1} \sum_{j=0}^{1} \sum_{k=0}^{1} c_x^i \left(\frac{\partial c_y^j}{\partial y} \right) c_z^k \phi_{ijk}, \tag{13}$$

$$\frac{\partial \hat{\phi}}{\partial z} = \sum_{i=0}^{1}\sum_{j=0}^{1}\sum_{k=0}^{1} c_x^i c_y^j \left(\frac{\partial c_z^k}{\partial z}\right) \phi_{ijk}, \qquad (14)$$

where the derivatives of the coefficients read

$$\frac{\partial c_l^i}{\partial x_l} = \mp \frac{1}{\Delta x_l}, \quad i = 0, 1, \quad l = x, y, z. \qquad (15)$$

The corresponding formulas for bilinear interpolation in two dimensions are obtained by letting the coefficient $c_z^k = 1$ and, thus, $\partial c_z^k/\partial z = 0$, and skipping the summation over k.

When interpolating to a Lagrangian node near fluid grid refinements the smallest local cell size is used in order to form the interpolation box. Furthermore, not all corners of the interpolation box coincide with a cell center. The velocities at such corners are obtained through least squares interpolation, by fitting a first order polynomial from the centers of the cells that intersect the interpolation box. The interpolation box is visualized in Figure 3.

(a) Uniform   (b) Refined

Figure 3. Basis for interpolating properties stored at the fluid grid to Lagrangian nodes in areas with uniform grid spacing (a) and near refinements (b).

The volume of fluid (VOF) method is used to model the presence of a viscoelastic fluid phase and a Newtonian phase. A color function ζ, is then defined, such that

$$\zeta = \begin{cases} 1 & \text{In the viscoelastic phase} \\ 0 & \text{In the Newtonian phase} \end{cases}. \qquad (16)$$

The discrete counterpart of ζ is the fluid volume fraction $\alpha \in [0,1]$, which is the local volume average of ζ in an Eulerian cell. The transport of α is described by the convection equation [34]

$$\frac{\partial \alpha}{\partial t} + \mathbf{u} \cdot \nabla \alpha = 0. \qquad (17)$$

By definition, α is the local volume fraction of the viscoelastic phase in a control volume. Hence, if $\alpha = 1$, the control volume is filled with the viscoelastic fluid and if $\alpha = 0$ with the Newtonian fluid. If $\alpha \in (0,1)$, this is interpreted as that the control volume intersects the interface between the fluid phases. The transport Equation (17) is discretized with the finite volume method on the Eulerian octree grid. In order to minimize the numerical diffusion of α and, thus, avoid smearing the fluid interface, the convection term in (17) is discretized with the compact CICSAM scheme [35].

A single set of the Equations (1)–(3) is solved for the whole computational domain. Fluid properties are locally averaged as

$$\phi = \alpha \phi_1 + (1-\alpha)\phi_2, \qquad (18)$$

where ϕ_v and ϕ_N are the properties of the viscoelastic and the Newtonian phase, respectively. The averaging is applied for ρ, μ, η, and λ, as well as for τ when calculating the contribution to the momentum equation. When α is needed at a Lagrangian node, it is interpolated with the same method as used for velocity.

Special care for the viscoelastic stress in the interface region $\alpha \in (0,1)$ is necessary due to the non-sharp interface nature of the VOF method. A brief motivation for this follows. In many examples of viscoelastic free surface flow, the viscosity of the Newtonian fluid may very well be on the order of 10^{-5} of that of the viscoelastic phase. Hence, the average velocity gradient in the Newtonian phase may be much larger than in the viscoelastic phase. Consequently, if the velocity gradient of the Newtonian phase is allowed to have large influence on the constitutive equation in the interface region; this can result in unphysically large stresses. This is especially the case if not all velocity and length scales in the Newtonian phase are resolved near the interface. Thus, a means of reducing this effect around the free surface is proposed.

A threshold volume fraction is used, such that Lagrangian nodes are only created inside the viscoelastic phase, which are defined by the condition $\alpha \geq \alpha_{\lim} = 0.1$. Thus, the constitutive equation is only solved in the viscoelastic part of the computational domain and the stress is assumed to vanish outside the viscoelastic phase. Furthermore, when the velocity gradient is calculated at a Lagrangian node, the corners of the interpolation box at which $\alpha < 0.1 \cdot \alpha_{\lim}$ are also considered to be outside the viscoelastic phase. The velocities at such corners are excluded from the gradient calculation and they are replaced by 0th order extrapolation from the corners which are inside the viscoelastic phase.

It is remarked that, apart from numerical stability, the described approach is advantageous in terms of computational efficiency, since the constitutive equation only needs to be solved in part of the domain.

When the stress field has been calculated, the term $\nabla \cdot (\alpha \tau)$ is integrated and then added to the right hand side of the discretized momentum equation. At this stage, the product rule is applied, such that

$$\nabla \cdot (\alpha \tau) = \alpha \nabla \cdot \tau + \tau \cdot \nabla \alpha, \qquad (19)$$

which can be seen as separating the pure interfacial contribution of the viscoelastic stress to the fluid momentum from the remainder part [36]. The formulation of (19) is used, since it was found to enhance numerical stability. The first term of (19) is integrated over the cells with Gauss' divergence theorem, as

$$\int_{c.v.} \nabla \cdot \tau \, dV = \int_{c.s.} \hat{n} \cdot \tau \, dS = \sum_f \int_{f.s.} \hat{n}_f \cdot \tau \, dA, \qquad (20)$$

where c.v. denotes the cell volume, c.s. the cell surface, f.s. the surface of cell face f, and the vector $\hat{n}$ denotes the surface normal pointing outwards from the cell. In the second step of (20) the surface integral is divided into a sum of the integrals over the respective surface faces of the cell, for which the normal vectors $\hat{n}_f$ are constant. The integral over each face is approximated with the trapezoidal rule while using the stresses at the Eulerian grid nodes. If a cell has neighbors that have a higher refinement level, e.g. in the case shown in Figure 1, the face to each smaller neighbor cell is individually integrated, such that the stress contribution from each grid node is included correctly.

The volume integral of the second term of (19) is approximated with the cell average stress and volume fraction gradient, as

$$\int_{c.v.} (\tau \cdot \nabla \alpha) dV \approx (\overline{\tau} \cdot \overline{\nabla \alpha}) \Delta V, \qquad (21)$$

where ΔV is the cell volume and $\overline{(\bullet)}$ denotes volume average. The cells where $\alpha < 0.1 \cdot \alpha_{\lim}$ are assumed to be outside the viscoelastic phase and the stress contribution is instead set to zero.

The full algorithm for viscoelastic free surface flow can be summarized, as follows,

1. Calculate and add the viscoelastic stress contribution to the discrete momentum equation.
2. Solve velocity and pressure fields $\mathbf{u}$, p from the momentum Equation (1) and continuity Equation (2) while using SIMPLEC iterations.
3. Solve the transport of the fluid volume fraction α from (17)
4. Solve the viscoelastic stress from the constitutive Equation (3) to the new time while using the backwards-tracking procedure.

When compared to the original version of the Lagrangian–Eulerian framework that was proposed in Ingelsten et al. [27], the new method is expected to be more robust due to the structured nature of the Lagrangian nodes as well as the staggered arrangement between the storage locations of the velocity and the viscoelastic stress. Although the computational efficiency is not assessed in detail in this work, the computational cost is, in fact, reduced. As a reference, the computational performance of the original Lagrangian–Eulerian method was studied for the flow of a four-mode PTT fluid over a confined cylinder [27]. Four Lagrangian nodes per fluid cell were initialized for the node set. Out of the total stress calculation time, approximately 50% was spent on solving the ODE systems, 30% on the unstructured interpolation, and 10% on the redistribution of the Lagrangian node set. In the new method, the unstructured interpolation, as well as the redistribution, is completely removed. Furthermore, the number of Lagrangian nodes is reduced to be on the same order as the number of Eulerian cells. Thus, a large reduction in computational time should be expected.

The proposed method is implemented in the software IBOFlow® [37], an in-house CFD code that was developed at the Fraunhofer–Chalmers Research Institute for Industrial Mathematics in Gothenburg, Sweden. In addition to viscoelastic flow, the solver has previously been employed to simulate conjugated heat transfer [38–40], and fluid-structure interaction [41], as well as free surface flow of shear-thinning fluids with applications for seam sealing [42,43], adhesive application [44], and 3D-bioprinting [45].

4. Results

In this section, the proposed method is evaluated for three different flow cases. First, a basic validation is carried out by comparing the numerical results with the analytic solution for a transient channel flow. The method is then compared to the numerical results from the literature for simulations of the viscoelastic die swell effect. The capability to simulate viscoelastic free surface with adaptive mesh refinement is demonstrated for simulations of viscoelastic jet buckling.

The test cases that are reported in this section are simulated while using a single-mode Oldroyd-B model, which has the constitutive equation

$$\lambda \overset{\triangledown}{\tau} + \tau = 2\eta \mathbf{S}, \tag{22}$$

corresponding to $F(\tau) = 1$ in (3). Furthermore, a viscosity ratio β is defined as

$$\beta = \frac{\mu}{\mu + \eta}. \tag{23}$$

By definition of the total viscosity $\eta_t = \mu + \eta$, it follows from (23) that $\mu = \beta \eta_t$ and $\eta = (1 - \beta)\eta_t$.

Note that the Lagrangian–Eulerian method that is proposed in this work is by no means limited to the Oldroyd-B model. All of the constitutive equations in the form of (3) are supported by the framework.

4.1. Planar Poiseuille Flow

A basic validation of the proposed method is performed for a viscoelastic planar Poiseuille flow. The numerical results for the transient as well as the steady flow are compared to the corresponding analytic solution.

The computational domain has height and length h and it is shown in Figure 4. The upper boundary is treated as a wall with the no-slip condition $\mathbf{u} = 0$ and the lower boundary has a symmetry condition. At the left and right boundaries, the Dirichlet conditions $p = \Delta p$ and $p = 0$ are imposed, respectively, while the cyclic conditions are used for the velocity and viscoelastic stress. Thus, the numerical model represents an infinitely long channel of height $2h$, which is subjected to a constant pressure drop Δp. The simulation starts from rest, with velocity and viscoelastic stress equal to zero.

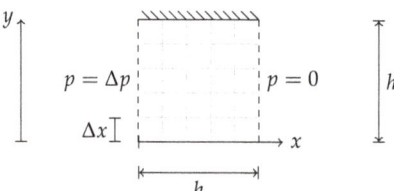

Figure 4. Schematic of the planar Poiseuille flow.

A Weissenberg number for the flow is defined as $\text{Wi} = \lambda U/h$, where U is the mean steady flow velocity, and a Reynolds number as $\text{Re} = 2\rho U h/(\mu + \eta)$. The flow is simulated for $\beta = 1/9, 1/18, 1/27$ and $\text{Wi} = 0.1, 1$. Here, $U = 0.1\,\text{m/s}$, $\rho = 1\,\text{kg/m}^3$ and $\eta = 1\,\text{Pa s}$ are constant, while Wi and β are varied, respectively, by changing λ and μ between simulations.

In order to ensure sufficient temporal resolution, the flow is simulated using different time step lengths. These simulations are performed for the lowest viscosity ratio $\beta = 1/27$, for which the transient variations are the largest. A uniform Eulerian grid with the cell size $\Delta x = H/20$ is used.

In Figure 5, the simulated centerline velocity, i.e., at $x = h/2, y = 0$, obtained with the time step lengths 10^{-4} s, 10^{-5} s and 10^{-6} s, are shown for $\text{Wi} = 0.1$ and $\text{Wi} = 1$. The velocity that si obtained with the longest step length is clearly different from those obtained with the two shorter step lengths. However, the two shorter lengths produce results that practically overlap on the scale of comparison. Thus, the step length $\Delta t = 10^{-5}$ s is considered to be sufficiently small and it is used to obtain the remaining results reported in this section.

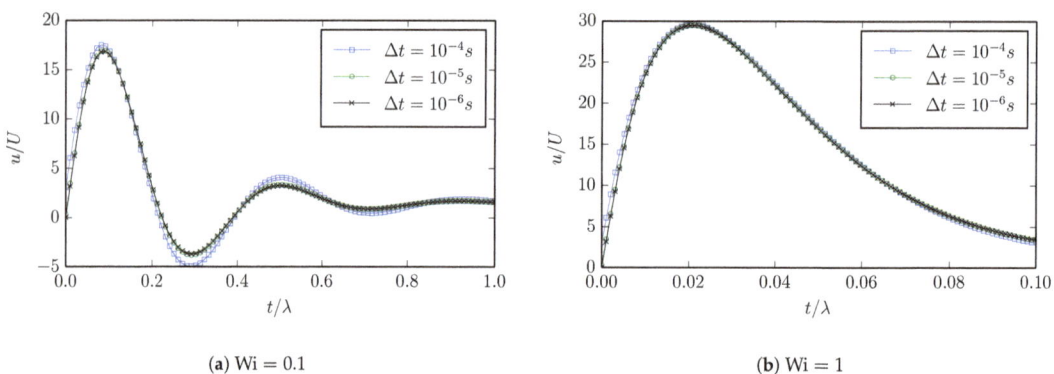

(a) Wi = 0.1   (b) Wi = 1

Figure 5. Centerline velocity in channel flow simulated using different time steps.

Xue et al. presented an analytic solution of the transient solution for the flow considered [46]. In Figure 6, the obtained centerline velocities for the $\beta = 1/9, 1/18, 1/27$ are shown for $\text{Wi} = 0.1$ and $\text{Wi} = 1$ and compared to the analytic solution. The results demonstrate the strong influence of the relationship between the solvent viscosity and polymeric viscosity on the transient flow. The numerical results overlap the analytic solution, showing that the proposed method treats the transient flow dynamics correctly. Furthermore, the results confirm that the spatial resolution $\Delta x/h = 20$ is sufficient for this case.

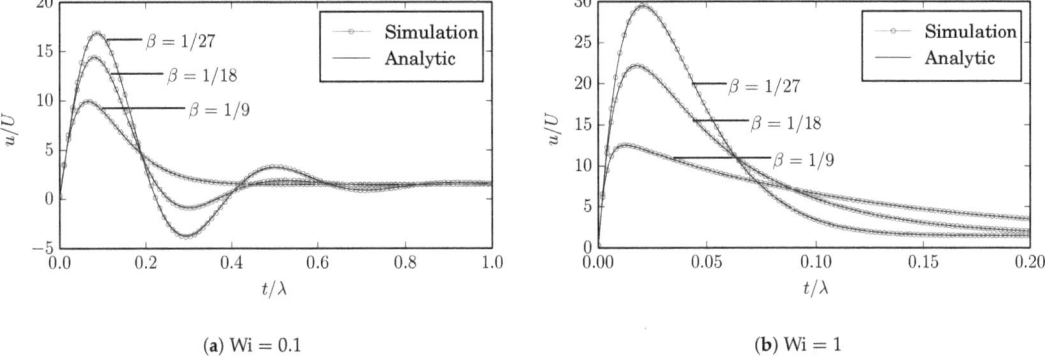

Figure 6. Centerline velocity in channel flow simulated using viscosity ratios β.

In order to assess grid dependency in more detail, the flows for $\beta = 1/27$ and $Wi = 0.1, 1$ are simulated until a steady flow is reached while using varying spatial resolution. The flow being steady is ensured through the condition

$$\frac{||\phi_{n-1} - \phi_n||_{L_2}}{||\phi_n||_{L_2}} < 10^{-10}, \qquad (24)$$

where $||\bullet||_{L_2}$ denotes the L_2-norm over the cells of the Eulerian grid, ϕ is velocity or viscoelastic stress, and the subscript n denotes the quantity at the nth time step.

The error measurements with respect to the analytic steady state solution are calculated as

$$E_\phi = \frac{||\phi_s - \phi_a||_{L_2}}{||\phi_a||_{L_2}}, \qquad (25)$$

where ϕ_s and ϕ_a denote the simulated and analytic solution, respectively. The computed errors for velocity and viscoelastic normal stress are shown as a function of grid size in Figure 7. The errors decrease with second order slope with grid refinements, which is coherent with the second order accuracy that was observed for the original Lagrangian–Eulerian method [27,28].

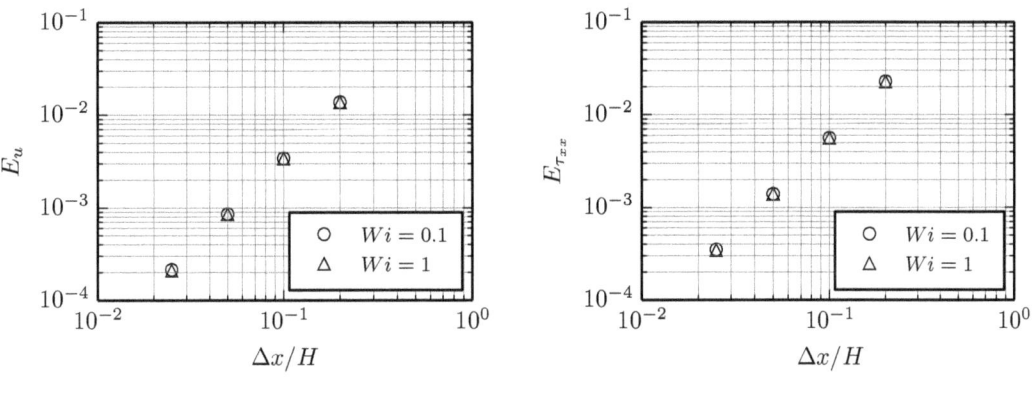

Figure 7. Computed errors with respect to the analytic solution for velocity (**a**) and viscoelastic normal stress (**b**).

4.2. Die Swell

The die swell effect is a phenomenon that occurs for the free surface flows of viscoelastic fluids flowing in a pipe or channel and exiting through circular or slit dies. Barnes et al. [47] described the origin of the die swell effect by viewing the viscoelastic fluid as a bundle of elastic threads. In the channel flow, the threads are stretched by the normal stress component τ_{xx}, where x is the flow direction. When the fluid exits the channel, the threads are allowed to relax and shorten in length. Consequently, the diameter of the emerging fluid increases.

Flows that exhibit the die swell effect are commonly used to benchmark and validate numerical methods [11,13,14,16–20,22,24]. In addition to the free surface, the flow has extensional characteristics around the channel exit as well as a stress singularity at the channel exit corner. These flow features make the flow suitable for testing robustness and accuracy of new numerical methods. Therefore, in this work, a planar die swell flow is simulated. The case parameters are selected to allow for a meaningful comparison with numerical data from the literature.

A schematic description of the computational domain can be found in Figure 8. A channel of height $2h$ and length $10h$ is considered. A symmetry boundary condition is used at $y = 0$, such that effectively half of the domain is simulated. The expansion zone is $5h$ high and $12h$ long, giving the domain a total length of $22h$. The length of the channel and size of the expansion zone are chosen to be sufficiently large for boundary effects not to affect the flow near the channel exit.

The exterior boundaries of the expansion zones are treated as outlets with the pressure condition $p = 0$, and an immersed boundary is used to impose the no-slip condition for the channel walls. Gravity is neglected.

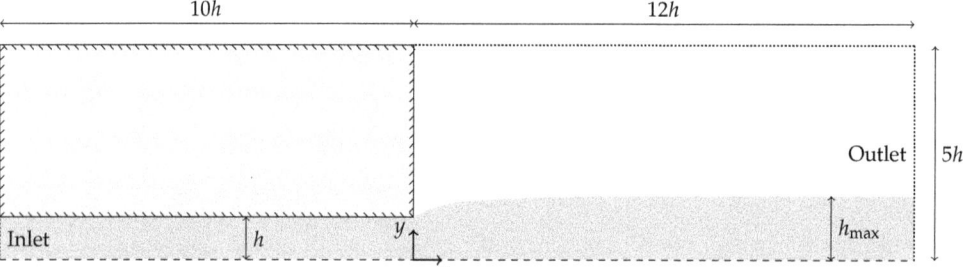

Figure 8. Die swell domain.

The viscoelastic fluid in the channel is an Oldroyd-B fluid with the viscosity ratio $\beta = 1/9$. The surrounding fluid is a Newtonian fluid with much lower viscosity and density than the viscoelastic fluid. In this work, these parameters are set to $\mu_{\text{air}} = 10^{-6}\mu$ and $\rho_{\text{air}} = 10^{-3}\rho$, respectively. Comminal et al. used a similar definition [24], but with $\rho_{\text{air}} = 10^{-2}\rho$.

The simulation of the flow starts from rest, with the channel filled with viscoelastic fluid. Transient flow is simulated for a sufficiently long physical time for the flow in the channel and the expansion region to fully develop, as well as for the free surface flow to exit through the outlet with a uniform velocity profile.

At the inlet, fully developed flows are imposed for the velocity and viscoelastic stress. For fully developed channel flow of an Oldroyd-B fluid, it can be shown that

$$u(y) = \frac{3}{2}U\left(1 - \frac{y^2}{h^2}\right), \qquad (26)$$

$$\frac{\partial u}{\partial y} = -\frac{3Uy}{h^2}, \qquad (27)$$

$$\tau_{xy}(y) = \eta \frac{\partial u}{\partial y}, \tag{28}$$

$$\tau_{xx}(y) = 2\lambda \tau_{xy}(y)\frac{\partial u}{\partial y} = 2\lambda\eta \left(\frac{\partial u}{\partial y}\right)^2, \tag{29}$$

where U is the mean velocity. A derivation of the expressions can be found in Appendix A.

The channel half height h and velocity U are taken as characteristic length and velocity scales for the flow. Thus, a Weissenberg number for the flow may be defined as

$$\text{Wi} = \frac{\lambda U}{h}, \tag{30}$$

and a Reynolds number as

$$\text{Re} = \frac{2\rho U h}{\mu + \eta}. \tag{31}$$

A dimensionless time is also defined as $t^* = tU/h$.

The swell ratio $S_r = h_{\max}/h$ is the main quantity of interest, in terms of validation by comparison. The swell ratio is a function of the so-called recoverable shear S_R, which is defined as [11]

$$S_R = \left|\frac{N_1}{2\tau_{xy}}\right|_{y=h} = \left|\frac{\tau_{xx} - \tau_{yy}}{\tau_{xy}}\right|_{y=h} = \frac{3\lambda U}{h} = 3\text{Wi}, \tag{32}$$

where $N_1 = \tau_{xx} - \tau_{yy}$ is the first normal stress difference. In the simulations that are reported in this section, the quantities U, η, μ, and ρ are constant, while S_R is varied by changing λ between simulations. The Reynolds number is $\text{Re} = 0.5$ for all simulations.

Tanner [48,49] developed a theoretical solution for the amplitude of the die swell as a function of the recoverable shear. For an Oldroyd-B channel emerging from a slit the solution reads

$$S_r = 0.19 + \left(1 + \frac{S_R^2}{3}\right)^{\frac{1}{4}}. \tag{33}$$

The theory does not take the stress singularity at the channel exit corner into account and is, therefore, only expected to yield a good approximation of the swell ratio when the influence of elasticity is small, i.e., for small Wi. Nevertheless, it is often included for comparison in numerical studies of the die swell effect and it is included in this study.

The influence of the spatial discretization on the numerical results is assessed through a grid dependence study and the flow is simulated for $S_R = 2.5$ with three different grids, denoted as M1, M2, and M3. A grid is defined by a base cell size Δx_{base} and a set of refinements, as described in Section 3. The grids M1, M2, and M3, respectively, have the base cell sizes $\Delta x_{\text{base}} = h/5, h/10, h/20$. The largest cells are located in the expansion zone, far from the channel exit. One level of refinement is used in the channel and around the channel opening and two levels of refinement near the exit corner. As an example, grid M1 is shown in Figure 9. The area around the channel exit corner has been zoomed for clarity. Table 1 summarizes the grids used. A constant time step length is used, such that $U\Delta t/\Delta x_{\min} = 0.01$, where $\Delta x_{\min}$ is the smallest cell size of the grid.

Table 1. Summary of grids that are used to study the grid dependence of the die swell simulations.

Grid	$h/\Delta x_{\text{base}}$	$h/\Delta x_{\text{channel}}$	$h/\Delta x_{\text{corner}}$	Num. Cells Total.
M1	5	10	20	4829
M2	10	20	40	19,685
M3	20	40	80	78,740

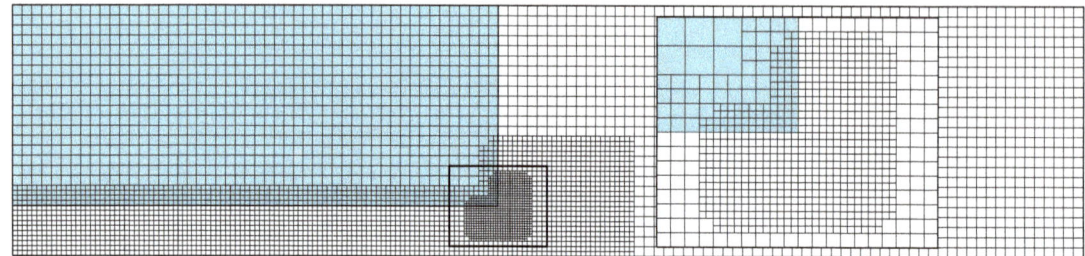

Figure 9. Grid M1 used in the grid dependence study, as defined in Table 1.

First, the velocity as well as the viscoelastic normal and shear stress are compared to the analytic steady solution inside the channel. In Figure 10, the corresponding profiles obtained using grid M1 across the channel at $x/h = 5$, halfway from the inlet to the channel exit. Already at the lowest grid resolution, the simulated profiles overlap the analytic solution. The results that are obtained with M2 and M3 are visually the same and the figures are, therefore, omitted. The results validate that the constitutive equations are solved correctly and that the immersed boundary method used to impose the boundary conditions at the channel wall works as intended for the solution algorithm for the viscoelastic stresses.

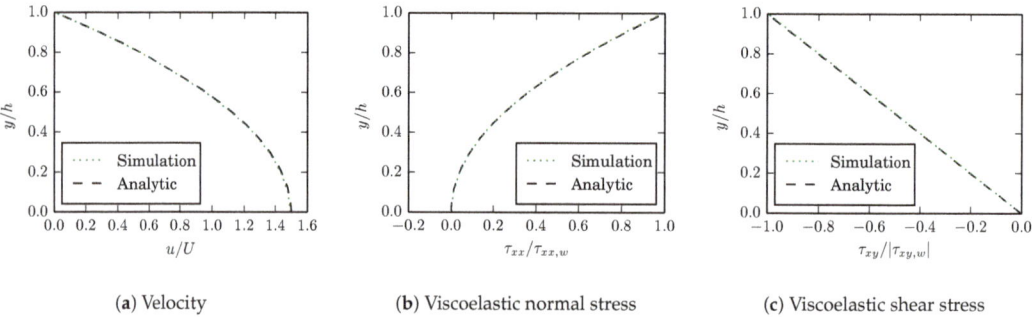

Figure 10. Simulated profiles for $S_R = 2.5$ across the channel at $x = 5h$, obtained with grid M1, compared to the analytic solution.

In Figure 11, the position of the free surface that is visualized by the contour $\alpha = 0.5$ is shown for $S_R = 2.5$ for the grids that are defined in Table 1. Some small surface oscillations are observed in the results from the two finer grids M2 and M3 in the region $x/h < 2$. As reference, Comminal et al. [24] reported similar behavior for their corresponding simulations. In their simulations, small self-sustained surface oscillations appeared at a certain level of recoverable shear, in their case $S_R > 1.5$ for the Oldroyd-B fluid, and for sufficiently high resolution. Similar observations were also made for simulations with the Giesekus model in their study. The oscillations were damped out further downstream and they did not influence the calculated swell ratio h_{max}/h. Similar characteristics are observed in the current work, as in Figure 11. Comminal et al. attributed the oscillations to numerical difficulties around the free surface due to the nature of the non-sharp interface in the VOF method.

In the light of this discussion, the results that are obtained on grid M2 and M3 are in relatively good agreement. This is particularly true far downstream of the channel exit, where the free surface produced by M2 and M3 are very close, while M1 produces a different result. The results indicate that M2 is of sufficient spatial resolution to predict the swell ratio S_r. Therefore, the results presented in the remainder of this section have been obtained while using grid M2. In addition, the simulation with grid M2 has been repeated with half the time step size, such that $U\Delta t/\Delta x_{\min} = 0.005$, which produced equivalent results.

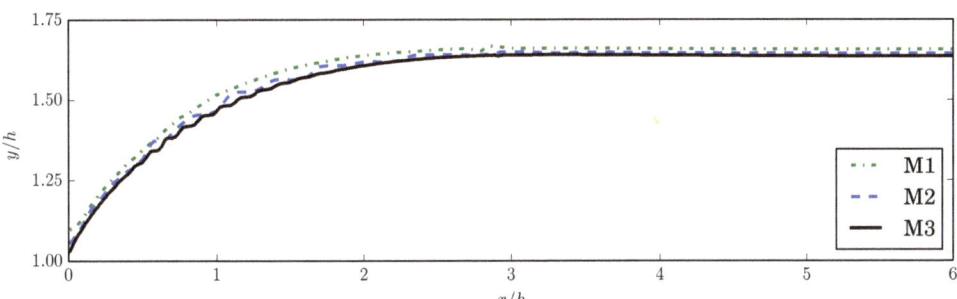

Figure 11. Free surface position in the expansion zone of the die swell geometry, simulated for grids M1, M2, and M3 for $S_R = 2.5$.

The simulations are repeated with grid M2 for $S_R \in \{1.0, 1.5, 2.0, 2.5\}$. In Figure 12, the contour $\alpha = 0.5$ is shown for $S_R = 1$ at different times. Figure 13 shows the corresponding results for $S_R = 2.5$. Similar behavior is observed in both cases. The emerging viscoelastic fluid increases in diameter directly upon exiting the channel. The magnitude of the increase is strongly influenced by the level of elasticity, as qualitatively predicted by Tanners theory. At a certain distance from the channel exit, approximately for $x/h > 3$, the emerging viscoelastic fluid reaches a terminal diameter.

The swell ratio is calculated for the simulations as $S_r = h_{max}/h$, where h_{max} is taken as the position of the free surface at $x/10$. The calculated swell ratios that were obtained with the proposed method are compared to available numerical data from the literature. The data used are the original FEM simulations reported by Crochet and Keunings [11], the marker-and-cell simulations by Tomé et al. [14], the pseudo-VOF simulation by Habla et al. [22], and the three different VOF methods used by Comminal et al. [24]. Table 2 briefly summarizes the data. For a full description of the cases and numerical methods, the reader is referred to the respective studies. However, a brief discussion concerning the differences between the results is given below.

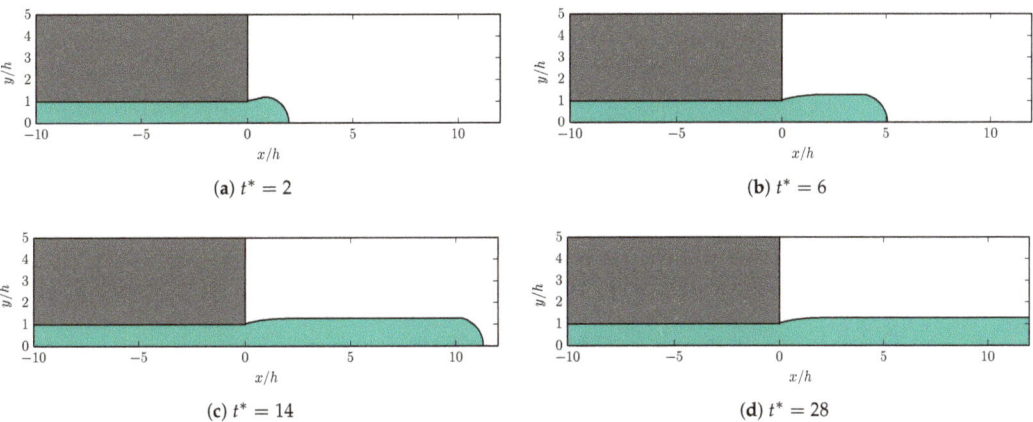

Figure 12. Snapshots of die swell simulation with $S_R = 1.0$, interface between viscoelastic phase (green) and Newtonian phase (white) visualized by $\alpha = 0.5$.

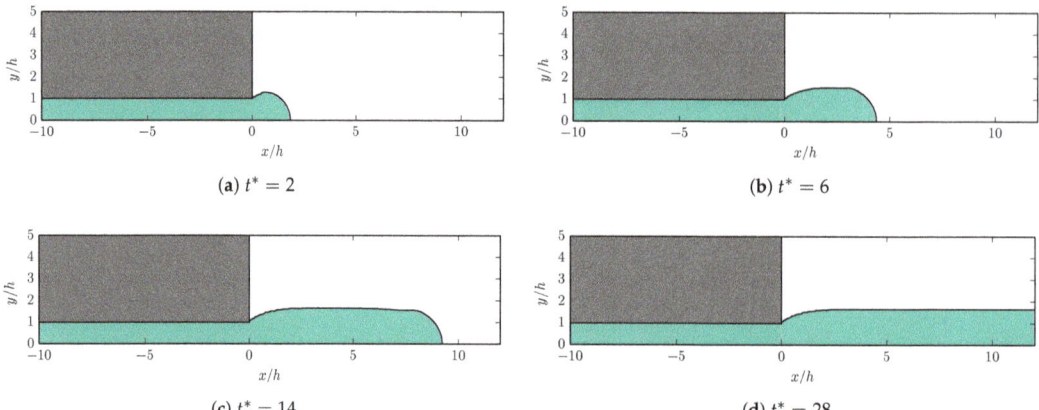

Figure 13. Snapshots of die swell simulation with $S_R = 2.5$, interface between the viscoelastic phase (green) and Newtonian phase (white) visualized by $\alpha = 0.5$.

Table 2. Summary of numerical data as compared for the die swell flow.

Work	Method	Re
Current work	VOF, Lagrangian-Eulerian	0.5
Crochet & Keunings [11]	Mixed FEM	0
Tomé et al. [14]	GENSMAC	0.5
Habla et al. [22]	pseudo-VOF	0.5
Comminal et al. [24] (CCU)	VOF, Geometric scheme	0
Comminal et al. [24] (HRIC)	VOF, Algebraic scheme	0
Comminal et al. [24] (RheoTool)	VOF, Algebraic scheme (MULES)	0.01

The calculated swell ratios as a function of the recoverable shear for the flows simulated with the proposed method are compared to the results from the literature in Figure 14. It is remarked that Comminal et al. employed a different definition of the Weissenberg number and swell ratio. However, all of the data that are presented in Figure 14 have been adopted to the definition used in the current work.

Indeed, a certain spread among the data is observed. The differences are not unexpected, though, as there are slight differences between how the different data have been obtained. This is in terms of the numerical method used as well as the simulation setup. Furthermore, Comminal et al. even obtained different results while using three slightly different numerical methods for exactly the same flow.

Crochet and Keunings and Tomé et al. did not consider a fluid surrounding the viscoelastic phase, as the per construction of their numerical methods. Comminal et al. used a similar definition of the surrounding fluid as in the current work, but with larger density. Habla et al. stated that the surrounding fluid is treated as air with $\mu_{\text{air}} \to 0$ and $\rho_{\text{air}} \to 0$, but the magnitudes were not reported. Furthermore, the value of viscosity ratio of the viscoelastic fluid β varied between their simulations. This is in contrast to a constant $\beta = 1/9$, which has been used in the other simulations included in this work.

Given the above discussion, a certain variance in the data is to be expected. However, the different results do follow the same general trend, including the swell ratios predicted with the proposed Lagrangian–Eulerian method.

Another aspect is that of grid dependence. The simulation performed by Crochet and Keunings had a grid resolution which, by today's standards, was very coarse. They used six triangular finite elements across the half-width of the channel. Therefore, the resulting swell ratios should be treated with caution when comparing to more recent results. Tomé et al. used a uniform grid of $\Delta x/h = 1/10$ which is on the order of the coarse grid M1.

Habla et al. reported the number of control volumes to be 4165, but did not specify the local cell sizes in detail. The number of control volumes is again comparable to grid M1 in the current work. Comminal et al. did conduct a grid dependence study with three different grids. Their finest cells were closer to the grids that were used in this work as compared to the other authors. Therefore, could be expected for the results reported in this work to be closest to the results of Comminal et al. given the similarities in grid resolution as well as the use of the VOF method. Their computed swell ratios are generally larger than the other literature results, with the exception of those from Crochet and Keunings. The same is also true for the swell ratios that were computed with the proposed method, particularly for increasing S_R.

In conclusion, while there is a spread between the data that are produced in different numerical studies, they follow the same general trend. The results that were obtained with the method proposed in this work follow the same trend and with the swell ratios comparable to the earlier published works.

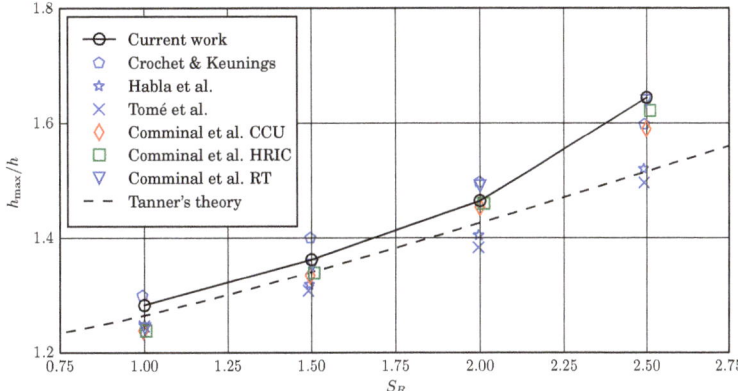

Figure 14. Simulated swell ratio compared to data from the literature.

4.3. Jet Buckling

The simulation of jet buckling is a common flow case for testing the capabilities of numerical methods for free surface flow. The phenomenon has been numerically investigated for different viscoelastic constitutive models by several authors [14–18,20,26].

The phenomenon occurs for a fluid jet that flows onto a rigid plate under certain conditions. Cruickshank [50] proposed a condition that is based on experimental and theoretical observations, stating that a planar Newtonian jet will buckle if $H/D > 3\pi$ and $Re < 0.56$, where H is the distance from the inlet to the plate, D the inlet width, and $Re = \rho U D / \mu$, where U is the inlet velocity. A modified yet approximate condition, based on numerical investigation, was later proposed by Tomé and McKee, stating that buckling should occur if

$$Re^2 \leq \frac{1}{\pi} \frac{(H/D)^{2.6} - 8.8^{2.6}}{(H/D)^{2.6}}. \tag{34}$$

It is remarked that, in simulations, numerical round-off errors are responsible for triggering the buckling, since the flow setup itself is actually symmetric.

Figure 15 shows a schematic of the fluid jet buckling domain that is used for simulation with the Lagrangian–Eulerian method proposed in this work. The domain is 50 mm wide and 100 mm high. An inlet of width $D = 5$ mm is located at the center of the upper boundary. At the inlet, a uniform velocity $U = 0.5$ m/s is imposed. The upper boundaries to the left and right, respectively, are outlets with the Dirichlet pressure condition $p = 0$. The remaining domain boundaries are treated as walls by imposing the no-slip condition. Gravity is acting in the negative y-direction with $g = 9.81$ m/s^2.

The viscoelastic fluid is a single-mode Oldroyd-B fluid with the viscosity ratio $\beta = 0.1$, density $\rho = 1000\,\text{kg/m}^3$, and the total viscosity $\eta_t = 10\,\text{Pa}\cdot\text{s}$. The surrounding Newtonian viscosity is air with viscosity $\mu_{\text{air}} = 1.8205 \times 10^{-5}\,\text{Pa}\cdot\text{s}$ and density $\rho_{\text{air}} = 1.204\,\text{kg/m}^3$.

The Reynolds number for the flow is defined as $\text{Re} = \rho U D/(\mu + \eta)$, which, for the chosen parameters, yields $\text{Re} = 0.25$. A Weissenberg number is defined as $\text{Wi} = \lambda U/D$. The flow is simulated for the Oldroyd-B fluid with $\lambda = 0.1\,\text{s}$, corresponding $\text{Wi} = 10$. As reference, a simulation is also performed with a Newtonian fluid, for which $\beta = \text{Wi} = 0$.

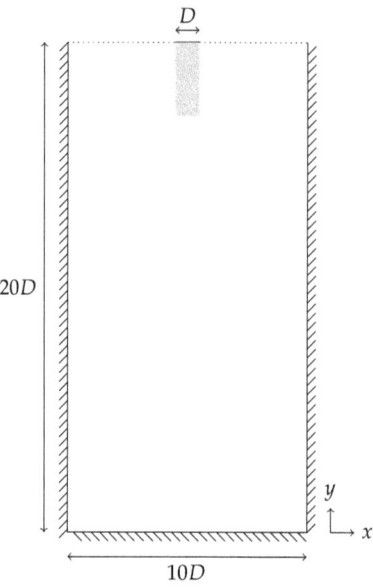

Figure 15. Viscoelastic fluid buckling domain.

In the viscoelastic jet buckling simulations that were performed in this work, the octree grid is adaptively refined with two levels around the viscoelastic fluid phase. As the simulation progresses, the grid is therefore updated and the cells are refined and coarsened, where needed. A grid dependence study is performed while using three adaptive grids of increasing resolution. The grids have the respective base cell sizes $\Delta x_{\text{base}} = D/2, D/4, D/8$, which correspond to the finest cell sizes $\Delta x_{\text{min}} = D/8, D/16, D/32$. The simulations are performed with a constant time step Δt satisfying $\Delta t U/\Delta x_{\text{min}} = 0.08$.

The grid dependence is evaluated with respect to two simulation properties. Firstly, the free surface $\alpha = 0.5$ is compared for the three adaptive grids. The first normal stress difference $N_1 = \tau_{yy} - \tau_{xx}$ along the jet at $x = 5D$ is also compared. Because the buckling itself is triggered by numerical round-off errors, the exact conditions at which the buckling starts in the simulations may vary between grids. Therefore, the grid study comparison is made at a simulation time before buckling is initiated.

In Figure 16, the free surface is compared for the three grids at the simulation times $t = 130\,\text{ms}$ and $t = 140\,\text{ms}$. The results obtained with the three grids are in relatively good agreement, particularly those that are obtained with the two finer grids. In Figure 17, the first normal stress differences along the jet are compared at the same simulation times. These results indicate even more strongly that the two finer grids produce stresses that are very close, while a slight discrepancy for the coarsest grid is observed. The remainder of the results reported in this section have been obtained while using $\Delta x_{\text{min}} = D/16$.

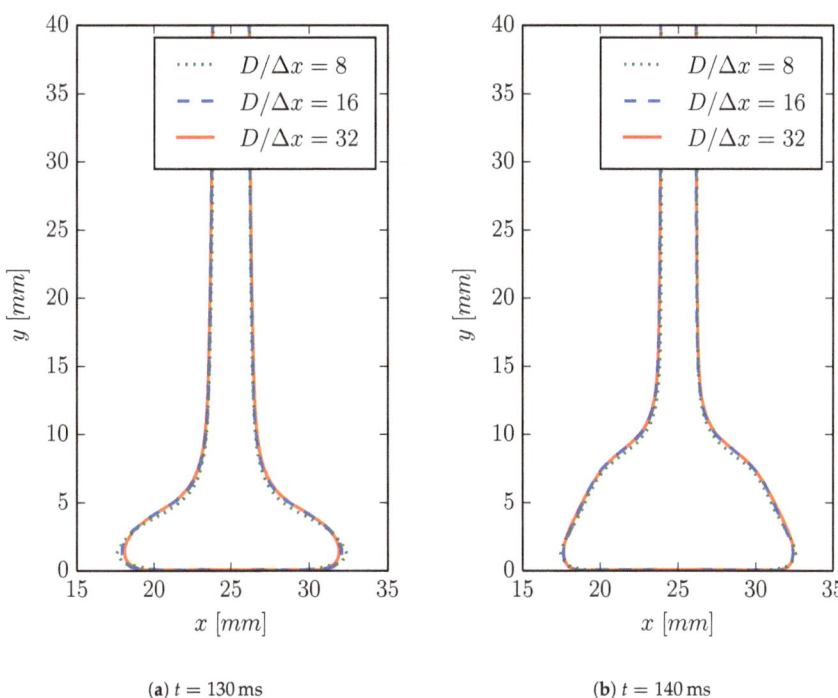

Figure 16. Isosurface $\alpha = 0.5$ in viscoelastic jet buckling simulation with $\lambda = 0.1$ s and Wi = 10 obtained with the finest cell sizes $\Delta x_{min} = D/8, D/16, D/32$.

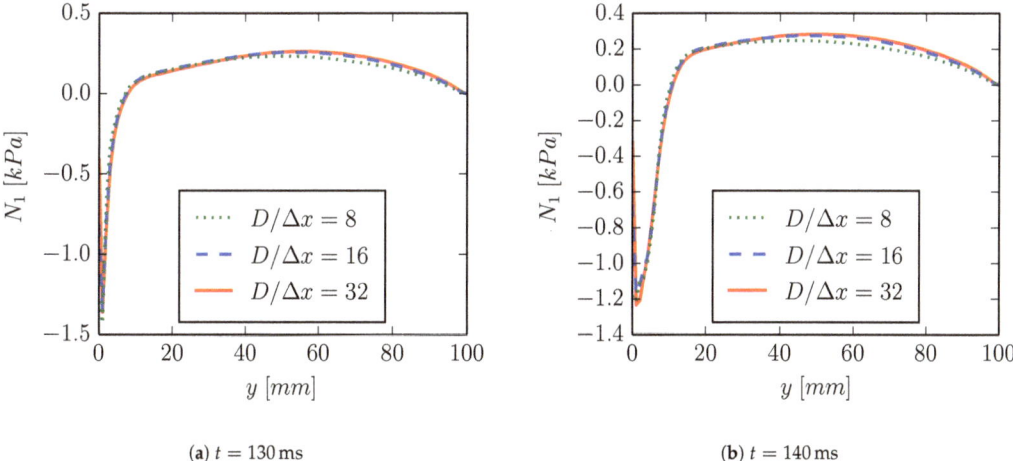

Figure 17. First normal stress difference $N_1 = \tau_{yy} - \tau_{xx}$ along the line $x/D = 5$ in viscoelastic jet buckling simulation with $\lambda = 0.1$ s and Wi = 10 obtained with the finest cell sizes $\Delta x_{min} = D/8, D/16, D/32$.

In Figures 18 and 19, a series of snapshots from the simulation of the Newtonian and the viscoelastic jets are shown, respectively. For the Newtonian case, only a very slight tendency to buckle is observed as a slight asymmetry in the fluid on the rigid surface. However, for the viscoelastic fluid, the buckling phenomenon is very apparent. When the viscoelastic jet impacts the rigid surface, the material first builds upwards until, inevitably, the buckling is initiated. The results demonstrate the strong influence of elasticity to the fluid jet buckling phenomenon. Furthermore, it confirms the capability of the proposed method to capture the viscoelastic effects and predict the phenomenon.

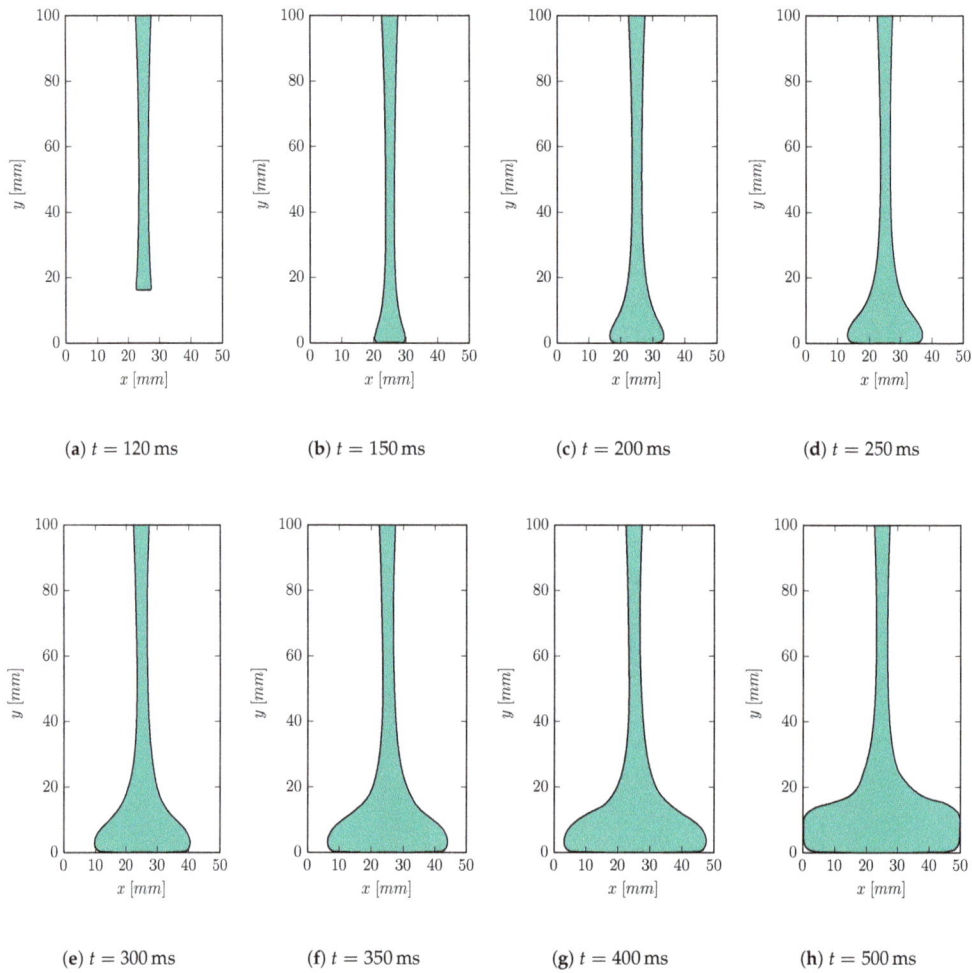

Figure 18. Newtonian jet buckling simulation.

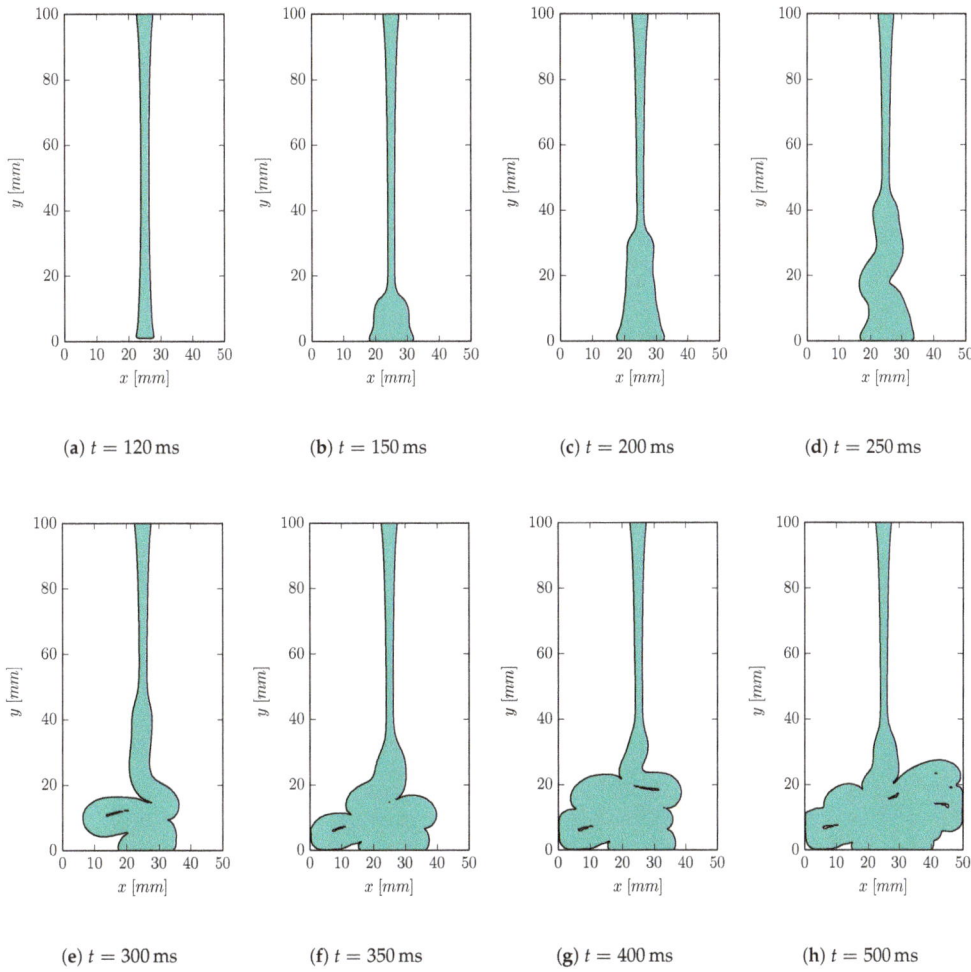

Figure 19. Viscoelastic jet buckling simulation with $\lambda = 0.1$ s.

5. Conclusions

In this work, a new Lagrangian–Eulerian method for the simulation of viscoelastic free surface flow has been proposed. The method was developed from a previously proposed method, in which the fluid momentum and continuity equations were solved on a stationary Eulerian octree grid, while the viscoelastic constitutive equation was solved along the trajectories of Lagrangian nodes which were convected by the flow. The main improvements in this work was the introduction of a backwards-tracking procedure for solving the viscoelastic constitutive equation, as well an extension to the simulation of viscoelastic free surface flow with the volume of fluid method.

The backwards-tracking procedure allowed for the storage of the viscoelastic stresses in a structured arrangement. Consequently, the unstructured interpolation of stresses as well as the addition and deletion of Lagrangian nodes, which was required in the previous method, was eliminated in favor of a significant reduction of the computational cost and increased robustness. Furthermore, no additional stabilization method, such as both sides diffusion or reformulation of the constitutive equation, was employed for any of the flows studied in this work.

The proposed method was validated with analytic solutions for a planar Poiseuille flow of Oldroyd-B fluids. The transient, as well steady flow, solutions were in excellent agreement with analytic solutions and the steady solution converged to the analytic solution with second order accuracy. Furthermore, the proposed method was evaluated for two types of viscoelastic free surface flow. The viscoelastic die swell effect was simulated with the Oldroyd-B model and the predicted swell ratios were in good agreement with the numerical data from the literature. Planar jet buckling of a highly viscoelastic fluid was also simulated, demonstrating the capability of the method for an additional flow case as well as the use of adaptive grid refinements.

The computational performance was not subject to detailed study in this work. However, as per the construction of the new method, the computational cost for the viscoelastic stress calculation was significantly reduced when compared to the previous Lagrangian–Eulerian method. Furthermore, it has previously been shown that the Lagrangian formulation of the constitutive equation is highly suitable for parallel calculations, including for GPU-acceleration. This also applies for the new method.

In conclusion, the new method is an important step in the development of efficient and useful tools for simulating industrial processes involving viscoelastic fluids. In future research, the proposed Lagrangian–Eulerian method will be used for the numerical investigation of viscoelastic free surface flows with moving objects, including for viscoelastic adhesive application and parts assembly, as well as for additive manufacturing processes.

Author Contributions: Conceptualization, S.I. and A.M.; Methodology, S.I. and A.M.; Software, S.I. and A.M.; Validation, S.I.; Writing—original draft preparation, S.I.; Writing—review and editing, A.M., R.K., F.E.; Supervision, A.M., R.K., F.E.; Funding acquisition, F.E. All authors have read and agreed to the published version of the manuscript.

Funding: This research has been partly carried out in a Centre for Additive Manufacturing—Metal (CAM2) in a joint project financed by Swedish Governmental Agency of Innovation Systems (Vinnova), coordinated by Chalmers University of Technology. The work has also been supported in part by Vinnova through the FFI Sustainable Production Technology program and in part by the Chalmers Area of Advance Production. The support is gratefully acknowledged.

Conflicts of Interest: The authors declare no conflict of interest. The funders had no role in the design of the study; in the collection, analyses, or interpretation of data; in the writing of the manuscript, or in the decision to publish the results.

Appendix A. Oldroyd-B Fluid Fully Developed Channel Flow

Assuming steady two-dimensional, fully developed channel flow, the constitutive equation reduces do

$$\tau_{xy} = \eta \frac{\partial u}{\partial y}, \tag{A1}$$

$$\tau_{xx} - 2\lambda \tau_{xy} \frac{\partial u}{\partial y} = 0, \tag{A2}$$

$$\tau_{yy} = 0. \tag{A3}$$

Insertion of τ_{xy} from (A1) into (A2) gives

$$\tau_{xx} = 2\lambda \eta \left(\frac{\partial u}{\partial y}\right)^2, \tag{A4}$$

The momentum equation in the flow direction reduces to

$$\frac{\partial p}{\partial x} = \mu \frac{\partial^2 u}{\partial y^2} + \frac{\partial \tau_{xy}}{\partial y}, \tag{A5}$$

which by insertion of τ_{xy} from (A1) becomes

$$\frac{\partial p}{\partial x} = \mu \frac{\partial^2 u}{\partial y^2} + \frac{\partial}{\partial y}\left(\eta \frac{\partial u}{\partial y}\right) = (\mu + \eta)\frac{\partial^2 u}{\partial y^2}, \tag{A6}$$

Integration of (A6) with respect to y once yields

$$y\frac{\partial p}{\partial x} = (\mu + \eta)\frac{\partial u}{\partial y} + C_1, \tag{A7}$$

where C_1 is an integration constant. Due to the symmetry condition $\frac{\partial u}{\partial y}|_{y=0} = 0$, which by insertion of $y = 0$ into (A7) leads to $C_1 = 0$. Inserting $C_1 = 0$ to (A7) and integrating in y again yields

$$\frac{y^2}{2}\frac{\partial p}{\partial x} = (\mu + \eta)u + C_2 \tag{A8}$$

where C_2 is an integration constant. Using the no-slip condition $u_{y=h} = 0$ gives

$$C_2 = \frac{h^2}{2}\frac{\partial p}{\partial x}, \tag{A9}$$

which inserted to (A8) after some rearrangement gives

$$u(y) = -\frac{1}{2(\mu + \eta)}\frac{\partial p}{\partial x}\left(h^2 - y^2\right), \tag{A10}$$

or, simply,

$$u(y) = A(h^2 - y^2), \tag{A11}$$

where a is a constant.

Now, the mean velocity U can be calculated as

$$U = \frac{1}{h}\int_0^h u(y)dy = \frac{A}{h}\int_0^h (h^2 - y^2)dy = \ldots = \frac{2}{3}Ah^2, \tag{A12}$$

such that $A = 3U/(2h^2)$. The velocity can then be expressed in terms of the mean velocity as

$$u(y) = \frac{3U}{2h^2}(h^2 - y^2) = \frac{3U}{2}\left(1 - \frac{y^2}{h^2}\right). \tag{A13}$$

Hence, the velocity gradient reads

$$\frac{\partial u}{\partial y} = -\frac{3Uy}{h^2}, \tag{A14}$$

which gives the final expressions for τ_{xy} and τ_{xx}.

References

1. Oliveira, P.; Pinho, F.; Pinto, G. Numerical simulation of non-linear elastic flows with a general collocated finite-volume method. *J. Non-Newton. Fluid Mech.* **1998**, *79*, 1–43. [CrossRef]
2. Alves, M.; Pinho, F.; Oliveira, P. The flow of viscoelastic fluids past a cylinder: Finite-volume high-resolution methods. *J. Non-Newton. Fluid Mech.* **2001**, *97*, 20–232. [CrossRef]
3. Alves, M.A.; Oliveira, P.J.; Pinho, F.T. Benchmark solutions for the flow of Oldroyd-B and PTT fluids in planar contractions. *J. Non-Newton. Fluid Mech.* **2003**, *110*, 45–75. [CrossRef]
4. Baaijens, H.P.; Peters, G.W.; Baaijens, F.P.; Meijer, H.E. Viscoelastic flow past a confined cylinder of a polyisobutylene solution. *J. Rheol.* **1995**, *39*, 1243–1277. [CrossRef]

5. Hulsen, M.A.; Fattal, R.; Kupferman, R. Flow of viscoelastic fluids past a cylinder at high Weissenberg number: Stabilized simulations using matrix logarithms. *J. Non-Newton. Fluid Mech.* **2005**, *127*, 27–39. [CrossRef]
6. Rasmussen, H.; Hassager, O. Simulation of transient viscoelastic flow with second order time integration. *J. Non-Newton. Fluid Mech.* **1995**, *56*, 65–84. [CrossRef]
7. Harlen, O.; Rallison, J.; Szabo, P. A split Lagrangian-Eulerian method for simulating transient viscoelastic flows. *J. Non-Newton. Fluid Mech.* **1995**, *60*, 81–104. [CrossRef]
8. Halin, P.; Lielens, G.; Keunings, R.; Legat, V. The Lagrangian particle method for macroscopic and micro–macro viscoelastic flow computations Dedicated to Professor Marcel J. Crochet on the occasion of his 60th birthday. *J. Non-Newton. Fluid Mech.* **1998**, *79*, 387–403. [CrossRef]
9. Gallez, X.; Halin, P.; Lielens, G.; Keunings, R.; Legat, V. The adaptive Lagrangian particle method for macroscopic and micro–macro computations of time-dependent viscoelastic flows. *Comput. Methods Appl. Mech. Eng.* **1999**, *180*, 345–364. [CrossRef]
10. Wapperom, P.; Keunings, R.; Legat, V. The backward-tracking Lagrangian particle method for transient viscoelastic flows. *J. Non-Newton. Fluid Mech.* **2000**, *91*, 273–295. [CrossRef]
11. Crochet, M.; Keunings, R. Finite element analysis of die swell of a highly elastic fluid. *J. Non-Newton. Fluid Mech.* **1982**, *10*, 339–356. [CrossRef]
12. Balemans, C.; Hulsen, M.; Anderson, P. Sintering of Two Viscoelastic Particles: A Computational Approach. *Appl. Sci.* **2017**, *7*, 516. [CrossRef]
13. Spanjaards, M.; Hulsen, M.; Anderson, P. Transient 3D finite element method for predicting extrudate swell of domains containing sharp edges. *J. Non-Newton. Fluid Mech.* **2019**, *270*, 79–95. [CrossRef]
14. Tomé, M.; Mangiavacchi, N.; Cuminato, J.; Castelo, A.; McKee, S. A finite difference technique for simulating unsteady viscoelastic free surface flows. *J. Non-Newton. Fluid Mech.* **2002**, *106*, 61–106. [CrossRef]
15. de Paulo, G.; Tomé, M.; McKee, S. A marker-and-cell approach to viscoelastic free surface flows using the PTT model. *J. Non-Newton. Fluid Mech.* **2007**, *147*, 149–174. [CrossRef]
16. Tomé, M.; Castelo, A.; Ferreira, V.; McKee, S. A finite difference technique for solving the Oldroyd-B model for 3D-unsteady free surface flows. *J. Non-Newton. Fluid Mech.* **2008**, *154*, 179–206. [CrossRef]
17. Oishi, C.M.; Tomé, M.F.; Cuminato, J.A.; McKee, S. An implicit technique for solving 3D low Reynolds number moving free surface flows. *J. Comput. Phys.* **2008**, *227*, 7446–7468. [CrossRef]
18. Tomé, M.; Paulo, G.; Pinho, F.; Alves, M. Numerical solution of the PTT constitutive equation for unsteady three-dimensional free surface flows. *J. Non-Newton. Fluid Mech.* **2010**, *165*, 247–262. [CrossRef]
19. Oishi, C.; Martins, F.; Tomé, M.; Cuminato, J.; McKee, S. Numerical solution of the eXtended Pom-Pom model for viscoelastic free surface flows. *J. Non-Newton. Fluid Mech.* **2011**, *166*, 165–179. [CrossRef]
20. Tomé, M.; Castelo, A.; Afonso, A.; Alves, M.; Pinho, F. Application of the log-conformation tensor to three-dimensional time-dependent free surface flows. *J. Non-Newton. Fluid Mech.* **2012**, *175–176*, 44–54. [CrossRef]
21. Izbassarov, D.; Muradoglu, M. A front-tracking method for computational modeling of viscoelastic two-phase flow systems. *J. Non-Newton. Fluid Mech.* **2015**, *223*, 122–140. [CrossRef]
22. Habla, F.; Marschall, H.; Hinrichsen, O.; Dietsche, L.; Jasak, H.; Favero, J.L. Numerical simulation of viscoelastic two-phase flows using openFOAM®. *Chem. Eng. Sci.* **2011**, *66*, 5487–5496. [CrossRef]
23. OpenFOAM. Available online: https://www.openfoam.org/ (accessed on 18 December 2020).
24. Comminal, R.; Pimenta, F.; Hattel, J.H.; Alves, M.A.; Spangenberg, J. Numerical simulation of the planar extrudate swell of pseudoplastic and viscoelastic fluids with the streamfunction and the VOF methods. *J. Non-Newton. Fluid Mech.* **2018**, *252*, 1–18. [CrossRef]
25. RheoTool. Available online: https://github.com/fppimenta/rheoTool (accessed on 18 December 2020).
26. Bonito, A.; Picasso, M.; Laso, M. Numerical simulation of 3D viscoelastic flows with free surfaces. *J. Comput. Phys.* **2006**, *215*, 691–716. [CrossRef]
27. Ingelsten, S.; Mark, A.; Edelvik, F. A Lagrangian-Eulerian framework for simulation of transient viscoelastic fluid flow. *J. Non-Newton. Fluid Mech.* **2019**, *266*, 20–32. [CrossRef]
28. Ingelsten, S.; Mark, A.; Jareteg, K.; Kádár, R.; Edelvik, F. Computationally efficient viscoelastic flow simulation using a Lagrangian-Eulerian method and GPU-acceleration. *J. Non-Newton. Fluid Mech.* **2020**, 104264. [CrossRef]
29. Versteeg, H.; Malalasekera, W. *An Introduction to Computational Fluid Dynamics: The Finite Volume Method*; Pearson Education Limited: Harlow, UK, 2007.
30. Mark, A.; van Wachem, B.G.M. Derivation and validation of a novel implicit second-order accurate immersed boundary method. *J. Comput. Phys.* **2008**, *227*, 6660–6680. [CrossRef]
31. Mark, A.; Rundqvist, R.; Edelvik, F. Comparison Between Different Immersed Boundary Conditions for Simulation of Complex Fluid Flows. *Fluid Dyn. Mater. Process.* **2011**, *7*, 241–258.
32. Doormaal, J.P.V.; Raithby, G.D. Enhancements of the simple method for predicting incompressible fluid flows. *Numer. Heat Transf.* **1984**, *7*, 147–163. [CrossRef]
33. Tahir-Kheli, R. *Ordinary Differential Equations. [Electronic Resource]: Mathematical Tools for Physicists*; Springer International Publishing: Cham, Switzerland, 2018.

34. Tryggvason, G.; Scardovelli, R.; Zaleski, S. *Direct Numerical Simulations of Gas-Liquid Multiphase Flows*; Cambridge University Press: Cambridge, UK 2011.
35. Ubbink, O.; Issa, R. A Method for Capturing Sharp Fluid Interfaces on Arbitrary Meshes. *J. Comput. Phys.* **1999**, *153*, 26–50. [CrossRef]
36. Niethammer, M.; Brenn, G.; Marschall, H.; Bothe, D. An extended volume of fluid method and its application to single bubbles rising in a viscoelastic liquid. *J. Comput. Phys.* **2019**, *387*, 326–355. [CrossRef]
37. IPS IBOFlow. Available online: http://ipsiboflow.com (accessed on 18 December 2020).
38. Mark, A.; Svenning, E.; Edelvik, F. An immersed boundary method for simulation of flow with heat transfer. *Int. J. Heat Mass Transf.* **2013**, *56*, 424–435. [CrossRef]
39. Andersson, T.; Nowak, D.; Johnson, T.; Mark, A.; Edelvik, F.; Küfer, K.H. Multiobjective Optimization of a Heat-Sink Design Using the Sandwiching Algorithm and an Immersed Boundary Conjugate Heat Transfer Solver. *J. Heat Transf.* **2018**, *140*. 102002. [CrossRef]
40. Nowak, D.; Johnson, T.; Mark, A.; Ireholm, C.; Pezzotti, F.; Erhardsson, L.; Ståhlberg, D.; Edelvik, F.; Küfer, K.H. Multicriteria Optimization of an Oven With a Novel ε-Constraint-Based Sandwiching Method. *J. Heat Transf.* **2020**, *143*. 012101. [CrossRef]
41. Svenning, E.; Mark, A.; Edelvik, F. Simulation of a highly elastic structure interacting with a two-phase flow. *J. Math. Ind.* **2014**, *4*, 7. [CrossRef]
42. Edelvik, F.; Mark, A.; Karlsson, N.; Johnson, T.; Carlson, J. Math-Based Algorithms and Software for Virtual Product Realization Implemented in Automotive Paint Shops. In *Math for the Digital Factory*; Ghezzi, L., Hömberg, D., Landry, C., Eds.; Springer: Berlin/Heidelberg, Germany, 2017; pp. 231–251.
43. Mark, A.; Bohlin, R.; Segerdahl, D.; Edelvik, F.; Carlson, J.S. Optimisation of robotised sealing stations in paint shops by process simulation and automatic path planning. *Int. J. Manuf. Res.* **2014**, *9*, 4–26. [CrossRef]
44. Svensson, M.; Mark, A.; Edelvik, F.; Kressin, J.; Bohlin, R.; Segerdahl, D.; Carlson, J.S.; Wahlborg, P.J.; Sundbäck, M. Process Simulation and Automatic Path Planning of Adhesive Joining. *Procedia CIRP* **2016**, *44*, 298–303. [CrossRef]
45. Göhl, J.; Markstedt, K.; Mark, A.; Håkansson, K.; Gatenholm, P.; Edelvik, F. Simulations of 3D bioprinting: Predicting bioprintability of nanofibrillar inks. *Biofabriaction* **2018**, *10*. [CrossRef]
46. Xue, S.C.; Tanner, R.; Phan-Thien, N. Numerical modelling of transient viscoelastic flows. *J. Non-Newton. Fluid Mech.* **2004**, *123*, 33–58. [CrossRef]
47. Barnes, H.A.; Hutton, J.F.; Walters, K. *An Introduction to Rheology*; Rheology Series; Elsevier: New York, NY, USA, 1989; Volume 3.
48. Tanner, R.I. A theory of die-swell. *J. Polym. Sci. Part—Polym. Phys.* **1970**, *8*, 2067–2078. [CrossRef]
49. Tanner, R.I. A theory of die-swell revisited. *J. Non-Newton. Fluid Mech.* **2005**, *129*, 85–87. [CrossRef]
50. Cruickshank, J.O. Low-Reynolds-number instabilities in stagnating jet flows. *J. Fluid Mech.* **1988**, *193*, 111–127. [CrossRef]

Article

Numerical Simulation of KBKZ Integral Constitutive Equations in Hierarchical Grids

Juliana Bertoco [1,*], Manoel S. B. de Araújo [2], Rosalía T. Leiva [1], Hugo A. C. Sánchez [1] and Antonio Castelo [1]

1 Instituto de Ciências Matemáticas e de Computação, Universidade de São Paulo-USP, São Carlos 13566-590, SP, Brazil; rosalia.taboada@usp.br (R.T.L.); hugo_acs@icmc.usp.br (H.A.C.S.); castelo@icmc.usp.br (A.C.)
2 Instituto de Ciências Exatas e Naturais, Universidade Federal do Pará-UFPA, Belém 66075-110, PA, Brazil; silvino@ufpa.br
* Correspondence: jubertoco@alumni.usp.br

Abstract: In this work, we present the implementation and verification of HiGTree-HiGFlow solver (see for numerical simulation of the KBKZ integral constitutive equation. The numerical method proposed herein is a finite difference technique using tree-based grids. The advantage of using hierarchical grids is that they allow us to achieve great accuracy in local mesh refinements. A moving least squares (MLS) interpolation technique is used to adapt the discretization stencil near the interfaces between grid elements of different sizes. The momentum and mass conservation equations are solved by an implicit method and the Chorin projection method is used for decoupling the velocity and pressure. The Finger tensor is calculated using the deformation fields method and a three-node quadrature formula is used to derive an expression for the integral tensor. The results of velocity and stress fields in channel and contraction-flow problems obtained in our simulations show good agreement with numerical and experimental results found in the literature.

Keywords: KBKZ integral constitutive equation; tree-based hierarchical grids; deformation fields

1. Introduction

Over the years, several software programs have been developed to solve problems involving complex viscoelastic fluid flows. Due to a lack of generality, some challenges can arise when trying to solve problems with specific characteristics. In most works that develop numerical methods for simulating viscoelastic flows, the constitutive equations are approximated by differential equations, such as the Oldroyd-B [1–3], Upper-Convected-Maxwell (UCM) [4,5], Phan–Thien–Tanner [6,7], eXtended Pom-Pom [8,9] models, among others. However, advances in computational resources have motivated researchers to consider more sophisticated rheological models that are expressed in integral form instead of differential equations. In this sense, integral models allow a better approximation of the behavior of viscoelastic fluids. However, they require a greater computational effort and this is because, at each moment of the simulation, it is necessary to store and access the history of the entire deformation of the fluid (since it previously started to be deformed). Among the integral models that we found in the literature, the constitutive equation KBKZ-PSM has been considered by many researchers who study numerical methods for this kind of fluid. A detailed discussion of the importance of the KBKZ-PSM integral constitutive model and the development of numerical techniques to approximate integral models can be found in the works of Tanner [10] and Mitsoulis [11]. The vast majority of problems using the KBKZ-PSM integral model involve confined flows, such as channel-flows [12,13] and flows in abrupt contractions [3,14,15]. Flow problems possessing free surface(s) have also been considered by some researchers. More interesting flow problems that involve transient free surface(s) and integral models are the filament stretching [16] and a numerical study of the die swell phenomenon [17–21].

On the other hand, a numerical solution of partial differential equations in general grids has been questioned by many researchers in recent decades. Many schemes try to combine efficiency and simplicity with the flexibility of unstructured mesh networks. A major advantage of using such meshes is the ability to refine the mesh locally, improving accuracy in specific regions without dramatically increasing the number of unknowns. Among all possible ways of discretizing the spatial domain (simplified meshes, curvilinear meshes, among others), hierarchical meshes based on Cartesian trees are a common choice. They allow the development of finite difference methods, without the hassle of mapping and transforming distorted elements or dealing with general and complicated stencils, as in non-Cartesian grids. Since flows are generally computed on facets aligned with the Cartesian axis, numerical schemes are generally simpler to derive. However, these facets are generally shared by different numbers of elements on each side, which is the main challenge in implementing numerical methods. Different techniques to deal with this problem have been developed in the literature, most of them restricted to quadtree meshes (in 2D) or octree (3D) meshes, which are special cases of hierarchical grids represented by data structures quadtree/octree. Despite this restriction, these tree-based data structures are generally good enough and still an adequate choice for adaptive grids and moving borders [22]. Thus, we intend to implement in the present work the transient KBKZ-PSM model through a method of finite differences in hierarchical meshes that employ interpolations using the moving least squares (MLS) method [23]. The developed numerical method is verified by using mesh refinement in channel flow and we show results from the simulation of the 4:1 contraction problem using a KBKZ fluid. Our results are compared to the ones obtained using the OpenFOAM system [24], which uses finite volumes in the discretization of Navier–Stokes equations. We used the OpenFOAM v2006 version to implement the equations with the finite volume method.

2. Governing Equations

The governing equations for transient, isothermal and incompressible flows are the mass conservation and the equation of motion, which, in dimensionless form, can be written as follows (for details, see Tomé et al. [21]):

$$\nabla \cdot \mathbf{v} = 0, \tag{1}$$

$$\frac{\partial \mathbf{v}}{\partial t} + \nabla \cdot (\mathbf{v}\mathbf{v}) = -\nabla p + \epsilon \nabla^2 \mathbf{v} + \nabla \cdot \mathbf{S} + \mathbf{F}. \tag{2}$$

Using the EVSS transformation [25], the extra-stress tensor τ is written as

$$\tau = \mathbf{S} + \epsilon\, \dot{\gamma}, \text{ where } \dot{\gamma} = \nabla \mathbf{v} + (\nabla \mathbf{v})^t \text{ and } \epsilon = \frac{c}{Re}; \; c > 0,$$

where S is a non-Newtonian tensor, $\mathbf{v}$ is the velocity field, p is the kinematic pressure and t is the time. In these equations, F represents the external forces, ϵ is a stability parameter (as shown in Araújo et al. [26]), $Re = \dfrac{\rho_0 UL}{\eta_0}$ is the Reynolds number, η_0 is the zero-shear-rate viscosity, ρ_0 is the fluid density and U and L are the velocity and length scales, respectively.

In this work, the rheological model that defines the behavior of fluid flow is the KBKZ-PSM [11] integral constitutive equation, which is shown below:

$$\tau(t) = \int_{-\infty}^{t} M(t-t') H(I_1, I_2) \mathbf{B}_{t'}(t) dt', \tag{3}$$

where $\mathbf{B}_{t'}(t)$ is the Finger tensor and M is the memory function, which adopts the following form:

$$M(t-t') = \sum_{k=1}^{m_1} \frac{a_k}{\lambda_k De} e^{-\frac{t-t'}{\lambda_k De}} \tag{4}$$

and H is the Papanastasiou–Scriven–Macosko [25] damping function, which is calculated using the following equation:

$$H(I_1, I_2) = \frac{\alpha}{\alpha - 3 + \beta I_1 + (1-\beta) I_2}. \tag{5}$$

The parameters λ_k, a_k, k are relaxation times, relaxation moduli and the number of relaxation modes, respectively. The quantities $I_1 = tr[\mathbf{B}_{t'}(t)]$ and $I_2 = \frac{1}{2}((I_1)^2 - tr[\mathbf{B}_{t'}^2(t)])$ are the first and second invariants of $\mathbf{B}_{t'}(t)$, respectively. The parameters $a_k, \lambda_k, \alpha, \beta$ are obtained from a curve fitting to the rheological properties of the fluid. $De = \lambda_{ref} \frac{U}{L}$ is the Deborah number, $\lambda_{ref} = \sum \frac{a_k \lambda_k^2}{a_k \lambda_k}$ is the average relaxation time and the zero-shear-rate viscosity is written as $\eta_0 = \sum a_k \lambda_k$.

3. Numerical Method

In this section, we present the methodology used in this work, the HiGTree/HiGFlow and the OpenFoam systems.

3.1. HiGTree/HiGFlow System

The HiGFlow system is a C language software, developed at ICMC-USP, which brings together a series of methods for the numerical simulation of flow of single-phase and multiphase fluids, using the finite difference technique. This system is being developed in a modular way, allowing new techniques and methods to be easily tested and added to the system. One feature is that the user chooses the dimension and the modules to be used in the program (such as single-phase, Newtonian, generalized Newtonian, viscoelastic) at compile time. In the same way, the user specifies the numerical techniques to be used in the input files: projection method, numerical scheme for the convective term, model of the constitutive equation for viscoelastic flows, in addition to the various parameters for simulation. In this work, all tests were performed in two dimensions (2D), and the following numerical techniques were chosen: an implicit Euler method to compute the velocity, the CUBISTA scheme to discretize the convective terms and an explicit Euler method for the convection of the Finger tensor.

On the other hand, the HiGTree system is responsible for creating the data structure, domains, linear and non-linear system solvers, as well as carrying out the interpolations schemes. Parallelization strategies are also implemented through the PETSc library (Portable, Extensible Toolkit for Scientific Computation), which contains a set of functions implementing the best-known methods for representing matrices, vectors and data in parallel, solution of linear systems with pre-conditioning, solution of linear and non-linear systems, ordinary differential equations, etc.

3.1.1. Hierarchical Grids

Equations (1) and (2) are approximated using finite differences in hierarchical Cartesian meshes. An illustrative representation of the mesh is given in Figure 1a and its structure of dependencies is illustrated in Figure 1b. In this data structure, each cell can be partitioned into distinct geometric shapes. Such generalization imposes difficulties in the numerical approximation in finite differences.

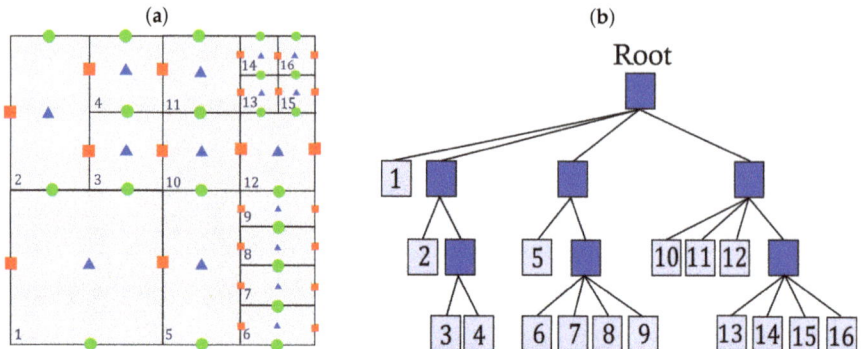

Figure 1. HiGTree data structure: (**a**) computational cell representation, (**b**) tree-based data structures.

For example, considering Figure 2, suppose that we are interested in approximating the second derivative in y direction centered on U_c. Using second-order finite differences, we have:

$$\frac{\partial^2 U_c}{\partial y^2} \approx \frac{1}{\delta y}(U_t - 2U_c + U_b); \tag{6}$$

We can notice that, for this case, U_b does not match known values in the mesh (recalling that the components of the velocity field are computed in the facet centers), but it can be calculated by interpolation using values of neighboring cells as shown in the following equation:

$$U_b = \sum_{k=1}^{V_b} w_k^b U_k; \tag{7}$$

The number of neighbors V_b is defined according to the imposed precision. The weights $w_k^b = w_k(\mathbf{x})$ are calculated using the moving least squares (MLS) method [23].

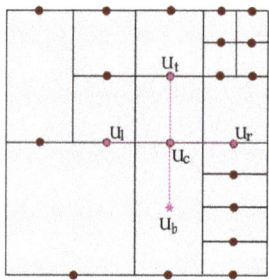

Figure 2. Finite difference 2nd-order stencil discretization.

3.1.2. Calculation of $\mathbf{v}(\mathbf{x}, t_{n+1})$ and $p(\mathbf{x}, t_{n+1})$

Upon discretizing Equation (2) in time using, for instance, a first-order explicit discretization, the idea of the incremental projection method is to use the newest previous pressure field, which yields an explicitly computed velocity field $\mathbf{v}^*$ that is not divergence-free, through the solution of the following equation:

$$\frac{\mathbf{v}^* - \mathbf{v}^n}{\delta t} + \mathbf{v}^n \cdot \nabla \mathbf{v}^n = -\nabla p^n + \epsilon \nabla^2 \mathbf{v}^n + \nabla \cdot \mathbf{S}^n + \mathbf{F}^n \tag{8}$$

The corrected velocity field can be computed from the decomposition itself:

$$\mathbf{v}^* = \mathbf{v}^{n+1} + \nabla \varphi, \tag{9}$$

where $\varphi = -\delta t(p^{n+1} - p^n)$, which is obtained by solving the Poisson equation:

$$\nabla^2 \varphi = \nabla \cdot \mathbf{v}^*, \tag{10}$$

with $n \cdot \nabla \varphi = 0$ on the boundaries $\partial \Omega$. Equation (10) can be easily derived by obtaining the divergence of Equation (9) (for more details, see [23,27]).

3.1.3. Calculation of the Extra-Stress Tensor $\tau(\mathbf{x}, t_{n+1})$

We follow the methodologies described in Tomé et al. [21] and Araújo et al. [26] to calculate the extra-stress tensor $\tau(\mathbf{x}, t_{n+1})$. The constitutive Equation (3) can be written as follows:

$$\tau(t_{n+1}) = \int_{-\infty}^{t-t_c} M(t_{n+1} - t') H(I_1, I_2) \mathbf{B}_{t'}(t_{n+1}) dt' \\ + \int_{t-t_c}^{t} M(t_{n+1} - t') H(I_1, I_2) \mathbf{B}_{t'}(t_{n+1}) dt', \tag{11}$$

where $t_c = t$ if $t < s_c$ or $t_c = s_c$ if $t \geq s_c$. The parameter s_c (s_c is a time interval) depends on the relaxation parameter λ_{ref}. This methodology is called s-approach and is described in more detail in Hulsen et al. [28].

Now, let $t'_j, j = 0, 1, \cdots, N$, be $(N+1)$-points in the interval $[t_{n+1} - t_c, t_{n+1}]$, where N is a fixed number. Then, the integral equation can be written as:

$$\tau(t_{n+1}) = \int_{-\infty}^{t-t_c} M(t_{n+1} - t') H(I_1, I_2) \mathbf{B}_{t'}(t_{n+1}) dt' \\ + \sum_{j=0}^{\frac{N-2}{2}} \int_{t'_{2j}}^{t'_{2j+2}} M(t_{n+1} - t') H(I_1, I_2) \mathbf{B}_{t'}(t_{n+1}) dt', \tag{12}$$

where $t'_0 = 0$ or $t'_0 = t_{n+1} - t_c$.

We consider $\mathbf{B}_{t'}(t_{n+1}) = \mathbf{B}_{t-t_c}(t_{n+1})$ for $t' < t_{n+1} - t_c$ and, therefore, the first integral becomes:

$$\int_{-\infty}^{t-t_c} M(t_{n+1}) H(I_1(\mathbf{B}_{t-t_c}(t_{n+1})), I_2(\mathbf{B}_{t-t_c}(t_{n+1}))) \mathbf{B}_{t-t_c}(t_{n+1}) dt' \tag{13}$$

which can be solved without any further issues.

Regarding the integrals within the summation operator in Equation (12), we use the method of undetermined coefficients (with a second-order quadrate formula) for their calculation (for details, see Tomé et al. [21]). In the following sections, we describe the method used to compute the tensor $\mathbf{B}_{t'(t_{n+1})}(t_{n+1})$ and how the points $t'_j(t_{n+1})$ are calculated.

- Discretization of the time interval $[t - t_c, t_{n+1}]$

  One of the key issues of the *deformation fields method* is how the integration nodes $t - t_c = t'_0 < t'_1 < \cdots < t'_N = t_{n+1}$ are distributed over the interval $[t - t_c, t_{n+1}]$, because such distribution can affect the accuracy of the results when solving complex flows. In Araújo et al. [26], the authors presented one discretization using a function that allowed them to determine the distribution of the time-integration points, which showed excellent results in some of the specific flow cases studied (such as extensional flows). However, care must be taken if we plan to generalize these results to more complex flows. In this work, we decide to use the more generic methodology presented in Tomé et al. [21]. We consider time-dependent flows so that the integration nodes are calculated using a *geometric progression* at time t_{n+1} as follows:

1. Set $t'_0 = t - t_c$ and $t'_N = t_{n+1}$;
2. Using t_c, where $t_c = t_{n+1}$ if $t < s_c$ or $t_c = s_c$ if $t \geq s_c$ make $t'_{N-j} = t'_N - \delta t \, q^j$, $j = 1, 2, \cdots, N - 1$, where $q = (t_c/\delta t)^{1/N}$, δt is the time-step.

- Computation of the Finger tensor $\mathbf{B}_{t'(t_{n+1})}(\mathbf{x}, t_{n+1})$
  One of the difficulties in the numerical simulation of viscoelastic flows using integral constitutive models is how to calculate accurately the strain history. In finite elements, this can be accomplished by a particle-tracking method based on the velocity field (see [12]), but, here, a different approach is taken. We follow the ideas of the deformation fields method [28] in which the Finger tensor is obtained by solving an appropriate evolution equation, where $\mathbf{B}_{t'(t)}(\mathbf{x}, t)$ is given by:

$$\frac{\partial}{\partial t} \mathbf{B}_{t'(t)}(\mathbf{x},t) + \mathbf{v}(\mathbf{x},t) \cdot \nabla \mathbf{B}_{t'(t)}(\mathbf{x},t) = [\nabla \mathbf{v}(\mathbf{x},t)]^T \cdot \mathbf{B}_{t'(t)}(\mathbf{x},t) + \mathbf{B}_{t'(t)}(\mathbf{x},t) \cdot \nabla \mathbf{v}(\mathbf{x},t), \quad (14)$$

with the condition $\mathbf{B}_{t'=t_{n+1}}(\mathbf{x}, t_{n+1}) = \mathbf{I}$.
The Finger tensor $\mathbf{B}_{t'(t)}(\mathbf{x}, t_{n+1})$ is calculated using the Euler method and the high-order upwind scheme CUBISTA [29] is used to discretize the convective terms. We point out that the Finger tensor $\mathbf{B}_{t'(t)}(\mathbf{x}, t_{n+1})$ is calculated at the past times $t'(t)$. The updated Finger tensor $\mathbf{B}_{t'(t_{n+1})}(\mathbf{x}, t_{n+1})$ is evaluated using a second-order interpolation method that is discussed in detail by Tomé et al. [3].

3.2. OpenFOAM System

All numerical experiments carried out in the present work will be compared with the results obtained using the OpenFOAM solver for integral models implemented by Araujo et al. [26]. The meshes were adapted in order to have simulations with similar conditions (and as close as possible) to the HiGFlow meshes. For the simulation of the contraction problem, for instance, the mesh shown in Figure 3 was used, where five regions with different refinements in the x direction can be observed. Notice that the upstream and downstream regions of the contraction geometry have volumes with exactly the same dimensions used in the HiGFlow simulations. On the other hand, a regular mesh was used for the channel-flow case. It is worth noting that the simulations were performed using the PISO method and half of the computational domain, considering the flow symmetry and the lower computational cost.

The coupling between stress and velocity was performed using the Improved Both Sides Diffusion (iBSD) [30] method, which adds a diffusive term on both sides of the momentum equation. For the solution of the linear systems resulting from the discretization of the velocity, the Bi-CGSTAB (BiConjugate Gradient Stabilized) method [31] was used with DILU (Simplified Diagonal-based Incomplete LU preconditioner) preconditioner, and, for the pressure, the conjugated preconditioned gradients (PCG) method was used with DIC (Simplified Diagonal-based Incomplete Cholesky) preconditioner.

In OpenFOAM, it is possible to choose the methods of discretization for some terms of an equation—for instance, diffusive or convective terms. Regarding this work, the numerical schemes used are described in Table 1.

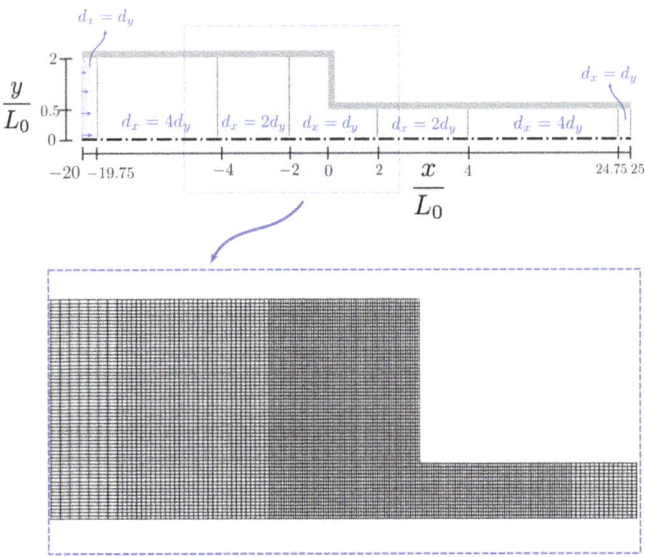

Figure 3. Computational domain used in OpenFOAM for the contraction problem.

Table 1. Numerical schemes in OpenFOAM.

Term	Scheme
$\nabla \cdot \mathbf{vv}$	Minmod
$\nabla \cdot \mathbf{vB}$	Minmod
$\nabla \cdot \tau$	Gauss linear
∇p	Gauss linear
$\nabla \mathbf{v}$	Gauss linear
$\nabla^2 \mathbf{v}$	Gauss linear corrected

Computation of the Finger Tensors

In OpenFOAM implementation, the Finger tensors, $\mathbf{B}(\mathbf{x}, t, t-s)$, are labeled by the elapsed time, s. The integration points, s_k, are distributed in the interval $[t - s_{max}, t]$, according to the following expression:

$$s_k = t_c \times \frac{e^{\zeta k} - 1}{e^{\zeta N} - 1} \tag{15}$$

where $t_c = \min\{t, s_{max}\}$, N is the number of integration points and ζ is a parameter that depends on the value of s_1 (for more details, see [26]). All the simulations were performed using $s_1 = \Delta t$ and $N = 51$.

The Finger tensors $\mathbf{B}(t_n, t_n - s_k)$ are convected according to Equation (14). We use a Euler explict scheme to obtain the fields $\mathbf{B}(t_{n+1}, t_n - s_k)$. These fields are then interpolated, allowing us to calculate the fields $\mathbf{B}(t_{n+1}, t_{n+1} - s_k)$.

4. Results

In this section, we present a verification of the methodology described in Section 3.1. Initially, the methodology is applied to the channel-flow problem. Using several meshes (uniform and non-uniform), the HiGFlow system showed good agreement with the solution obtained with the OpenFOAM system. Results of meshes' orders and errors are also shown. Lastly, the numerical simulation of contraction flows is presented. The results are compared with solutions of the OpenFOAM system [26], Freeflow system [3], Mitsoulis [14] and Quinzani [15].

4.1. Mesh Independence in Channel-Flow

The numerical method described in Section 3 was applied to simulate the flow of a KBKZ fluid in a 2D planar channel (see Figure 4) of length $10L$ and height L, where $L = 0.006$ m. At the channel entrance, a dimensionless parabolic velocity profile given by $u(y) = 4y(1 - y)$ was used. The scaling parameters were the centerline velocity, $U = 0.1$ ms^{-1}, and the fluid simulated was FLUID S1, whose parameters are described in Table 2. In this flow, we had $Re = 0.34$, $De = 1$, $\epsilon = 0.1$, $s_c = 0.1s$ and the number of deformation fields was $N = 100$.

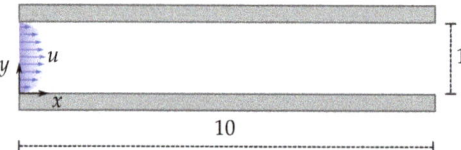

Figure 4. Dimensionless representation of the channel domain.

Table 2. Fluid parameters used in this work. Adapted from [15].

FLUID S1			
$\rho_0 = 801.5$ kg/m^3, $\alpha = 10$, $\beta = 0.7$, $\lambda_{ref} = 0.06$ s $\eta_0 = 1.424$ Pa·s			
k	λ_k (s)	a_k (Pa)	η_k (Pa·s)
1	0.6855	0.058352	0.0400
2	0.1396	1.664756	0.2324
3	0.0389	14.560411	0.5664
4	0.0059	99.152542	0.5850

In order to verify the mesh convergence of the results, the flow was simulated using several meshes (see Tables 3 and 4 and Figure 5).

Table 3. Uniform meshes.

Meshes	$d_x = d_y$
MI (8 × 80)	0.125
MII (16 × 160)	0.0625
MIII (32 × 320)	0.03125
MIV (64 × 640)	0.015625
MV (128 × 1280)	0.0078125

Table 4. Non-uniform meshes used in the 2D planar channel.

	Refined Meshes—Two Levels		
Meshes	Larger d_x	Smaller d_x	
MRI	0.125	0.0625	
MRII	0.0625	0.03125	
MRIII	0.03125	0.015625	
	Refined Meshes—Three Levels		
Meshes	Larger d_x	Middle d_x	Smaller d_x
MRVI	0.125	0.0625	0.03125
MRV	0.0625	0.03125	0.015625

Figure 5 shows the non-uniform meshes, where we can see the structure of the mesh. In Figure 5a, the mesh MRI (with two levels of refinement) is depicted, and in Figure 5b, we can see the mesh $MRVI$ (with three levels of refinement).

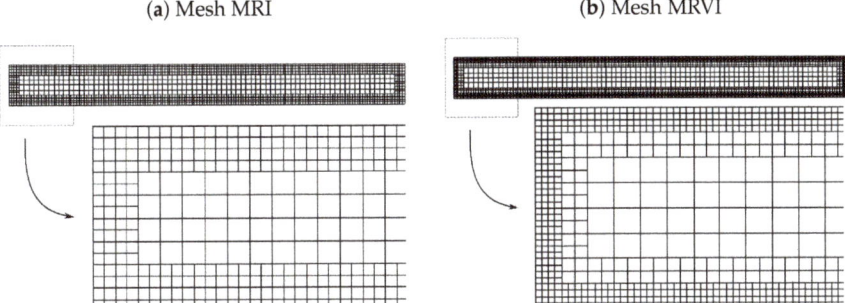

Figure 5. Refined meshes: (**a**) two levels and (**b**) three levels.

The u-profiles are illustrated in Figure 6, where the mesh convergence can be seen. We adopted mesh MV as a reference mesh (black line) and the solutions of the refined meshes (full symbols) and uniform meshes (empty symbols) are shown. In this figure, we also show the OpenFoam system profile using the mesh MV. We saw good agreement between the solutions obtained in both systems.

Our results for the tensor components τ_{xx} and τ_{yy} are presented in Figure 7, where we can also see good agreement between the numerical solutions of the HiGFlow and the OpenFOAM systems.

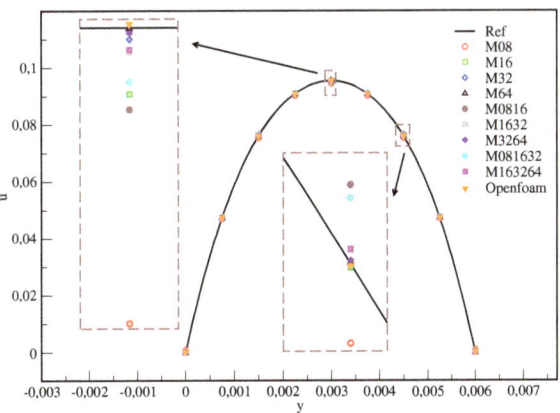

Figure 6. The u-profiles used by simulation of channel problem.

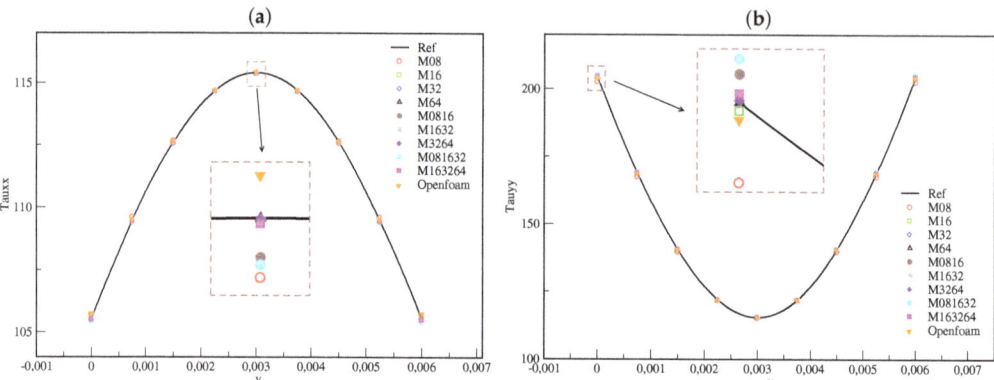

Figure 7. Comparison between the sumerical solutions of the tensor components obtained using HiGFlow (with different meshes) and OpenFOAM (mesh MV). (a) τ_{xx} and (b) τ_{yy} tensor components.

To verify the convergence, we show the errors (L_1, L_2, L_∞) and the orders in Table 5. The errors are calculated using the following equations:

$$L_1 = \frac{\sum_0^n |u(i)^{MV} - u(i)^*|}{\sum_0^n |u(i)^*|}, \quad L_2 = \sqrt{\frac{\sum_0^n (u(i)^{MV} - u(i)^*)^2}{\sum_0^n u(i)^{*2}}} \quad \text{and} \quad L_\infty = \frac{max|u(i)^{MV} - u(i)^*|}{max|u(i)^*|}$$

where $u(i)^{MV}$ is the solution in mesh MV and $u(i)^*$ is the solution in meshes $MI - MIV$ and $MRI - MRV$, $u(i)$ is the u_x profile in points $i = (x(i), y(i))$ in which $x(i) = 5$ and $y(i) = i * 0.125$, $i = 0, 1, \cdots, 8$. The orders Q for uniform meshes $Q = \frac{\log(E_{M2} \backslash E_{M1})}{\log(h_2 \backslash h_1)}$ show values close to 2 ($Q \approx 2$), which is the correct value that we expected to observe, since the velocity is calculated using an implicit Euler method. The values E_{M2} and E_{M1} are the errors (in the norms L_1, L_2 or L_∞) for two consecutive meshes (the d_x value in M_2 is lower than in M_1) and h_2 and h_1 are the d_x values in their respective meshes.

Table 5. Errors and orders for u-velocity. The mesh MV was assumed as a reference solution.

	u_x Errors		
Mesh	L_1	L_2	L_∞
MI	1.046×10^{-3}	1.012×10^{-3}	1.022×10^{-3}
MII	1.718×10^{-4}	1.815×10^{-4}	2.327×10^{-4}
MIII	3.602×10^{-5}	3.657×10^{-5}	4.493×10^{-5}
MIV	5.619×10^{-6}	5.897×10^{-6}	7.583×10^{-6}
RMI	4.809×10^{-4}	5.598×10^{-4}	7.115×10^{-4}
RMII	1.007×10^{-4}	1.018×10^{-4}	1.110×10^{-4}
RMIII	2.253×10^{-5}	2.238×10^{-5}	2.382×10^{-5}
RMVI	5.306×10^{-4}	5.452×10^{-4}	5.918×10^{-4}
RMV	1.077×10^{-4}	1.129×10^{-4}	1.250×10^{-4}
	u_x Orders		
Mesh	L_1	L_2	L_∞
MI-MII	2.606	2.479	2.134
MII-MIII	2.254	2.311	2.373
MIII-MIV	2.680	2.633	2.567

4.2. Numerical Simulation of 4:1 Abrupt Planar Contraction Problem

In this section, we show the simulations for the 4:1 abrupt planar contraction flow. This problem is interesting because, for instance, the flow near the contraction is a complex mixture of shear and elongation, and secondary fluid motions might exist, even in the Newtonian limit (see [15]). For this reason, contraction flows have been extensively studied previously in the literature (see [3,14,15,26]).

Figure 8 shows the domain representation, where we adopted a dimensionless parabolic inlet velocity profile $u(y) = \frac{3}{8}\frac{1}{4}(2-y)(2+y)$. The scaling parameter $L = 0.0064$ m is the height of the small channel, and we used $N = 50$ deformation fields, $\epsilon = 0.1$ and the time interval $s_c = 0.1$ s. In Table 6, we report the scaling parameters of average velocity $\bar{u}$ used in all simulations and the dimensionless parameters $Re = \frac{\rho L \bar{u}}{\eta_0}$, $De = \frac{\lambda \bar{u}}{L}$, $De(\dot{\gamma})$ (see [15]) and the characteristic shear rate $\dot{\gamma}$.

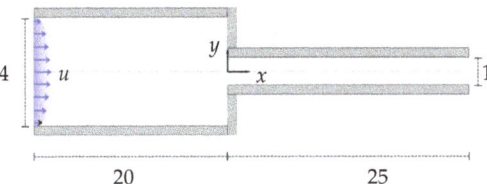

Figure 8. Domain representation.

Table 6. Flow parameter values used in the contraction problem.

	Planar Contraction Flows			
$\bar{u}\left[\frac{m}{s}\right]$	$\dot{\gamma}\left[\frac{1}{s}\right]$	$De(\dot{\gamma})$	Re	De
0.044	13.9	0.38	0.16	0.41
0.100	31.3	0.55	0.36	0.94
0.150	48.4	0.66	0.56	1.45
0.221	69.1	0.77	0.80	2.07

The mesh M_1 used in these simulations is shown in Figure 9. We used three levels of refinement, with the most refined part near the contraction region. We also used one uniform mesh M_2 for $De = 0.94$ in order to check the convergence solutions in two meshes. In M_2, $d_x = d_y = 0.03125$ m, and in M_1, we use small values of $dx = dy = 0.03125$ m as well as larger values, $dx = dy = 0.125$ m.

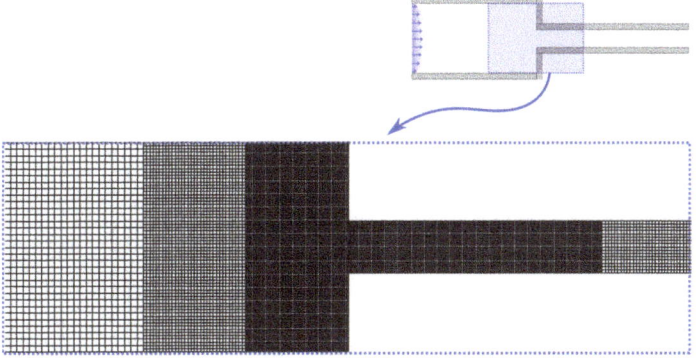

Figure 9. Graphical representation of the mesh M_1 used in the contraction simulation.

In Figure 10, we illustrate the centerline axial profile velocity $u_x(y)$ solutions for different values of the Deborah number $De = 0.41, 0.94, 1.45$ and 2.07 for the HiGFlow (green lines) and OpenFOAM (black lines) systems using mesh $M1$. The experimental results of Quinzani et al. [15] and numerical results of Mitsoulis [14] and Tomé et al. [3] (for $De = 1.45$) are also shown for comparison purposes. For the case with $De = 0.94$, we show two solutions using our methodology in mesh M_1 (non-uniform mesh) and mesh M_2, where we can see good agreement between the solutions. For this reason, we adopted the M_1 mesh to simulate the other cases with different values of number De. For the profile $u_x(y)$, the HiGFlow system showed good agreement with the OpenFOAM solutions in the regions before and after the contraction. Near to contraction (see Figure 10), we have a region of instability and the methodology behaves differently, but our results showed similar behavior to the instabilities presented in the works of Quinzani et al. (orange triangles) [15], Mitsoulis [14] (blue bullet) and Tomé et al. (violet square) [3].

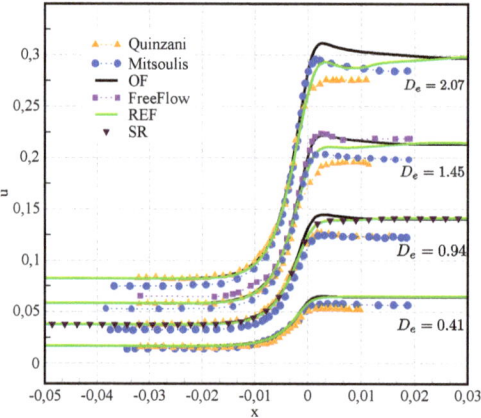

Figure 10. Centerline axial velocity profiles obtained using HiGFlow (green lines) and OpenFOAM (black lines). Experimental [15] and numerical results [3,14] found in the literature are also shown for comparison purposes.

In Figure 11, we show the numerical solution using HiGFlow (green lines) and OpenFOAM (black lines) systems for the tensor components τ_{xx} and τ_{yy} with the same flow parameter values reported in Table 6 using M_1 and M_2 for the case with $De = 0.94$. The methodologies presented in this work are different. OpenFOAM uses the finite volume method while HiGFlow approximates the equations using finite differences. Therefore, the solutions obtained will not be equal but should be comparable. Outside the contraction region, the OpenFOAM and HiGFlow solutions are very similar for all values of De. However, in the region close to the contraction, the solutions obtained using OpenFOAM showed a higher peak ($x \approx 0$) but this is mostly seen in the cases with the highest values of De.

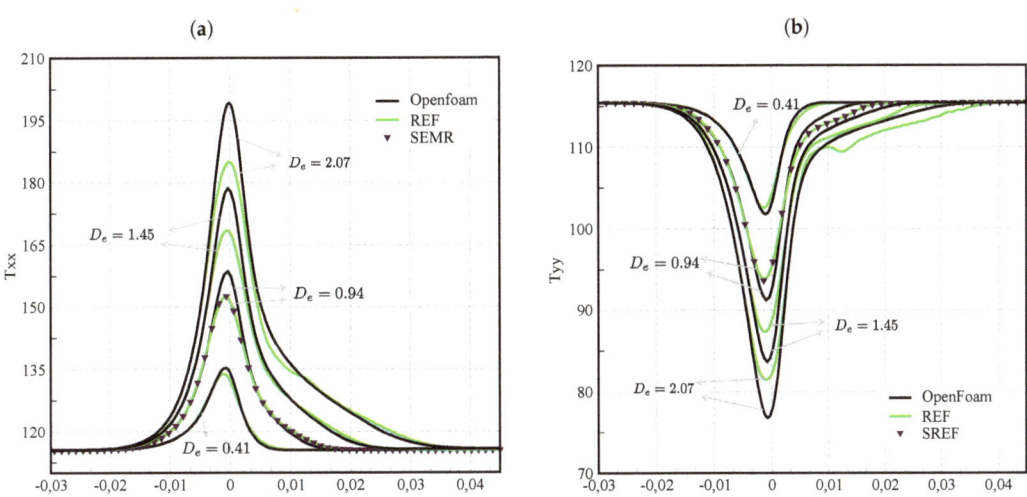

Figure 11. Numerical results for (**a**) τ_{xx} and (**b**) τ_{yy} using HiGFlow system (HF-M1 e HF-M2) and OpenFOAM system (OF-M1).

In Figure 12a, we show the comparison between the first normal stress difference values $N1 = \tau_{xx} - \tau_{yy}$ for two different cases of Deborah number, $De = 0.41$ and $De = 1.45$. For better visualization, the other two values of De (see Table 6) are illustrated in Figure 12b, where we can see that the values of HiGFlow (green lines) have good agreement with experimental data (orange triangles) reported by Quinzani [15], while the solution using OpenFOAM was similar to the solution presented by Mitsoulis [14].

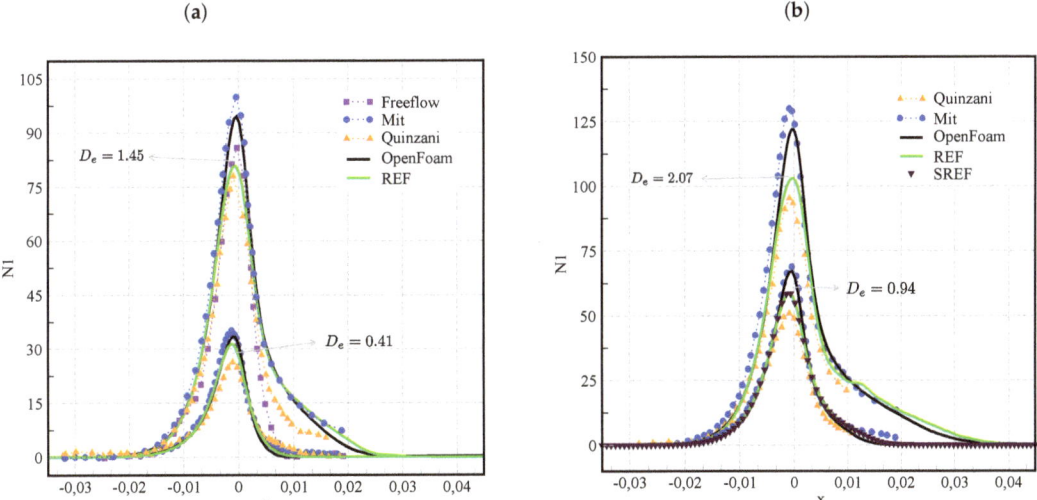

Figure 12. The first normal stress differences $N1$: (**a**) for $De = 0.41$ and $De = 1.45$; and (**b**) for $De = 0.94$ and $De = 2.07$.

In Figure 13, we compare the streamlines obtained using OpenFOAM (a) and HiGFlow (b) with a fixed value of $De = 2.07$. We can see that there is vortex formation in both cases and that the solutions are relatively comparable.

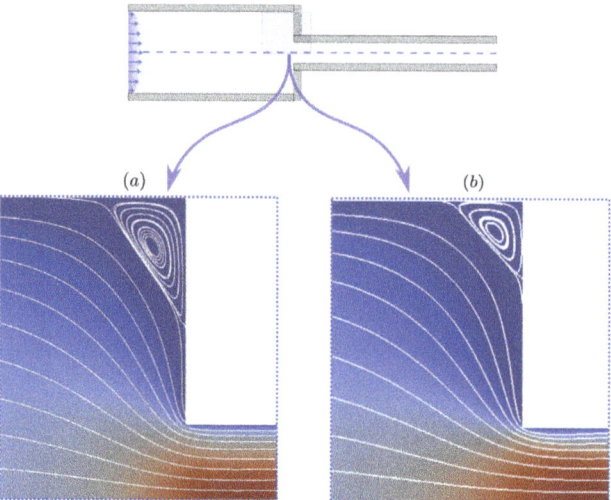

Figure 13. Streamlines of the contraction flow problem with $De = 2.07$. (**a**) OpenFOAM and (**b**) HiGFlow.

5. Discussion

The work aims to present the numerical simulation of the KBKZ integral constitutive equations for incompressible and transient complex flows. We used a new solver HiGTree/HiGFlow lately developed by Souza et al. [23]. In this solver, we have implemented the methodology described in Section 3.1 to simulate viscoelastic flows modeled by integral constitutive equations. Initially, the numerical technique was verified by refined mesh in channel flows. Using the FLUID S1 (see Table 2), we performed nine simulations using non-refined and refined meshes. For comparison purposes, a mesh was chosen and the simulation using the OpenFOAM solver [26] was performed. In these simulations, we can see that, although the methodologies used in HiGFlow and OpenFOAM are quite different (the first uses finite differences and the second uses finite volume), we obtained very similar results in both systems. We also verified that the errors decrease with the mesh refinement and that the order of convergence of the velocity was around two, as expected.

A classic problem in the simulation of integral viscoelastic flows is known as 4:1 abrupt contraction and, thus, the literature for this problem is extensive. We chose to check our methodology for the four values of De presented in Mitsoulis [14]. In addition to the comparison with the results of this author, we performed the simulations using the OpenFOAM solver [26] and also compared our results with the experimental ones from Quinzani [15] and with the numerical results from FreeFlow [3]. We know that, in the contraction region, there are singularities and numerical techniques that might exhibit different behaviors. Although the values obtained by us in this work differ somewhat from the values obtained by Mitsoulis [14] or OpenFOAM [26], they were quite comparable to the experimental values of Quinzani [15]. Thus, we verified that the methodology presented here is capable of simulating complex flows in transient fluid regimes governed by integral constitutive models using the rapid technique of finite differences in hierarchical meshes with local refinement.

The computational efficiency of the models has been previously studied [3,21,26]. However, it is worth mentioning that the methods used in the present work used fewer integration points (deformation fields) compared to the points used in the early work of Hulsen et al. [28] and are similar to those used more recently by Hulsen and Anderson [32]. This improvement is due mainly to the distinct methodologies adopted to obtain the integration points, which allow efficient distribution of the elapsed time. Although the results presented here are two-dimensional, our methodology is still able to simulate flows in higher dimensions (the user just has to specify the dimension (2, 3 or higher) in the input data file). Our future work will be to present simulations in three dimensions for classic problems. In three dimensions, for example, integral models are still computationally expensive, as there is a need to store and connect a fixed N number of fields to each cell. Therefore, we will also work on ways to improve or modify the integral calculation—for instance, as was done in the recent work of Hulsen [32].

Author Contributions: Funding acquisition, A.C.; Methodology, J.B., M.S.B.d.A., R.T.L. and A.C.; Project administration, A.C.; Software, J.B., M.S.B.d.A. and A.C.; Supervision, A.C.; Validation, J.B. and M.S.B.d.A.; Writing—original draft, J.B., M.S.B.d.A., R.T.L. and A.C.; Writing—review & editing, H.A.C.S. All authors have read and agreed to the published version of the manuscript.

Funding: All authors are grateful for the financial support from the Brazilian Petroleum Agency (ANP)/Petrobras, grant 0050.0075367.12.9, and from the São Paulo Research Foundation (FAPESP), grants 2013/07375-0, 2017/21105-6 and 2020/02990-1. A. Castelo is also grateful for the financial support from the National Council for Scientific and Technological Development (CNPq), grant 307483/2017-7. Research was carried out using the computational resources of the Center for Mathematical Sciences Applied to Industry (CeMEAI), funded by FAPESP grant 2013/07375-0.

Institutional Review Board Statement: Not applicable.

Informed Consent Statement: Not applicable.

Data Availability Statement: Not applicable.

Acknowledgments: Bertoco, J.; Leiva, R.T.; Sánchez, H. A. C. and Castelo, A. acknowledge the support of the ICMC-Instituto de Ciências Matemáticas e de Computação, Departamento de Matemática e Estatística, USP, São Carlos, SP. Araújo, M. S. B. acknowledges the support of the UFPA-Universidade Federal do Pará, Instituto de Ciências Exatas e Naturais.

Conflicts of Interest: The authors declare no conflict of interest.

Abbreviations

The following abbreviations are used in this manuscript:

HF   HiGTree-HiGFlow
OF   OpenFOAM
Re Reynolds number
De Deborah number

References

1. Aboubacar, M.; Matallah, H.; Tamaddon-Jahromi, H.R.; Webster, M.F. Highly elastic solutions for Oldroyd-B and Phan-Thien Tanner fluids with a finite volume element method: Planar contraction flows. *J. Non-Newt. Fluid Mech.* **2020**, *103*, 65–103. [CrossRef]
2. Clermont, J.R.; Normandin, M. Numerical simulation of extrudate swell for Oldroyd-B fluid using the stream-tube analysis and a streamline approximation. *J. Non-Newt. Fluid Mech.* **1993**, *50*, 193–215. [CrossRef]
3. Tomé, M.F.; Araujo, M.S.B.; Alves, M.A.; Pinho, F.T. Numerical simulation of viscoelastic flows using integral constitutive equations: A finite difference approach. *J. Comput. Phys.* **2008**, *227*, 4207–4243. [CrossRef]
4. Tomé, M.F.; Castelo, A.; Afonso, A.M.; Alves, M.A.; Pinho, F.T. Application of the log-conformation tensor to three-dimensional time-dependent free surface flows. *J. Non-Newt. Fluid Mech.* **2012**, *175–176*, 44–54. [CrossRef]
5. Mompean, G.; Thais, L.; Tomé, M.F.; Castelo, A. Numerical prediction of three-dimensional time-dependent viscoelastic extrudate swell using differential and algebraic models. *Comput. Fluids* **2011**, *44*, 68–78. [CrossRef]
6. Berauto, C.; Fortin, A.; Coupez, T.; Demay, Y.; Vergnes, B.; Agassant, J.F. A finite element method for computing the flow of multi-mode viscoelastic fluids: Comparison with experiments. *J. Non-Newt. Fluid Mech.* **1998**, *75*, 1–23 [CrossRef]
7. Tomé, M.F.; Paulo, G.S.; Alves, M.A.; Pinho, F.T. Numerical Solution of the PTT constitutive equation for three-dimensional free surface flows. *J. Non-Newt. Fluid Mech.* **2010**, *165*, 247–262. [CrossRef]
8. Oishi, C.M.; Martins, F.P.; Tomé, M.F.; Alves, M.A. Numerical simulation of drop impact and jet buckling problems using the eXtended Pom Pom model. *J. Non-Newt. Fluid Mech.* **2012**, *169–170*, 91–103. [CrossRef]
9. Oishi, C.M.; Martins, F.P.; Tomé, M.F.; Cuminato, J.A.; Mckee, S. Numerical solution of the eXtended Pom-Pom model for viscoelastic free surface flows. *J. Non-Newt. Fluid Mech.* **2011**, *166*, 165–179. [CrossRef]
10. Tanner, R.I. From A to (BK)Z in constitutive relations. *J. Rheol.* **1988**, *32*, 673–702. [CrossRef]

11. Mitsoulis, E. 50 Years of the K-BKZ constitutive relation for polymers. *Polym. Sci.* **2013**. [CrossRef]
12. Luo, X.L.; Mitsoulis, E. An efficient algorithm for strain history tracking in finite element computation of non-Newtonian fluids with integral constitutive equations. *Int. J. Num. Meth. Fluids* **1990**, *11*, 1015–1031. [CrossRef]
13. Ansari, M.; Alabbas, A.; Hatzikiriakos, S.G.; Mitsoulis, E. Entry flow of polyethylene melts in tapered dies. *Int. Polym. Proc.* **2010**, *25*, 287–296. [CrossRef]
14. Mitsoulis, E. Numerical simulation of planar entry flow for a polyisobutylene solution using an integral constitutive equation. *J. Rheol.* **1993**, *37*, 1029. [CrossRef]
15. Quinzani, L.M.; Brown, R.A.; Armstrong, R.C. Birefringence and laser-Doppler velocimetry (LDV) studies of viscoelastic flow through a planar contraction. *J. Non-Newt. Fluid Mech.* **1994**, *52*, 1–36. [CrossRef]
16. McKinley, G.H.; Sridhar, T. Filament-stretching rheometry of complex fluids. *Annu. Rev. Fluid Mech.* **2002**, *34*, 375–415. [CrossRef]
17. Chai, M.S.; Yeow, Y.L. Modelling of fluid M1 using multiple-relaxation-time constitutive equations. *J. Non-Newt. Fluid Mech.* **1990**, *35*, 459–470. [CrossRef]
18. Mitsoulis, E. Extrudate swell of Boger fluids. *J. Non-Newt. Fluid Mech.* **2010**, *165*, 812–824. [CrossRef]
19. Ganvir, V.; Lele, A.; Thaokar, R.; Gautham, B.P. Prediction of extrudate swell in polymer melt extrusion using an arbitrary Langrangian Eulerian (ALE) based finite element method. *J. Non-Newt. Fluid Mech.* **2009**, *156*, 21–28. [CrossRef]
20. Ahmed, R.; Liang, R.F.; Mackley, M.R. The experimental observation and numerical prediction of planar entry flow and die swell for molten polyethylenes. *J. Non-Newt. Fluid Mech.* **1995**, *59*, 129–153. [CrossRef]
21. Tomé, M.F.; Bertoco, J.; Oishi, C.M.; Araujo, M.S.B.; Cruz, D.; Pinho, F.T.; Vynnycky, M. A finite difference technique for solving a time strain separable k-bkz constitutive equation for two-dimensional moving free surface flows. *J. Comput. Phys.* **2016**, *311*, 114–141. [CrossRef]
22. Losasso, F.; Gibou, F.; Fedkiw, R. Simulating water and smoke with an octree data structure. *ACM Trans. Grap. (TOG)* **2004**, *3*, 457–462. [CrossRef]
23. Sousa, F.S.; Lages, C.F.; Ansoni, J.L.; Castelo, A.; Simão, A. A finite difference method with meshless interpolation for incompressible flows in non-graded tree-based grids. *J. Comput. Phys.* **2019**, *396*, 848–866. [CrossRef]
24. Weller, H.G.; Tabor, G.; Jasak, H.; Fureby, C. A tensorial approach to computational continuum mechanics using object-oriented techniques. *Comp. Phys.* **1998**, *12*, 620–631. [CrossRef]
25. Papanastasiou, A.C.; Scriven, L.E.; Macosko, C.W. An integral constitutive equation for mixed flows: Viscoelastic characterization. *J. Rheol.* **1993**, *27*, 387. [CrossRef]
26. Araújo, M.S.B.; Fernandes, C.; Ferrás, L.L.; Tukovic, Z.; Jasakc, H.; Nóbrega, J.M. A stable numerical implementation of integral viscoelastic models in the OpenFOAM computational library. *Comput. Fluids* **2018**, *172*, 1–13. [CrossRef]
27. Chorin, A.J. Numerical solution of the Navier-Stokes equations. *Math. Comp.* **1968**, *22*, 745–762. [CrossRef]
28. Hulsen, M.A.; Peters, E.; Brule, B.V.D. A new approach to the deformation fields method for solving complex flows using integral constitutive equations. *J. Non-Newt. Fluid Mech.* **2001**, *98*, 201–221. [CrossRef]
29. Alves, M.; Oliveira, P.; Pinho, F.T. A convergent and universally bounded interpolation scheme for the treatment of advection. *Int. J. Num. Meth. Fluids* **2003**, *41*, 47–75. [CrossRef]
30. Fernandes, C.; Araujo, M.S.B.; Ferrás, L.L.; Nóbrega, J.M. Improved both sides diffusion (iBSD): A new and straightforward stabilization approach for viscoelastic fluid flows. *J. Non-Newt. Fluid Mech.* **2017**, *49*, 63–78. [CrossRef]
31. Van der Vorst, H.A. Bi-CGSTAB: A Fast and Smoothly Converging Variant of Bi-CG for the Solution of Nonsymmetric Linear Systems. *SIAM J. Sci. Stat. Comput.* **1992**, *13*, 631–644. [CrossRef]
32. Hulsen, M.A.; Anderson, P.D. The deformation fields method revisited: Stable simulation of instationary viscoelastic fluid flow using integral models. *J. Non-Newt. Fluid Mech.* **2018**, *262*, 68–78. [CrossRef]

Article

Analytical Investigation of Viscoelastic Stagnation-Point Flows with Regard to Their Singularity

Jie Liu *, Martin Oberlack and Yongqi Wang

Chair of Fluid Dynamics, Department of Mechanical Engineering, Technical University of Darmstadt, Otto-Berndt-Str. 2, 64287 Darmstadt, Germany; oberlack@fdy.tu-darmstadt.de (M.O.); wang@fdy.tu-darmstadt.de (Y.W.)
* Correspondence: liu@fdy.tu-darmstadt.de

Citation: Liu, J.; Oberlack, M.; Wang, Y. Analytical Investigation of Viscoelastic Stagnation-Point Flows with Regard to Their Singularity. Appl. Sci. 2021, 11, 6931. https://doi.org/10.3390/app11156931

Academic Editors: Luís L. Ferrás and Alexandre M. Afonso

Received: 6 July 2021
Accepted: 24 July 2021
Published: 28 July 2021

Publisher's Note: MDPI stays neutral with regard to jurisdictional claims in published maps and institutional affiliations.

Copyright: © 2021 by the authors. Licensee MDPI, Basel, Switzerland. This article is an open access article distributed under the terms and conditions of the Creative Commons Attribution (CC BY) license (https://creativecommons.org/licenses/by/4.0/).

Abstract: Singularities in the stress field of the stagnation-point flow of a viscoelastic fluid have been studied for various viscoelastic constitutive models. Analyzing the analytical solutions of these models is the most effective way to study this problem. In this paper, exact analytical solutions of two-dimensional steady wall-free stagnation-point flows for the generic Oldroyd 8-constant model are obtained for the stress field using different material parameter relations. For all solutions, compatibility with the conservation of momentum is considered in our analysis. The resulting solutions usually contain arbitrary functions, whose choice has a crucial effect on the stress distribution. The corresponding singularities are discussed in detail according to the choices of the arbitrary functions. The results can be used to analyze the stress distribution and singularity behavior of a wide spectrum of viscoelastic models derived from the Oldroyd 8-constant model. Many previous results obtained for simple viscoelastic models are reproduced as special cases. Some previous conclusions are amended and new conclusions are drawn. In particular, we find that all models have singularities near the stagnation point and most of them can be avoided by appropriately choosing the model parameters and free functions. In addition, the analytical solution for the stress tensor of a near-wall stagnation-point flow for the Oldroyd-B model is also obtained. Its compatibility with the momentum conservation is discussed and the parameters are identified, which allow for a non-singular solution.

Keywords: viscoelastic models; stagnation-point flow; stress singularities; Weissenberg numbers

1. Introduction

The working fluids encountered in practical applications and industry are often non-Newtonian, and research on this type of fluids has been conducted for decades. In theoretical research, a variety of non-Newtonian fluid models has developed [1]. One group of these models is of the rate type which involves differential transport equations for the stress tensor. As these models are highly complex and mostly non-linear, exact analytical solutions can only be obtained for very special flow cases. Many theoretical works limit themselves to investigating the distribution of stress tensor in simple canonical viscoelastic flows by means of relatively simple models, such as the Oldroyd-B and Maxwell-B model, see, e.g., [2–4]. A classical model problem in this context, and also to be considered presently, is the similarity solution for the velocity field in a stagnation-point flow described, e.g., in [5]. Therein, the velocity distribution is usually assumed in the form $(u,v) = (xf'(y), -f(y))$, though for a wall-free stagnation-point flow or a creeping stagnation-point flow far away from the wall, the velocity profile reduces to $f(y) = ay$, i.e., $(u,v) = (ax, -ay)$, where a is constant rate of the strain. Under this assumption, Phan-Thien [6,7] obtained exact solutions to the plane and axisymmetric stagnation-point flows for both Maxwellian and Oldroyd-B fluids, respectively, where the governing equations were reduced to a system of ordinary differential equations. It was shown that in a stagnation-point flow with a certain Weissenberg number, there is a singularity in the stress field. Renardy [8] analyzed a steady creeping flow of the upper convected Maxwell (UCM fluid, in which the shear stress was assumed to be zero, and the normal stress depended only on the transverse coordinate of the outflow. His solution demonstrated that a singularity exists not only at the stagnation point, but also along the entire streamline passing the stagnation point downstream. Thomases et al. [9] used the same ansatz for the velocity field and solved the stress field of an Oldroyd-B fluid by using the method of characteristics. Their results showed that the behavior of the solutions is very sensitive to the Weissenberg number. However, the exact analytical solutions were only constructed for the model equations of the stress field without considering the compatibility with the momentum conservation equation. Cruz et al. [10] obtained a general analytical solution for a steady planar extensional wall-free stagnation-point flow of a viscoelastic fluid described by the UCM model. This solution depends on both space coordinates, and represents an extension of the previous solutions considering only the dependence of the transverse coordinate. Recently, Meleshko et al. [11] extended the analysis of the stress distribution in a wall-free stagnation-point flow from the UCM model to the Johnson–Segalman model and also took the momentum conservation equation into account. Their solution demonstrates that the Johnson–Segalman model has a non-removable logarithmic singularity.

In a near-wall stagnation-point flow, the velocity must satisfy the no-slip condition at the wall, so the velocity profile for the wall-free stagnation-point flow, i.e., $(u, v) = (xf'(y), -f(y))$ with $f(y) = ay$, is here no longer suitable. In this case, the function $f(y)$ could take the form of $f(y) = ay^n$ with $n > 1$ for an impermeable wall located at $y = 0$. Under the simplest assumption $f(y) = ay^2$, Becherer et al. [12] and Van Gorder et al. [13], respectively, considered an UCM fluid and presented the exact solutions to the coupled PDEs of the viscoelastic stress. However, they did not analyze the compatibility of their solutions with the momentum conservation equation. Van Gorder [14] analyzed the same problem and showed that the solution for the stress components fails to satisfy the momentum conservation equation except in the linear case $f(y) = -ay$, corresponding to a wall-free stagnation-point flow. Therefore, to investigate near-wall stagnation-point flows, a more general velocity profile obeying the momentum conservation has to be proposed.

All the models mentioned above can be derived from the Oldroyd 8-constant model, and many other models are also special cases of the Oldroyd 8-constant model [15]. It is interesting to investigate whether the other viscoelastic models have similar singularities as the Oldroyd-B, Maxwell-B and Johnson–Segalman models. For this purpose, we will search for exact analytical solutions for the Oldroyd 8-constant constitutive framework of a wall-free stagnation-point flow satisfying the compatibility with the conservation of momentum by means of the method of characteristics and analyze the effect of various model parameters on the solutions with regard to their singularity. Many conclusions previously obtained by other researchers for simple viscoelastic constitutive models are either confirmed or rectified. Some new conclusions are drawn. In particular, we find that all models have singularities near the stagnation point and most of them can effectively be avoided by appropriately choosing the model parameters and free functions. For the near-wall stagnation-point flow satisfying the no-slip conditions, it is impossible to analytically solve the constitutive equations of the Oldroyd 8-constant model. Instead, we focus on the Oldroyd-B model and analyze the compatibility of the solution with the momentum conservation equation.

2. Model Equations
2.1. Conservation Equations

The balance equation of momentum and mass conservation for an incompressible fluid take the form:

$$\rho \frac{D\mathbf{u}}{Dt} = \nabla \cdot \sigma + \rho \mathbf{f}, \tag{1}$$

$$\nabla \cdot \mathbf{u} = 0, \tag{2}$$

where ρ is the fluid density, $\mathbf{u}$ the flow velocity, σ is the Cauchy stress tensor, and $\mathbf{f}$ the volume force which is neglected in the present study. For a Maxwell fluid, σ can be split in two parts:

$$\sigma = -p\mathbf{I} + \mathbf{T}_p, \tag{3}$$

where $\mathbf{T}_p$ is the polymetric stress contribution and p the dynamic pressure. For an Oldroyd-type fluid, a Newtonian stress part with a viscosity η_s is added to the total stress and the Cauchy stress tensor is given by

$$\sigma = -p\mathbf{I} + 2\eta_s \mathbf{D} + \mathbf{T}_p = -p\mathbf{I} + \mathbf{T}. \tag{4}$$

here $\mathbf{D} = \frac{1}{2}(\nabla \mathbf{u}^T + \nabla \mathbf{u})$ is the symmetric rate-of-strain tensor. $\mathbf{T}$ is called the deviatoric stress tensor with $\mathbf{T} = 2\eta_s \mathbf{D} + \mathbf{T}_p$.

2.2. The Oldroyd 8-Constant Model

The most general linear viscoelastic model is the Oldroyd 8-constant model [16] described by

$$\mathbf{T} + \lambda_1 \overset{\circ}{\mathbf{T}} + \mu_0 (tr\mathbf{T})\mathbf{D} - \mu_1(\mathbf{TD} + \mathbf{DT}) + \nu_1[tr(\mathbf{TD})]\mathbf{I} \\ = 2\eta_0 [\mathbf{D} + \lambda_2 \overset{\circ}{\mathbf{D}} - 2\mu_2 \mathbf{D}^2 + \nu_2 tr(\mathbf{D}^2)\mathbf{I}], \tag{5}$$

where $\lambda_1, \lambda_2, \mu_0, \mu_1, \mu_2, \nu_1, \nu_2$ are material constants with the dimension of time. η_0 is the total viscosity split by $\eta_0 = \eta_p + \eta_s$ and η_p is the polymer contribution to the viscosity. $\overset{\circ}{\mathbf{T}}$ is the corotational objective time derivative of $\mathbf{T}$ defined as

$$\overset{\circ}{\mathbf{T}} := \frac{d\mathbf{T}}{dt} - \mathbf{WT} + \mathbf{TW}, \tag{6}$$

where $\mathbf{W} = \frac{1}{2}(\nabla \mathbf{u}^T - \nabla \mathbf{u})$ is the skew-symmetric vorticity tensor. Giesekus [1] has extended the Oldroyd 8-constant model by adding the term $\nu_0(tr\mathbf{T})\mathbf{I}$ to the left-hand side of (5), but this extension will not be considered in the present investigation.

Most viscoelastic models can be derived from the Oldroyd 8-constant model [15]. For example, setting $\mu_1 = \lambda_1$, $\mu_2 = \lambda_2 = \frac{\eta_s}{\eta_0}\lambda_1$, and $\mu_0 = \nu_1 = \nu_2 = 0$ in (5) gives the Oldroyd-B model:

$$\mathbf{T} + \lambda_1 \overset{\triangledown}{\mathbf{T}} = 2\eta_0 \left(\mathbf{D} + \lambda_2 \overset{\triangledown}{\mathbf{D}} \right) \tag{7}$$

where the symbol $\triangledown$ is the upper-convected time derivative, defined by

$$\overset{\triangledown}{\mathbf{T}} := \frac{d\mathbf{T}}{dt} - \nabla \mathbf{u}^T \cdot \mathbf{T} - \mathbf{T} \cdot \nabla \mathbf{u}. \tag{8}$$

Usually, a retardation parameter is defined by $\beta = \frac{\eta_p}{\eta_0} \in [0,1]$. $\beta = 0$ corresponds to a Newtonian fluid, $\beta = 1$ to a Maxwell fluid, and in between a Oldroyd fluid. If $\frac{\eta_s}{\eta_0} = 1$, hence $\lambda_2 = \lambda_1$, the Oldroyd-B fluid reduces to the Newtonian fluid with the constitutive equation $\mathbf{T} = 2\eta_s \mathbf{D}$. If $\eta_s = 0$, hence $\lambda_2 = 0$, $\eta_0 = \eta_p$, $\mathbf{T} = \mathbf{T}_p$, and the Maxwell-B model is obtained:

$$\mathbf{T}_p + \lambda_1 \overset{\triangledown}{\mathbf{T}}_p = 2\eta_p \mathbf{D}. \tag{9}$$

Combining the constitutive Equation (7) and the relation $\mathbf{T} = 2\eta_s \mathbf{D} + \mathbf{T}_p$ for an Oldroyd fluid results in a constitutive relation in terms of $\mathbf{T}_p$, which is completely consistent with the expression of (9). However, in contrast to the Maxwell fluid, the Newtonian viscosity η_s in the momentum equation of an Oldroyd-type fluid is not zero. More examples of viscoelastic models derived from the Oldroyd 8-constant model can be found in the Table II of [15].

We employed the dimensionless variables, which are scaled by the characteristic length L_0, strain rate a and viscosity η_0 as following:

$$\tilde{x} = \frac{x}{L_0}, \quad \tilde{u} = \frac{u}{aL_0}, \quad \tilde{p} = \frac{p}{a\eta_0}, \quad \tilde{T} = \frac{T}{a\eta_0}, \quad Re = \frac{\rho a L_0^2}{\eta_0}, \quad W_1 = a\lambda_1,$$
$$W_2 = a\mu_0, \; W_3 = a\mu_1, \; W_4 = a\nu_1, \; W_5 = a\lambda_2, \; W_6 = a\mu_2, \; W_7 = a\nu_2,$$
(10)

where $W_1, \ldots, W_7$ are Weissenberg numbers and Re the Reynolds number. Omitting the tilde symbol, the constitutive Equation (5) can be rewritten in the following dimensionless form:

$$\mathbf{T} + W_1 \overset{\circ}{\mathbf{T}} + W_2(tr\mathbf{T})\mathbf{D} - W_3(\mathbf{TD} + \mathbf{DT}) + W_4[tr(\mathbf{TD})]\mathbf{I}$$
$$= 2[\mathbf{D} + W_5 \overset{\circ}{\mathbf{D}} - 2W_6 \mathbf{D}^2 + W_7 tr(\mathbf{D}^2)\mathbf{I}].$$
(11)

Then, the dimensionless Oldroyd 8-constant model (11) will be utilized for investigating the two-dimensional steady stagnation-point flows with regard to their singularities under different material parameter relations.

3. Analytical Solutions of a Wall-Free Stagnation-Point Flow and Their Singularities

For the velocity field of a two-dimensional steady stagnation-point flow, there exists a similarity solution described as

$$\mathbf{u} = (xf'(y), -f(y)),$$
(12)

where $f(y)$ is an arbitrary function depending only on y [5,6]. In the case of a wall-free stagnation-point flow, the flow is a potential flow, which means $\nabla \times \mathbf{u} = 0$. This leads to $f''(y) = 0$, and further $f(y) = ay$. The parameter a stands for the constant rate of the strain and is employed in the non-dimensionalization, as shown in (10). Hence, the dimensionless velocity field is given by

$$\mathbf{u} = (x, -y).$$
(13)

The velocity field is symmetrical for both the x and the y axis, as shown in Figure 1, so we only need to consider the case with $x, y > 0$ in the following analysis.

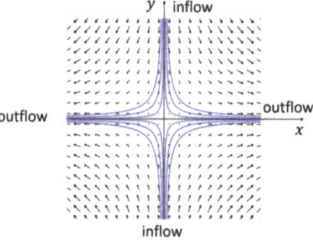

Figure 1. Sketch of the velocity field in a wall-free stagnation-point flow.

The stress tensor is symmetric and takes a two-dimensional form of:

$$\mathbf{T} = \begin{pmatrix} T_{11} & T_{12} \\ T_{12} & T_{22} \end{pmatrix}.$$
(14)

Substituting (13) and (14) into (11), we obtain the following three scalar constitutive equations:

$$W_1 x \frac{\partial T_{11}}{\partial x} - W_1 y \frac{\partial T_{11}}{\partial y} = (k_1 - 1)T_{11} - k_2 T_{22} - k_3 + 2, \tag{15}$$

$$W_1 x \frac{\partial T_{22}}{\partial x} - W_1 y \frac{\partial T_{22}}{\partial y} = k_2 T_{11} - (k_1 + 1)T_{22} - k_3 - 2, \tag{16}$$

$$W_1 x \frac{\partial T_{12}}{\partial x} - W_1 y \frac{\partial T_{12}}{\partial y} = -T_{12}, \tag{17}$$

where $k_1 = 2W_3 - W_2 - W_4$, $k_2 = W_2 - W_4$ and $k_3 = 4(W_6 - W_7)$. It is to notice that $\overset{\circ}{\mathbf{D}} = 0$ under the assumption of the velocity (13), so the Weissenberg number W_5 has not any influence on the stress field.

In the PDE system (15)–(17), Equation (17) for T_{12} is uncoupled with (15), (16), and thus can be solved independently. The coupling of (15) and (16) for T_{11} and T_{22} depends on the value of k_2. If $k_2 = 0$, both the equations are decoupled and can be solved separately for T_{11} and T_{22}. If $k_2 \neq 0$, Equations (15) and (16) constitute a coupled PDE system with non-constant coefficients. In addition, the solutions of the PDE system (15)–(17) must also satisfy the compatibility condition derived by substituting (4), (13) and (14) into the rotation of the momentum Equation (1):

$$\frac{\partial^2 T_{11}}{\partial x \partial y} - \frac{\partial^2 T_{22}}{\partial x \partial y} - \frac{\partial^2 T_{12}}{\partial x^2} + \frac{\partial^2 T_{12}}{\partial y^2} = 0. \tag{18}$$

In the following, the analytic solutions of the constitutive Equations (15)–(17) with consideration of the compatibility condition (18) will be derived, respectively, for the two cases $k_2 = 0$ and $k_2 \neq 0$.

3.1. Case $k_2 = 0$

Many of the viscoelastic models derived from the Oldroyd 8-constant model (11) are characterized by $W_2 = W_4$, i.e., $k_2 = 0$. Examples are the Oldroyd-B, Maxwell-B, Johnson–Segalman models. Their solutions are all contained in this part.

In the case of $k_2 = 0$, the three components of the stress tensor can be separately solved by using the method of characteristics. The characteristic equations of (15)–(17) are given by

$$\frac{dx}{W_1 x} = \frac{dy}{-W_1 y} = \frac{dT_{11}}{(k_1 - 1)T_{11} - k_3 + 2}, \tag{19}$$

$$\frac{dx}{W_1 x} = \frac{dy}{-W_1 y} = \frac{dT_{22}}{-(k_1 + 1)T_{22} - k_3 - 2}, \tag{20}$$

$$\frac{dx}{W_1 x} = \frac{dy}{-W_1 y} = \frac{dT_{12}}{-1}. \tag{21}$$

The values of $k_1 - 1$ and $k_1 + 1$, arising in (19) and (20) as the coefficients of T_{11} and T_{22}, have a crucial effect on the structure of their solutions. Therefore, the solutions will be further investigated for the following three cases, respectively.

3.1.1. Case $(k_1 - 1)(k_1 + 1) \neq 0$

In this case, $k_1 \neq \pm 1$, we can obtain the following solutions by solving the Equations (19)–(21):

$$T_{11} = F_1(\psi) \cdot y^{\frac{1-k_1}{W_1}} + \frac{k_3 - 2}{k_1 - 1}, \tag{22}$$

$$T_{22} = F_2(\psi) \cdot y^{\frac{1+k_1}{W_1}} - \frac{k_3 + 2}{k_1 + 1}, \tag{23}$$

$$T_{12} = F_3(\psi) \cdot y^{\frac{1}{W_1}}. \tag{24}$$

$F_1(\psi)$, $F_2(\psi)$ and $F_3(\psi)$ are arbitrary functions of $\psi = xy$, and $\psi = const.$ represents the characteristic lines of the solutions. This result is consistent with that obtained by Van Gorder [14] for the stagnation-point flow of an upper convected Maxwell fluid, in which $k_1 = 2W_1$ and $k_3 = 0$.

Further satisfying the compatibility condition (18) yields the following restriction on functions F_1, F_2, and F_3:

$$\left[\psi F_1'' + \left(\frac{1-k_1}{W_1} + 1\right) F_1'\right] y^{-\frac{k_1}{W_1}} - \left[\psi F_2'' + \left(\frac{1+k_1}{W_1} + 1\right) F_2'\right] y^{\frac{k_1}{W_1}} \\ + \left[\psi^2 F_3'' + \frac{2}{W_1}\psi F_3' + \frac{1}{W_1}\left(\frac{1}{W_1} - 1\right) F_3\right] y^{-2} - F_3'' y^2 = 0. \tag{25}$$

the only way to satisfy this condition for any value of y is that all coefficients of y^k vanish. This induces three different cases depending on the value of $\frac{k_1}{W_1}$. The case of $\frac{k_1}{W_1} = \pm 2$ corresponds to the Oldroyd and Maxwell fluids. Their solutions for Weissenberg number $W_1 \neq \frac{1}{2}$ are covered in this investigation.

(i) If $\frac{k_1}{W_1} = 2$, only two restriction conditions for F_1, F_2, and F_3 can be obtained from (25), meaning that:

$$\psi F_1'' + \left(\frac{1}{W_1} - 1\right) F_1' + \psi^2 F_3'' + \frac{2}{W_1}\psi F_3' + \frac{1}{W_1}\left(\frac{1}{W_1} - 1\right) F_3 = 0, \tag{26}$$

$$\psi F_2'' + \left(\frac{1}{W_1} + 3\right) F_2' + F_3'' = 0. \tag{27}$$

Hence, F_1 and F_2 can be related to F_3 by

$$F_1 = -\psi F_3 + \left(1 - \frac{1}{W_1}\right) \int F_3 d\psi + C_1 \psi^{2-\frac{1}{W_1}} + C_2, \tag{28}$$

$$F_2 = -\psi^{-1} F_3 + \left(\frac{1}{W_1} + 1\right) \psi^{-\frac{1}{W_1} - 2} \int \psi^{\frac{1}{W_1}} F_3 d\psi + C_3 \psi^{-\frac{1}{W_1} - 2} + C_4, \tag{29}$$

where C_1, C_2, C_3 and C_4 are arbitrary constants, and F_3 is still an arbitrary function of $\psi = xy$. To avoid the integral terms emerging in the above relations, we replace F_3 by another arbitrary function $F(\psi)$ as follows:

$$F_3 = \left(\psi^{1-\frac{1}{W_1}} F'\right)' = \psi^{-\frac{1}{W_1}}\left(\psi F' - \frac{1}{W_1} F\right)'. \tag{30}$$

Substituting (30) into (28), (29) and then into (22)–(24), results in the final solutions of the stress tensor:

$$T_{11} = -x^{2-\frac{1}{W_1}} F'' + C_1 x^{2-\frac{1}{W_1}} + C_2 y^{\frac{1}{W_1}-2} + \frac{k_3-2}{k_1-1}, \tag{31}$$

$$T_{22} = -x^{-\frac{1}{W_1}} y^2 F'' + \frac{2}{W_1} x^{-\frac{1}{W_1}-1} y F' - \frac{1}{W_1}\left(\frac{1}{W_1}+1\right) x^{-\frac{1}{W_1}-2} F$$

$$+ C_3 x^{-\frac{1}{W_1}-2} + C_4 y^{\frac{1}{W_1}+2} - \frac{k_3+2}{k_1+1}, \tag{32}$$

$$T_{12} = x^{1-\frac{1}{W_1}} y F'' + \left(1 - \frac{1}{W_1}\right) x^{-\frac{1}{W_1}} F'. \tag{33}$$

In a recent investigation by [17], the solution of a wall-free stagnation-point flow for the Maxwell fluid with $W_1 \neq \frac{1}{2}$ satisfying the momentum equation was discovered. This corresponds to our solutions (31)–(33) with $k_1 = 2W_1$ and $k_3 = 0$. Physically, the components of the stress tensor should be limited everywhere, including at the stagnation point $(x, y) = (0, 0)$ and at infinity. Hence, further restrictions on the arbitrary constants C_1, C_2, C_3, C_4 and the arbitrary function $F(xy)$ in the solutions are needed. To avoid the singularity of the stress tensor, Ref. [17] suggested the choice:

$$F(xy) = (xy)^{2+\frac{1}{W_1}+\delta} e^{-b(xy)^2}, \quad C_1 = C_2 = C_3 = C_4 = 0, \tag{34}$$

where $\delta > 0$ and $b > 0$. However, this choice (34) cannot prevent the singularity as was claimed. To demonstrate this, as an example, we substitute (34) into (33) resulting in:

$$T_{12} = \left[a_1 + a_2(xy)^2 + a_3(xy)^4\right] e^{-b(xy)^2} \cdot (xy)^{1+\delta} \cdot y^{\frac{1}{W_1}} = G(xy) \cdot y^{\frac{1}{W_1}}, \tag{35}$$

where $a_1 = (2+\delta)(2+\delta+\frac{1}{W_1})$, $a_2 = -2b(6+2\delta+\frac{1}{W_1})$ and $a_3 = 4b^2$ are constant coefficients. For $W_1 > 0$, the particular solution (33) has the following properties:

- Along the characteristic curves $xy = 0$ and $xy \to \infty$, $T_{12} \to 0$, no singularity occurs.
- Along the characteristic curve $xy = c_0$, where c_0 is a non-zero finite constant, the value of $G(xy)$ is a bounded constant $G(c_0)$, while $y^{\frac{1}{W_1}}$ is singular at $y \to \infty$. An infinite shear stress T_{12} arises in the region far away from the stagnation point ($y \to \infty$) and near the y-axis ($x \to 0$).

With the choice (34), similar singularities also appear in the stress tensor components T_{11} and T_{22} far away from the stagnation point, for $y \to \infty$ (but $x \to 0$) or $x \to \infty$ (but $y \to 0$). As an example, the corresponding stress components T_{11} and T_{12} excluding the constant part $\frac{k_3-2}{k_1-1}$ are presented in Figures 2 and 3 for the case $W_1 = 0.25$, $\delta = 2$ and $b = 12$ with a much finer spatial resolution than [17] used. The present figures show that a singularity in the stress field may arise in the region far away from the stagnation point ($y \to \infty$) and near the y axis ($x \to 0$), as previously analytically recognized, while in [17], this tendency was invisible due to the rather coarse resolution employed.

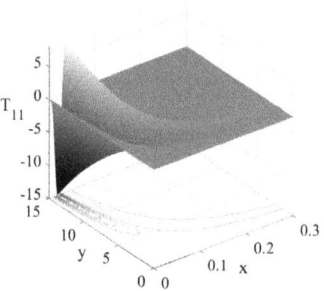

Figure 2. Stress component T_{11} with a spatial resolution of $x \in [0:0.001:0.3], y \in [0:0.001:15]$.

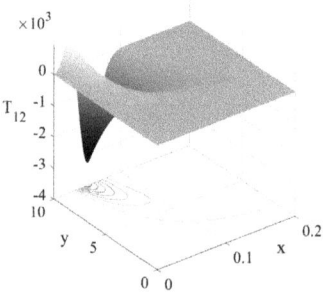

Figure 3. Stress component T_{12} with a spatial resolution of $x \in [0:0.001:0.2], y \in [0:0.001:10]$.

Actually, the appearance of the singularity at $(x,y) \to (0,\infty)$ or $(x,y) \to (\infty,0)$ is independent from the choice of the arbitrary function $F(xy)$. This can be easily recognized by observing the distribution of the stresses (31)–(33) along an arbitrary characteristic curve $xy = c_0$, where c_0 is non-zero finite constant. As an example, the term $x^{-\frac{1}{W_1}} F'(xy)$ tends to be infinite at $(x,y) \to (0,\infty)$ for $W_1 > 0$ or at $(x,y) \to (\infty,0)$ for $W_1 < 0$. These singularities are independent from the choice of $F(xy)$ and thus non-removable. However, possible singularity near the stagnation point $(x,y) \to (0,0)$ can be effectively prevented by choosing a reasonable function $F(xy)$, e.g., (34). This ensures that no singularity occurs in the stress field near the stagnation point. For a stagnation-point flow with the velocity field given by (13), the velocity becomes unbounded at $x \to \infty$ or $y \to \infty$ and is thus physically no longer meaningful. The singularity arising in the far field at $x \to \infty$ or $y \to \infty$ may not be relevant when a bounded stagnation-point flow is investigated. We should then focus on the analysis of singularity in a bounded area near the stagnation point.

(ii) If $\frac{k_1}{W_1} = -2$, performing the similar steps as for the above case gives the solutions:

$$T_{11} = x^{-\frac{1}{W_1}} y^2 F'' - \frac{2}{W_1} x^{-\frac{1}{W_1}-1} y F' + \frac{1}{W_1}\left(\frac{1}{W_1}+1\right) x^{-\frac{1}{W_1}-2} F$$
$$+ C_1 x^{-\frac{1}{W_1}-2} + C_2 y^{\frac{1}{W_1}+2} + \frac{k_3 - 2}{k_1 - 1}, \tag{36}$$

$$T_{22} = x^{2-\frac{1}{W_1}} F'' + C_3 x^{2-\frac{1}{W_1}} + C_4 y^{\frac{1}{W_1}-2} - \frac{k_3+2}{k_1+1}, \tag{37}$$

$$T_{12} = x^{1-\frac{1}{W_1}} y F'' + \left(1 - \frac{1}{W_1}\right) x^{-\frac{1}{W_1}} F'. \tag{38}$$

Similarly, by reasonably choosing of C_i ($i = 1, 2, 3, 4$) and $F(xy)$, singularity near the stagnation point can be avoided.

(iii) If $\frac{k_1}{W_1} \neq \pm 2$, we obtain four equations for F_1, F_2 and F_3 from (25). Solving them and then substituting them into (22)–(24) yield:

$$T_{11} = C_1 x^{-\frac{1-k_1}{W_1}} + C_2 y^{\frac{1-k_1}{W_1}} - \frac{k_3 - 2}{1 - k_1}, \tag{39}$$

$$T_{22} = C_3 x^{-\frac{1+k_1}{W_1}} + C_4 y^{\frac{1+k_1}{W_1}} - \frac{k_3 + 2}{1 + k_1}, \tag{40}$$

$$T_{12} = 0. \tag{41}$$

To prevent the singularity near the stagnation point $(x, y) = (0, 0)$, only one of C_1, C_2 and one of C_3, C_4 need to be zero according to the relative relationship between k_1 and W_1. This results in a stress distribution that depends on only one coordinate x or y.

3.1.2. Case $k_1 - 1 = 0$

In this case, the coefficient of T_{11} in (19) vanishes. The corresponding solution of T_{11} is simply given by

$$T_{11} = \frac{k_3 - 2}{W_1} \ln y + F_1, \tag{42}$$

where F_1 is an arbitrary function of $\psi = xy$. The solutions of T_{22} and T_{12} remain the same as given in (22)–(24). Again, considering the compatibility condition, the final solutions are obtained depending on the value of W_1. All intermediate steps, which are similar to those in the last Section 3.1.1, are omitted.

(i) If $W_1 = \frac{1}{2}$, we obtain the solutions:

$$T_{11} = 2(k_3 - 2) \ln y - F'' + C_1 \ln(xy) + C_2, \tag{43}$$

$$T_{22} = -x^{-2} y^2 F'' + 4 x^{-3} y F' - 6 x^{-4} F + C_3 x^{-4} + C_4 y^4 - \frac{k_3}{2} - 1, \tag{44}$$

$$T_{12} = x^{-1} y F'' - x^{-2} F'; \tag{45}$$

The logarithmic singularity in T_{11} caused by $\ln y$ at $y \to 0$ can only be avoided when $k_3 = 2$, i.e., $W_6 - W_7 = \frac{1}{2}$. In this case, the choice (34) for the arbitrary constants C_i ($i = 1, 2, 3, 4$) and arbitrary function $F(xy)$ is still suitable to prevent the singularity near the stagnation point.

However, for the Oldroyd model, i.e., the case of $W_6 - W_7 = 4(1 - \beta) W_1$ with $0 < \beta < 1$ and for the Maxwell model, i.e., the case of $W_6 - W_7 = 0$, the logarithmic singularity at the Weissenberg number $W_1 = \frac{1}{2}$ is unavoidable. The similar conclusion was also drawn by [11] for Maxwell fluid.

(ii) If $W_1 \neq \frac{1}{2}$:

$$T_{11} = \frac{k_3 - 2}{W_1} \ln y + C_1 \ln(xy) + C_2, \tag{46}$$

$$T_{22} = C_3 x^{-\frac{2}{W_1}} + C_4 y^{\frac{2}{W_1}} - \frac{k_3}{2} - 1, \tag{47}$$

$$T_{12} = 0. \tag{48}$$

The singularity caused by the term $\ln y$ can only be avoided if $k_3 = 2$. In addition, C_3 also has to be zero to prevent the singularity at $xy \to 0$. This choice will cause a uniform stress field, which may be physically disputable.

3.1.3. Case $k_1 + 1 = 0$

Analogous to the last two subsections, the solutions in this case are distinguished for the following cases.

(i) If $W_1 = \frac{1}{2}$:

$$T_{11} = x^{-2}y^2 F'' - 4x^{-3}yF' + 6x^{-4}F + 6C_1 x^{-4} + C_2 y^4 - \frac{k_3}{2} + 1,$$
$$T_{22} = 2(k_3 + 2)\ln y + F'' + C_3 \ln(xy) + C_4, \qquad (49)$$
$$T_{12} = x^{-1}yF'' - x^{-2}F'.$$

where the logarithmic singularity caused by $\ln y$ in T_{22} could only be avoided with the extra restriction $k_3 + 2 = 0$. However, for the Oldroyd model there is $k_3 = 2(1 - \beta)$ at $W_1 = \frac{1}{2}$ with $< 0\beta < 1$ and for the Maxwell model $k_3 = 0$, so that logarithmic singularity is non-removable. The similar conclusion was also drawn by [11] for a Maxwell fluid.

(ii) For $W_1 \neq \frac{1}{2}$:

$$T_{11} = C_1 x^{-\frac{2}{W_1}} + C_2 y^{\frac{2}{W_1}} - \frac{k_3}{2} + 1,$$
$$T_{22} = \frac{k_3 + 2}{W_1} \ln y + C_3 \ln(xy) + C_4, \qquad (50)$$
$$T_{12} = 0.$$

Similar to the last cases, the singularity in T_{11} and T_{22} can only be avoided at $k_3 = -2$ with $C_i = 0$ ($i = 1, 2, 3$). This again corresponds to a uniform stress field and thus may be physically disputable.

3.2. Case $k_2 \neq 0$

For some of the viscoelastic models derived from the Oldroyd 8-constant model k_2 is not zero, such as the Williams 3-constant Oldroyd model, Oldroyd-4-constant model, etc., details see [15]. For this case, the solution of T_{12} remains the same as in (24). T_{11} and T_{22} need to be solved from the coupled PDE system consisting of (15) and (16) with non-constant coefficients.

Applying the method of characteristics, the PDE system reduces to the following ODE system:

$$W_1 y \frac{dT_{11}}{dy} = (1 - k_1)T_{11} + k_2 T_{22} + k_3 - 2, \qquad (51)$$

$$W_1 y \frac{dT_{22}}{dy} = -k_2 T_{11} + (1 + k_1)T_{22} + k_3 + 2, \qquad (52)$$

along the characteristic curves $xy = \text{const}$. Furthermore, using the transformation:

$$y = e^{W_1 \tilde{y}} \qquad (53)$$

Equations (51) and (52) can be transformed into an ODE system with constant coefficients:

$$\frac{dT_{11}}{d\tilde{y}} = (1 - k_1)T_{11} + k_2 T_{22} + k_3 - 2, \qquad (54)$$

$$\frac{dT_{22}}{d\tilde{y}} = -k_2 T_{11} + (1 + k_1)T_{22} + k_3 + 2, \qquad (55)$$

or in matrix form:

$$\frac{d\mathbf{t}}{d\tilde{y}} = \mathbf{A}\mathbf{t} + \mathbf{b} \qquad (56)$$

with:

$$\mathbf{t} = \begin{pmatrix} T_{11} \\ T_{22} \end{pmatrix}, \quad \mathbf{A} = \begin{pmatrix} 1 - k_1 & k_2 \\ -k_2 & 1 + k_1 \end{pmatrix}, \quad \mathbf{b} = \begin{pmatrix} k_3 - 2 \\ k_3 + 2 \end{pmatrix}. \qquad (57)$$

Eliminating T_{22} by combining (54) and (55), gives an second-order ODE with constant coefficients for T_{11}:

$$\frac{d^2 T_{11}}{d\tilde{y}^2} - 2\frac{dT_{11}}{d\tilde{y}} + (1 - k_1^2 + k_2^2)T_{11} = k_2(k_3 - 2) - (k_1 + 1)(k_3 + 2). \qquad (58)$$

The homogeneous solution of this ODE depends on the type of solutions of the corresponding characteristic equation:

$$\lambda^2 - 2\lambda + (1 - k_1^2 + k_2^2) = 0. \qquad (59)$$

Its eigenvalues are:

$$\lambda_{1,2} = 1 \pm \sqrt{k_1^2 - k_2^2}. \qquad (60)$$

Depending on the value of $k_1^2 - k_2^2$, the homogeneous solution of (58) has the following three different cases:

(i) If $k_1^2 - k_2^2 > 0$, λ_1 and λ_2 are real numbers with $\lambda_1 \neq \lambda_2$, we obtain:

$$T_{11,h} = C_1 \cdot e^{\lambda_1 \tilde{y}} + C_2 \cdot e^{\lambda_2 \tilde{y}}; \qquad (61)$$

(ii) If $k_1^2 - k_2^2 = 0$, i.e., $\lambda_1 = \lambda_2 = 1$, we obtain:

$$T_{11,h} = (C_1 + C_2 \tilde{y}) \cdot e^{\tilde{y}}; \qquad (62)$$

(iii) If $k_1^2 - k_2^2 < 0$, $\lambda_{1/2} = 1 \pm i\omega$ with $\omega = \sqrt{k_2^2 - k_1^2}$ are conjugates complex numbers, the homogeneous solution of $T_{11,h}$ is given by

$$T_{11,h} = e^{\tilde{y}}[C_1 \sin(\omega \tilde{y}) + C_2 \cos(\omega \tilde{y})]. \qquad (63)$$

The particular solution of the ODE (58), depending on the value of the coefficient $1 - k_1^2 + k_2^2$, includes two cases as follows:

(i) If $1 - k_1^2 + k_2^2 \neq 0$, we obtain:

$$T_{11,p} = \frac{k_2(k_3 + 2) - (k_1 + 1)(k_3 - 2)}{1 - k_1^2 + k_2^2}; \qquad (64)$$

(ii) If $1 - k_1^2 + k_2^2 = 0$, the particular solution takes the form:

$$T_{11,p} = -\frac{1}{2}[k_2(k_3 + 2) - (k_1 + 1)(k_3 - 2)]\tilde{y}. \qquad (65)$$

The general solution of the ODE (58) is expressed as

$$T_{11} = T_{11,h} + T_{11,p}. \qquad (66)$$

Obviously, the second case of the particular solution, (65), can only be combined with the first case of the homogeneous solution, (61).

The solution of T_{22} can be directly determined by inserting the solution for T_{11}, (66), into (54):

$$T_{22} = \frac{1}{k_2}\left(\frac{dT_{11}}{d\tilde{y}} + (k_1 - 1)T_{11} - k_3 + 2\right). \qquad (67)$$

To obtain the final solutions of the original PDE system consisting of (15) and (16), we only need to replace the variable $\tilde{y}$ with y by the transformation $\tilde{y} = \ln y^{\frac{1}{W_1}}$, and all arbitrary constants C_i with arbitrary functions $F_i(xy)$. Finally, together with the solution of T_{12} in (24), we obtain the complete solutions of the PDE system (15)–(17) as follows.

(i) If $k_1^2 - k_2^2 > 0$ and $k_1^2 - k_2^2 \neq 1$:

$$T_{11} = k_2 \left(F_1 \cdot y^{\frac{1+\omega}{W_1}} + F_2 \cdot y^{\frac{1-\omega}{W_1}} \right) + \frac{k_2(k_3+2) - (k_1+1)(k_3-2)}{1 - k_1^2 + k_2^2},$$

$$T_{22} = (k_1+\omega)F_1 \cdot y^{\frac{1+\omega}{W_1}} + (k_1-\omega)F_2 \cdot y^{\frac{1-\omega}{W_1}} + \frac{(k_1-1)(k_3+2) - k_2(k_3-2)}{1 - k_1^2 + k_2^2}, \quad (68)$$

$$T_{12} = F_3 \cdot y^{\frac{1}{W_1}}.$$

(ii) If $k_1^2 - k_2^2 = 1$, the solutions take the form:

$$T_{11} = k_2 \left(F_1 \cdot y^{\frac{2}{W_1}} + F_2 \right) - \frac{1}{2W_1} [k_2(k_3+2) - (k_1+1)(k_3-2)] \ln y,$$

$$T_{22} = (k_1+1)F_1 \cdot y^{\frac{2}{W_1}} + (k_1-1)F_2 - \frac{k_1-1}{2k_2 W_1} [k_2(k_3+2) - (k_1+1)(k_3-2)] \ln y \quad (69)$$

$$- \frac{1}{2k_2} [k_2(k_3+2) - (k_1-1)(k_3-2)],$$

$$T_{12} = F_3 \cdot y^{\frac{1}{W_1}}.$$

(iii) If $k_1^2 - k_2^2 = 0$:

$$T_{11} = k_2 F_1 \cdot y^{\frac{1}{W_1}} + \frac{k_2}{W_1} F_2 \cdot \ln y \cdot y^{\frac{1}{W_1}} + \frac{k_2(k_3+2) - (k_1+1)(k_3-2)}{1 - k_1^2 + k_2^2},$$

$$T_{22} = k_1 F_1 \cdot y^{\frac{1}{W_1}} + F_2 \cdot y^{\frac{1}{W_1}} + \frac{k_1}{W_1} F_2 \cdot \ln y \cdot y^{\frac{1}{W_1}} + \frac{(k_1-1)(k_3+2) - k_2(k_3-2)}{1 - k_1^2 + k_2^2}, \quad (70)$$

$$T_{12} = F_3 \cdot y^{\frac{1}{W_1}}.$$

(iv) If $k_1^2 - k_2^2 < 0$:

$$T_{11} = k_2 \left[F_1 \cdot \cos\left(\ln y^{\frac{\omega}{W_1}} \right) + F_2 \cdot \sin\left(\ln y^{\frac{\omega}{W_1}} \right) \right] \cdot y^{\frac{1}{W_1}}$$

$$+ \frac{k_2(k_3+2) - (k_1+1)(k_3-2)}{1 - k_1^2 + k_2^2},$$

$$T_{22} = \left[(k_1 F_1 + \omega F_2) \cdot \cos\left(\ln y^{\frac{\omega}{W_1}} \right) + (k_1 F_2 - \omega F_1) \cdot \sin\left(\ln y^{\frac{\omega}{W_1}} \right) \right] \cdot y^{\frac{1}{W_1}} \quad (71)$$

$$+ \frac{(k_1-1)(k_3+2) - k_2(k_3-2)}{1 - k_1^2 + k_2^2},$$

$$T_{12} = F_3 \cdot y^{\frac{1}{W_1}}.$$

Here, $\omega = \sqrt{k_1^2 - k_2^2}$, F_1, F_2 and F_3 are arbitrary functions of xy.

The logarithmic singularity caused by the term $\ln y$ in the solution (69) can only be avoided under the special circumstance $k_2(k_3+2) - (k_1+1)(k_3-2) = 0$. Similar to the cases in the last subsection, satisfying the compatibility condition yields three to four restriction equations on the functions F_1, F_2, and F_3 according to the relationship between ω and k_1. Furthermore, to avoid the singularity near the stagnation point $(x, y) = (0, 0)$, the obtained final stress distribution either depends on only one variable or is again meaninglessly uniform. These tedious derivations are not given.

4. Analytical Solutions of a Near-Wall Stagnation-Point Flow

Analytic investigations of near-wall stagnation-point flows of viscoelastic fluids have rarely been dealt with. These investigations only treat the constitutive stress equations, however, the momentum equation is not satisfied, as can be seen, e.g., in [12,13]. The difficulty is the accessibility of analytic solutions. In contrast to a wall-free stagnation-point flow, the similarity solution of the velocity field (12) in a near-wall stagnation-point flow must satisfy the no-slip conditions at the wall $y = 0$. This requires $f'(0) = 0$ in addition to $f(0) = 0$. To satisfy this condition, the simplest choice is $f(y) = y^2$ as have been employed in [12,13]. Here, we investigate a more general form $f(y) = y^\alpha$ with $\alpha > 1$ and $y > 0$. The corresponding dimensionless velocity field takes the form:

$$\mathbf{u} = n(\alpha x y^{\alpha-1}, -y^\alpha), \tag{72}$$

where $n = \pm 1$ denotes the direction of the flow. As displayed in Figure 4, $n = 1$ corresponds to the inflow toward the stagnation point and $n = -1$ indicates the outflow away from the stagnation point.

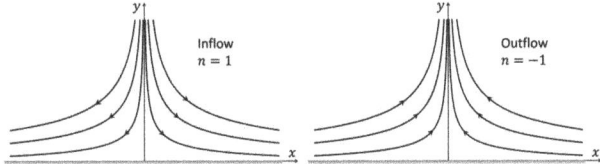

Figure 4. Sketch of velocity field in a near-wall stagnation-point flow.

For such a velocity field, an analytical solution is only attainable for less complex viscoelastic models. In this paper, we consider the most commonly employed Oldroyd-B model. As described in Section 2, the constitutive model equation for an Oldroyd-B fluid can be expressed for $\mathbf{T}_p$, which is the polymetric contribution to the stress tensor. Omitting the index p, the dimensionless model equation takes the form:

$$\mathbf{T} + W\overset{\triangledown}{\mathbf{T}} = 2\beta \mathbf{D}, \tag{73}$$

where W is the only Weissenberg number in this model, and $\beta = \frac{\eta_p}{\eta_p + \eta_s} \in [0,1]$ is the retardation parameter.

As has been shown, the Reynolds number does not appear in the constitutive model equations, so its value does not affect the singularities of the models. Therefore, it is convenient and common to assume the Reynolds number $Re = 1$. In this case, the corresponding momentum equation is given as

$$\mathbf{u} \cdot \nabla \mathbf{u} = -\nabla p + (1-\beta) \nabla \cdot \mathbf{D} + \nabla \cdot \mathbf{T}. \tag{74}$$

Inserting the velocity field (72) into the constitutive Equation (73), we obtain a PDE system, consisting of three equations as follows:

$$W\alpha x y^{\alpha-1} \frac{\partial T_{11}}{\partial x} - W y^\alpha \frac{\partial T_{11}}{\partial y} = (2W\alpha y^{\alpha-1} - n)T_{11} + 2W\alpha(\alpha-1)xy^{\alpha-2}T_{12} + 2\beta \alpha y^{\alpha-1}, \tag{75}$$

$$W\alpha x y^{\alpha-1} \frac{\partial T_{22}}{\partial x} - W y^\alpha \frac{\partial T_{22}}{\partial y} = -(2W\alpha y^{\alpha-1} + n)T_{22} - 2\beta \alpha y^{\alpha-1}, \tag{76}$$

$$W\alpha x y^{\alpha-1} \frac{\partial T_{12}}{\partial x} - W y^\alpha \frac{\partial T_{12}}{\partial y} = -nT_{12} + W\alpha(\alpha-1)xy^{\alpha-2}T_{22} + \beta\alpha(\alpha-1)xy^{\alpha-2}. \tag{77}$$

The solutions of this PDE system must also satisfy the momentum Equation (74). Applying the curl operator to (74) in order to eliminate the pressure p results in the compatibility condition:

$$\frac{\partial^2 T_{11}}{\partial x \partial y} - \frac{\partial^2 T_{22}}{\partial x \partial y} - \frac{\partial^2 T_{12}}{\partial x^2} + \frac{\partial^2 T_{12}}{\partial y^2}$$
$$= 2\alpha(\alpha-1)xy^{2\alpha-3} - n(1-\beta)\alpha(\alpha-1)(\alpha-2)(\alpha-3)xy^{\alpha-4}. \tag{78}$$

Then, we will give the analytical solution of the PDE system (75)–(77) and discuss its compatibility with Equation (78).

4.1. Analytical Solutions of the Model Equation

We firstly solve (76) for T_{22}, which is uncoupled from (75) and (77). Then, Equations (75) and (77) can be successively solved. Using the method of characteristic for (76) yields the characteristic equations:

$$\frac{dx}{W \epsilon xy^{\alpha-1}} = \frac{dy}{-Wy^\alpha} = \frac{dT_{22}}{-(2W\alpha y^{\alpha-1}+n)T_{22} - 2\beta\alpha y^{\alpha-1}}. \tag{79}$$

Its solution takes the form:

$$T_{22} = F_1(xy^\alpha) \cdot y^{2\alpha} e^{-\frac{y^{1-\alpha}}{nW(\alpha-1)}} + \frac{2\beta}{W}\alpha y^{2\alpha} e^{-\frac{y^{1-\alpha}}{nW(\alpha-1)}} \int y^{-2\alpha-1} e^{\frac{y^{1-\alpha}}{nW(\alpha-1)}} dy, \tag{80}$$

where $F_1(xy^\alpha)$ is an arbitrary function of xy^α, and $xy^\alpha = const.$ represents the characteristic curves of the solution. Substituting (80) into (77) and solving the resulting PDE with the same method as above yields:

$$T_{12} = F_2(xy^\alpha) \cdot e^{-\frac{y^{1-\alpha}}{nW(\alpha-1)}} - F_1(xy^\alpha) \cdot \alpha xy^{2\alpha-1} e^{-\frac{y^{1-\alpha}}{nW(\alpha-1)}} + \frac{\beta}{2nW^2}xy^{-\alpha}$$
$$+ \frac{\beta}{nW^2}xy^\alpha \left(\alpha y^{\alpha-1} + \frac{1}{2nW}\right) e^{-\frac{y^{1-\alpha}}{nW(\alpha-1)}} \int y^{-3\alpha} e^{\frac{y^{1-\alpha}}{nW(\alpha-1)}} dy, \tag{81}$$

with the arbitrary functions F_1, F_2 of xy^α. Similarly, substituting (81) into (75) and solving the consequent PDE result in the solution for T_{11}:

$$T_{11} = F_3(xy^\alpha) \cdot y^{-2\alpha} e^{-\frac{y^{1-\alpha}}{nW(\alpha-1)}} - 2\alpha F_2(xy^\alpha) \cdot xy^{-1} e^{-\frac{y^{1-\zeta}}{nW(\alpha-1)}}$$
$$+ \zeta^2 F_1(xy^\alpha) \cdot x^2 y^{2\alpha-2} e^{-\frac{y^{1-\alpha}}{nW(\alpha-1)}} - \frac{2\alpha\beta}{W} y^{-2\alpha} e^{-\frac{y^{1-\alpha}}{nW(\alpha-1)}} \int y^{2\alpha-1} e^{\frac{y^{1-\alpha}}{nW(\alpha-1)}} dy$$
$$- \frac{\alpha\beta}{W^2} x^2 y^{-\zeta-1} - \frac{\alpha\beta}{nW^2} x^2 y^{\alpha-1} \left(\alpha y^{\alpha-1} + \frac{1}{nW}\right) e^{-\frac{y^{1-\alpha}}{nW(\alpha-1)}} \int y^{-3\alpha} e^{\frac{y^{1-\alpha}}{nW(\alpha-1)}} dy \tag{82}$$
$$- \frac{\alpha^2\beta}{nW^2} x^2 e^{-\frac{y^{1-\alpha}}{nW(\alpha-1)}} \int y^{-\alpha-2} e^{\frac{y^{1-\alpha}}{nW(\alpha-1)}} dy,$$

where F_1, F_2, and F_3, again, are arbitrary functions of xy^α.

The solutions in the case $\alpha = 2$ are consistent with that obtained by [12,13].

4.2. Compatibility Condition

Substituting the solutions of the stress components (80)–(82) into the compatibility condition (78) yields the following restriction on the functions F_1, F_2, and F_3:

$$\begin{aligned}
&-\alpha F_1' \cdot y^{3\alpha-1} - \left(F_2'' + \frac{1}{nW}F_1'\right) \cdot y^{2\alpha} + \alpha\left(\alpha F_1'\psi + 2(\alpha-1)F_1\right)\psi y^{\alpha-3} \\
&+ \alpha\left(-\alpha F_2''\psi^2 - (3\alpha-1)F_2'\psi + 2F_2 - \frac{\alpha}{nW}F_1'\psi^2 - \frac{\alpha-2}{nW}F_1\psi\right) \cdot y^{-2} \\
&+ \alpha\left(F_3''\psi - F_3' - \frac{3}{nW}F_2 - \frac{1}{W^2}F_1\psi\right) \cdot y^{-\alpha-1} + \frac{1}{nW}\left(\frac{1}{nW}F_2 + F_3'\right) \cdot y^{-2\alpha} \\
&+ \psi \cdot G(y) = 0,
\end{aligned} \quad (83)$$

where $\psi = xy^\alpha$ and:

$$\begin{aligned}
G(y) =& 2\beta\Big[\frac{1}{nW^2}\alpha(\alpha-1)y^{-2\alpha} + \frac{1}{W}\alpha^2(\alpha+1)y^{-\alpha-1} + n\alpha^2(\alpha+1)^2 y^{-2} \\
&+ 4W\alpha^2(\alpha+1)(\alpha-1)y^{\alpha-3}\Big] \cdot \int y^{-3} e^{\frac{y^{1-\alpha}}{nW(\alpha-1)}} dy \\
&+ \left[n\alpha\left(\alpha(\alpha^2 - 2\alpha + 5) - 2(1-\beta)(2\alpha^2 - 3\alpha + 3)\right)y^{-4} - 2\alpha(\alpha-1)y^{\alpha-3}\right. \\
&\left.+ 4\beta W\alpha^2(\alpha+1)(\alpha-1)y^{\alpha-5} + \frac{2}{W}(1-\beta)\alpha(\alpha-1)y^{-\alpha-3}\right] \cdot e^{\frac{y^{1-\alpha}}{nW(\alpha-1)}}.
\end{aligned} \quad (84)$$

The condition (83) could only be satisfied for any value of y along the characteristic line $\psi = 0$, which denotes the symmetric line $x = 0$. Along this line, the compatibility Equation (83) will be greatly simplified to:

$$\begin{aligned}
&-\alpha F_1' \cdot y^{3\alpha-1} - \left(F_2'' + \frac{1}{nW}F_1'\right) \cdot y^{2\alpha} + 2\alpha F_2 \cdot y^{-2} \\
&- \alpha\left(F_3' + \frac{3}{nW}F_2\right) \cdot y^{-\alpha-1} + \frac{1}{nW}\left(\frac{1}{nW}F_2 + F_3'\right) \cdot y^{-2\alpha} = 0,
\end{aligned} \quad (85)$$

which leads to:

$$F_1(\psi) = F_1(\psi=0) = C_1, \qquad F_2 = 0, \qquad F_3(\psi) = F(\psi=0) = C_3, \quad (86)$$

where C_1 and C_3 are arbitrary constants. The corresponding stress components along the y-axis are given by

$$\begin{aligned}
T_{11} &= C_3 y^{-2\alpha} e^{-\frac{y^{1-\alpha}}{nW(\alpha-1)}} - \frac{2\beta}{W}\alpha y^{-2\alpha} e^{-\frac{y^{1-\alpha}}{nW(\alpha-1)}} \int y^{2\alpha-1} e^{\frac{y^{1-\alpha}}{nW(\alpha-1)}} dy, \\
T_{22} &= C_1 y^{2\alpha} e^{-\frac{y^{1-\alpha}}{nW(\alpha-1)}} + \frac{2\beta}{W}\alpha y^{2\alpha} e^{-\frac{y^{1-\alpha}}{nW(\alpha-1)}} \int y^{-2\alpha-1} e^{\frac{y^{1-\alpha}}{nW(\alpha-1)}} dy, \\
T_{12} &= 0.
\end{aligned} \quad (87)$$

Furthermore, applying the L'Hospital's rule, we can analyze the behavior of the stress components near the stagnation point, respectively. Since $\alpha > 1$ and $y > 0$, we obtain:

$$\lim_{y \to 0} y^{\pm 2\alpha} e^{-\frac{y^{1-\alpha}}{nW(\alpha-1)}} = \begin{cases} 0 & \text{for } n = 1, \\ \infty & \text{for } n = -1, \end{cases} \quad (88)$$

$$\lim_{y \to 0} y^{\pm 2\alpha} e^{-\frac{y^{1-\alpha}}{nW(\alpha-1)}} \int y^{\mp 2\alpha - 1} e^{\frac{y^{1-\alpha}}{nW(\alpha-1)}} dy = 0. \quad (89)$$

To avoid the singularity for the case of outflow, we can choose $C_1 = C_3 = 0$. Hence, for both the cases of inflow and outflow, regular solutions exist near the stagnation point.

As mentioned above, the solutions under the velocity assumption (72) can satisfy the conservation of the momentum equation only on the symmetric axis. In order to obtain a solution that satisfies the compatibility equation, a more suitable function $f(y)$ to describe the velocity field has to be proposed. This is an interesting topic for future study. Actually, in many previous investigations for both wall-free and near-wall stagnation-point flows, the singularity of the stress field was analyzed by forcing the solution to be independent of x, see, e.g., [8,12,13].

5. Conclusions

In this paper, we obtained the analytical solutions of the stress distributions of a wall-free steady stagnation-point flow with the proposed velocity profile $\mathbf{u} = (ax, -ay)$ for the Oldroyd 8-constant model under different material relations. All solutions here are compatible with the momentum conservation equation, and their singularities are discussed in detail. The results show that all models have singularities near the stagnation point. Most of these can be effectively avoided by appropriately choosing the model parameters and free functions. However, the singularity in Oldroyd-B and Maxwell-B models at the Weissenberg number $Wi = \frac{1}{2}$ is non-removable. The results in this investigation can be directly used to analyze the stress contributions and their singularities of a wide spectrum of viscoelastic models derived from the Oldroyd 8-constant model. Furthermore, for the Oldroyd-B model, we obtained the analytical solutions of the stress tensor in a near-wall stagnation-point flow with the proposed velocity profile $\mathbf{u} = \pm(\alpha x y^{\alpha-1}, -y^\alpha)$ with $\alpha > 1$ and show that the solutions can satisfy the momentum conservation along the streamline passing the stagnation point. To the best of our knowledge, such an analysis of the compatibility of model solutions with momentum conservation in a near-wall stagnation-point flow is absent in the literature. To further investigate near-wall stagnation-point flows, a more general velocity profile has to be proposed.

Furthermore, nowadays, there is a tendency to make comparisons, at least qualitative, between complete theoretical models that are not computer-implementable and experimental models in order to understand which terms of the theoretical models are not covered by the experimental models, and moreover, to understand the correspondences between terms of the theoretical and experimental models, see, e.g., [18]. For the next step, it will be an interesting and valuable work to further compare the models discussed in this article under different material parameter relations with the experimental models.

Author Contributions: Conceptualization, M.O. and Y.W.; validation, J.L.; formal analysis, J.L., M.O. and Y.W.; investigation, J.L.; writing—original draft preparation, J.L.; writing—review and editing, M.O. and Y.W.; visualization, J.L.; supervisions, M.O. and Y.W.; funding acquisition, M.O. and Y.W. All authors have read and agreed to the published version of the manuscript.

Funding: This research was funded by the Deutsche Forschungsgemeinschaft (DFG—German Research Foundation). The OAP was funded by Technical University of Darmstadt.

Institutional Review Board Statement: Not applicable.

Informed Consent Statement: Not applicable.

Data Availability Statement: Not applicable.

Acknowledgments: We acknowledge support by the Deutsche Forschungsgemeinschaft (DFG— German Research Foundation) and the Open Access Publishing Fund of Technical University of Darmstadt, and further partial funding by the DFG as part of the OB 96/46-1 and OB 96/52-1 projects.

Conflicts of Interest: The authors declare no conflict of interest.

References

1. Giesekus, H. *Phänomenologische Rheologie: Eine Einführung*; Springer: Berlin/Heidelberg, Germany, 1994.
2. Dillen, S.; Oberlack, M.; Wang, Y. Analytical investigation of rotationally symmetrical oscillating flows of viscoelastic fluids. *J. Non–Newton. Fluid Mech.* **2019**, *272*, 104168. [CrossRef]

3. Ma, B.; Wang, Y.; Kikker, A. Analytical solutions of oscillating Couette-Poiseuille flows for the viscoelastic Oldroyd B fluid. *Acta Mech.* **2019**, *230*, 2249–2266. [CrossRef]
4. Saengow, C.; Giacomin, A.J.; Kolitawong, C. Exact analytical solution for large-amplitude oscillatory shear flow from Oldroyd 8-constant framework: Shear stress. *Phys. Fluids* **2017**, *29*, 043101. [CrossRef]
5. Schlichting, H. *Boundary-Layer Theory*, 6th ed.; McGraw-Hill: New York, NY, USA, 1968; p. 96.
6. Phan-Thien, N. Plane and axi-symmetric stagnation flow of a Maxwellian fluid. *Rheol. Acta* **1983**, *22*, 127–130. [CrossRef]
7. Phan-Thien, N. Stagnation flows for the Oldoryd-B fluid. *Rheol. Acta* **1984**, *23*, 172–176. [CrossRef]
8. Renardy, M. A comment on smoothness of viscoelastic stresses. *J. Non–Newton. Fluid Mech.* **2006**, *138*, 204–205. [CrossRef]
9. Thomas, B.; Shelley, M. Emergence of singular structures in Oldroyd-B fluids. *Phys. Fluids* **2007**, *19*, 103103. [CrossRef]
10. Cruz, D.O.A.; Pinho, F.T. Analytical solution of steady 2D wall-free extensional flows of UCM fluids. *J. Non–Newton. Fluid Mech.* **2015**, *223*, 157–164. [CrossRef]
11. Meleshko, S.V.; Moshkin, N.P.; Pukhnachev, V.V.; Samatova,V. On steady two-dimensional analytical solutions of the viscoelastic Maxwell equation. *J. Non–Newton. Fluid Mech.* **2019**, *270*, 1–7. [CrossRef]
12. Becherer, P.; van Saarloosa, W.; Morozovb, A.N. Stress singularities and the formation of birefringent strands in stagnation flows of dilute polymer. *J. Non–Newton. Fluid Mech.* **2008**, *157*, 126–132. [CrossRef]
13. Van Gorder, R.A.; Vajravelu, K.; Akyildiz, F.T. Viscoelastic stresses in the stagnation flow of a dilute polymer solution. *J. Non–Newton. Fluid Mech.* **2009**, *161*, 94–100. [CrossRef]
14. Van Gorder, R.A. Do general viscoelastic stresses for the flow of an upper convected Maxwell fluid satisfy the momentum equation? *Maccanica* **2012**, *47*, 1977–1985. [CrossRef]
15. Saengow, C.; Giacomin, A.J.; Grizzuti, N.; Pasquino, R. Startup steady shear flow from the Oldroyd 8-constant framework. *Phys. Fluids* **2019**, *31*, 063101. [CrossRef]
16. Oldroyd, J.G. Non–Newtonien effects in steady motion of some idealized elastico-viscous liquids. *Proc. R. Soc. Lond. Ser. Math. Phys. Sci.* **1958**, *245*, 278–297.
17. Meleshko, S.V.; Moshkin, N.P.; Pukhnachev,V.V. On exact analytical solutions of equations of Maxwell incompressible. *Int. J. Non-Linear Mech.* **2018**, *105*, 152–157. [CrossRef]
18. Versaci, M.; Palumbo, A. Magnetorheological Fluids: Qualitative comparison between a mixture model in the Extended Irreversible Thermodynamics framework and an Herschel–Bulkley experimental elastoviscoplastic model. *Int. J. Non-Linear Mech.* **2020**, *118*, 103288. [CrossRef]

Article

Different Formulations to Solve the Giesekus Model for Flow between Two Parallel Plates

Laison Junio da Silva Furlan [1,†], Matheus Tozo de Araujo [1,†], Analice Costacurta Brandi [2], Daniel Onofre de Almeida Cruz [3] and Leandro Franco de Souza [1,*]

1. Department of Applied Mathematics and Statistics, University of Sao Paulo, Sao Carlos 13566-590, Brazil; laisonfurlan@usp.br (L.J.d.S.F.); mtaraujo@icmc.usp.br (M.T.d.A.)
2. Department of Mathematics and Computer Science, Sao Paulo State University, Presidente Prudente 19060-900, Brazil; analice.brandi@unesp.br
3. Department of Mechanical Engineering, Federal University of Rio de Janeiro, Rio de Janeiro 21941-972, Brazil; doac@mecanica.coppe.ufrj.br
* Correspondence: lefraso@icmc.usp.br
† These authors contributed equally to this work.

Abstract: This work presents different formulations to obtain the solution for the Giesekus constitutive model for a flow between two parallel plates. The first one is the formulation based on work by Schleiniger, G; Weinacht, R.J., [*Journal of Non-Newtonian Fluid Mechanics*, **40**, 79–102 (1991)]. The second formulation is based on the concept of changing the independent variable to obtain the solution of the fluid flow components in terms of this variable. This change allows the flow components to be obtained analytically, with the exception of the velocity profile, which is obtained using a high-order numerical integration method. The last formulation is based on the numerical simulation of the governing equations using high-order approximations. The results show that each formulation presented has advantages and disadvantages, and it was investigated different viscoelastic fluid flows by varying the dimensionless parameters, considering purely polymeric fluid flow, closer to purely polymeric fluid flow, solvent contribution on the mixture of fluid, and high Weissenberg numbers.

Keywords: Giesekus model; flow between two parallel plates; exact solution; numerical solution; high-order approximations; high Weissenberg number

1. Introduction

The solution for the velocity and extra-stress tensor distribution in a viscoelastic fluid flow using a specific model can be obtained numerically and, sometimes, analytically. Each specific model has its own complexity and limitations, compared to the real viscoelastic fluids. The numerical solution of a laminar viscoelastic fluid flow is necessary for many flow analyses, for instance, in laminar-turbulent transition flow studies. The fluid flow components of this laminar flow can be obtained easily with some viscoelastic models, for instance, the velocity and extra-stress tensor field solutions for the Oldroyd-B model for the flow between two parallel plates. For some other models, the solution can require more profound analysis and mathematical and numerical tools to be obtained, even for simplified geometries.

In general, practical problems do not allow for analytical solutions due to their complexity. For this reason, numerical methods for simulating non-Newtonian fluid flows have been part of a very active area of research. Techniques for simulating viscoelastic flows have been used to solve different constitutive models such as Oldroyd-B [1,2], FENE-P [3,4], FENE-CR [5,6], PTT [4,7,8], and Giesekus [9,10]. The fluids that are elastic and have a constant viscosity are known as Boger fluids and the Maxwell, Oldroyd-B, and Giesekus models are suitable to simulate these fluids type [11].

For the Giesekus constitutive model [12], Yoo and Choi [9] studied the analytical solution for Couette and Poiseuille flows. For the Poiseuille flow, they obtained a semi-analytical solution for the mean velocity. The values were obtained by the integration (numerical, due to the complexity of the resulting equation) of the expression obtained for the derivative of the mean velocity.

Schleiniger and Weinacht [10] continued Yoo and Choi's [9] studies, presenting a weak and classical solution for Poiseuille flow, with and without solvent viscosity contribution. However, just like Yoo and Choi [9], the presented solution for the average velocity is obtained implicitly.

More recently, Raisi [13] presented an approximation for the solution of the Couette–Poiseuille flow for the Giesekus model. However, its results depend on the numerical solution of the shear stress at the stationary plate by the Newton–Raphson method.

Ferrás [8] also carried out studies for the solution of the Poiseuille flow using the Giesekus model, and showed an agreement between the results obtained by the mentioned works and the guarantees of the solution existence for the branch $1/2 < \alpha_G \leq 1$.

Tomé et al. [14] presented a solution method for the Giesekus viscoelastic fluid flow based on work by Schleiniger and Weinacht [10], where an analytical solution for the flow between two parallel plates problem was proposed. In their work, the authors considered a purely polymeric fluid flow.

Among the differential constitutive models, the laminar flow solution with the Giesekus model can not be obtained directly, because the extra-stress tensor appears non-linearly through the quadratic term. On the other hand, the model is considered to better approximate the rheology of polymers [15,16] and has the advantage of simplicity where only three parameters are involved: the temporal relaxation λ, the mobility parameter α_G, and the viscosity of the polymer η_p. Furthermore, the Giesekus model is able to predict the first and second normal-stress differences.

The current study presents different formulations to obtain the solution for the Giesekus constitutive model considering the flow between two parallel plates (Poiseuille flow for Newtonian fluid flow). The first one is the formulation proposed by Schleiniger and Weinacht [10]. The second formulation is the independent variable change, a new method proposed here, and the last method is the numerical simulation of the governing equations using high-order approximations. Each formulation to solve the laminar flow between two parallel plates has its advantages and disadvantages, and these features are explored here.

The paper is divided as follows: Section 2 presents the governing equations; the different formulations to obtain the laminar solution are presented in Section 3, including a semi-analytical solution obtained through the results presented by Schleiniger and Weinacht [10], a formulation to solve considering the independent variable change, and a numerical formulation through the high-order numerical approximation. Section 4 shows the results obtained with each formulation to solve the governing equations, investigating the limitations of each solver. The main conclusions are presented in Section 5.

2. Mathematical Formulation

In this paper, we consider a non-Newtonian, two-dimensional, and incompressible fluid flow, which is assumed to be unsteady and without body forces. The dimensionless governing equations are given by the continuity equation:

$$\nabla \cdot \mathbf{u} = 0, \tag{1}$$

and the momentum equation:

$$\frac{\partial \mathbf{u}}{\partial t} + \nabla \cdot (\mathbf{u}\mathbf{u}) = -\nabla p + \frac{\beta}{Re}\nabla^2 \mathbf{u} + \nabla \cdot \mathbf{T}, \quad (2)$$

where $\mathbf{u}$ is the velocity field, t is the time, p is the pressure, $\beta = \frac{\eta_s}{\eta_0}$ is the coefficient that controls the solvent viscosity contribution (where $\eta_0 = \eta_s + \eta_p$, with η_0 being the total viscosity, η_s and η_p being the solvent and polymer viscosity, respectively), $Re = \frac{\rho U L}{\eta_0}$ is the Reynolds number (ρ is the fluid density) and $\mathbf{T}$ is the non-Newtonian extra-stress tensor that must obey an appropriate constitutive equation.

The Giesekus constitutive model [12–15] is given by the following equation,

$$\mathbf{T} + Wi\overset{\nabla}{\mathbf{T}} + \frac{\alpha_G Wi Re}{(1-\beta)}(\mathbf{T} \cdot \mathbf{T}) = \frac{2(1-\beta)}{Re}\mathbf{D}, \quad (3)$$

where $\mathbf{D}$ is the rate of deformation tensor, $Wi = \frac{\lambda U}{L}$ is the Weissenberg number, λ is the relaxation-time of the fluid, L is the channel half-width, U is the velocity scale, α_G is the so-called mobility parameter ($0 \leq \alpha_G \leq 1$) and $\overset{\nabla}{\mathbf{T}}$ is the upper-convected derivative of $\mathbf{T}$. This model is based on molecular concepts and it reproduces the characteristics of polymeric fluids [8].

The system of Equations (1) and (2) with the Giesekus constitutive Equation (3) in two-dimensional Cartesian coordinates (x, y) are adopted. For the formulation where the solution is analytic or semi-analytic, some simplifications are carried out in the governing equations. The assumptions of such flow for these formulations are: steady-state flow $\left(\frac{\partial(\cdot)}{\partial t} = 0\right)$, no variation of the velocity and tensor in the streamwise direction $\left(\frac{\partial(\cdot)}{\partial x} = 0, \mathbf{u} = \mathbf{u}(y), \mathbf{T} = \mathbf{T}(y)\right)$, normal velocity equal zero, and a constant streamwise pressure gradient $\left(\frac{\partial p(x,y)}{\partial x} = p_x < 0\right)$. The value of the streamwise pressure gradient is achieved considering the integral $\int_{-1}^{1} u\, dy = 4/3$. This value is obtained for the Newtonian velocity profile with a maximum value equal to 1 in the channel center. According to these assumptions, it is considered a horizontal channel where the fluid flows in the streamwise direction x; hence, the following system is taken into account:

$$p_x = \frac{\beta}{Re}u'' + T'_{xy}, \quad (4)$$

$$\frac{\partial p(x,y)}{\partial y} = T'_{yy}, \quad (5)$$

$$T_{xx} - 2Wi T_{xy}u' + \frac{\alpha_G Re Wi}{(1-\beta)}\left(T_{xx}^2 + T_{xy}^2\right) = 0, \quad (6)$$

$$T_{xy} - Wi T_{yy}u' + \frac{\alpha_G Re Wi}{(1-\beta)}T_{xy}(T_{xx} + T_{yy}) = \frac{(1-\beta)}{Re}u', \quad (7)$$

$$T_{yy} + \frac{\alpha_G Re Wi}{(1-\beta)}\left(T_{xy}^2 + T_{yy}^2\right) = 0. \quad (8)$$

In addition, it will be considered $-1 \leq y \leq 1$ and, thus, $T_{xx}(0) = T_{xy}(0) = T_{yy}(0) = u'(0) = 0$ at the center ($y = 0$).

3. Different Formulations to Obtain the Fully Developed Laminar Flow with the Giesekus Model

This section presents different formulations to find the fully developed viscoelastic fluid flow between parallel plates using the Giesekus model. The first formulation is based on Schleiniger and Weinacht [10], the second one is a new formulation that solves the Giesekus equation based on the independent variable change (y for the tensor T_{xy}). With this change in the system, it is possible to obtain a restriction condition for the pressure gradient, showing that this model has some restrictions on the previous formulation. The third formulation is a high-order simulation (HOS) code using vorticity–velocity formulation and a log-conformation formulation adopted for the extra-stress tensor calculation to overcome the high Weissenberg number problem (HWNP).

3.1. Schleiniger and Weinacht Formulation

Here, we present the steps to solve analytically the non-linear system of equations that represent the steady-state of the isothermal, incompressible flows of a Giesekus fluid with a Newtonian solvent between two parallel plates. The formulation is based on [9,10], who presented them using different dimensionless forms among each other. Schleiniger and Weinacht (SW) [10] solved and discussed the solutions mathematically, considering the Giesekus fluid with and without Newtonian solvent, and they commented about the Giesekus fluid without the Newtonian solvent for the axisymmetric case. In their paper, the solution is not obtained explicitly, i.e., the derivative of the solution may not be checked out directly by the reader. Hence, this section will provide a detailed explanation to achieve the analytical solution and a numerical algorithm for solving the implicit equation, which is named herein the "semi-analytical solution".

The system of equations—Equations (4)–(8) is analogous to the system (2.1)–(2.5), solved by Schleiniger and Weinacht [10], with some differences in the dimensional system.

Rewriting Equation (8) in an equivalent form, we get:

$$\left(T_{yy} + \frac{(1-\beta)}{2\alpha_G ReWi}\right)^2 + T_{xy}^2 = \frac{(1-\beta)^2}{4\alpha_G^2 Re^2 Wi^2}, \tag{9}$$

which provides two expressions to T_{yy} as a function of T_{xy},

$$T_{yy} = \frac{-(1-\beta) \pm \sqrt{(1-\beta)^2 - 4\alpha_G^2 Re^2 Wi^2 T_{xy}^2}}{2\alpha_G ReWi}. \tag{10}$$

Since the extra-stress tensor should be equal to zero along the centerline of the channel, the best choice in Equation (10) is the plus sign:

$$T_{yy} = \frac{-(1-\beta) + \sqrt{(1-\beta)^2 - 4\alpha_G^2 Re^2 Wi^2 T_{xy}^2}}{2\alpha_G ReWi}. \tag{11}$$

Adding Equations (6) and (8), one can obtain:

$$(T_{xx} + T_{yy}) + \frac{\alpha_G ReWi}{(1-\beta)}[(T_{xx} + T_{yy})^2 - 2T_{xx}T_{yy} + 2T_{xy}^2] - 2WiT_{xy}u' = 0, \tag{12}$$

or equivalently,

$$\frac{\alpha_G ReWi}{(1-\beta)}(T_{xx} + T_{yy})^2 + (T_{xx} + T_{yy}) - \frac{2\alpha_G ReWi}{(1-\beta)}T_{xx}T_{yy} + \frac{2\alpha_G ReWi}{(1-\beta)}T_{xy}^2 = 2WiT_{xy}u'. \tag{13}$$

The value of $(T_{xx} + T_{yy})$ can be obtained from Equation (7):

$$(T_{xx} + T_{yy}) = \frac{\frac{(1-\beta)}{Re}u' + WiT_{yy}u' - T_{xy}}{\frac{\alpha_G ReWi}{(1-\beta)}T_{xy}}. \tag{14}$$

Moreover, Equation (14) implies that:

$$T_{xx} = \frac{(1-\beta)[(1-\beta) + ReWiT_{yy}]u'}{\alpha_G Re^2 WiT_{xy}} - \frac{(1-\beta) + \alpha_G ReWiT_{yy}}{\alpha_G ReWi}. \tag{15}$$

Equations (14) and (15) are valid for all $y \neq 0$. The values of the extra-stress tensor components T_{xx}, T_{yy}, T_{xy} and the streamwise velocity component u' are known at centerline $y = 0$.

Substituting Equations (14) and (15) into Equation (13) and carrying on some algebraic manipulations, and with the use of Equation (8), one can obtain:

$$u' = \frac{[\frac{(1-\beta)}{Re} + bWiT_{yy}]T_{xy}}{[\frac{(1-\beta)}{Re} + WiT_{yy}]^2}, \tag{16}$$

where $b = 2\alpha_G - 1$.

Equation (16) shows that u' is a function of the extra-stress tensor component T_{yy} and T_{xy}. As Equation (11) shows that the extra-stress tensor component T_{yy} is a function of the extra-stress component T_{xy}, it is possible to use Equation (11) in Equation (16); after some algebraic manipulations, again, a equation for u' can be obtained:

$$u' = \frac{2\alpha_G ReT_{xy}[(1-\beta) + b\sqrt{(1-\beta)^2 - 4\alpha_G^2 Re^2 Wi^2 T_{xy}^2}]}{[b(1-\beta) + \sqrt{(1-\beta)^2 - 4\alpha_G^2 Re^2 Wi^2 T_{xy}^2}]^2}. \tag{17}$$

We should comment about the sign of the square root term presented in Equation (17). The solution of the Giesekus model needs to satisfy all equations, in particular Equation (8), which was rewritten as Equation (9). The Equation (9) means that:

$$\left(T_{yy} + \frac{(1-\beta)}{2\alpha_G ReWi}\right)^2 \leq \frac{(1-\beta)^2}{4\alpha_G^2 Re^2 Wi^2}, \tag{18}$$

and also

$$T_{xy}^2 \leq \frac{(1-\beta)^2}{4\alpha_G^2 Re^2 Wi^2}. \tag{19}$$

The restriction given by Equation (19) leads to $(1-\beta)^2 - 4\alpha_G^2 Re^2 Wi^2 \geq 0$, i.e., the square root term presented in Equation (17) is always non-negative since Equation (8) has to be taken into account. Therefore, this restriction must be respected in this paper.

Integrating Equation (4) with respect to y and using that $T_{xy} = u' = 0$ at the centerline $y = 0$, one can arrive at:

$$T_{xy} = \frac{-\beta}{Re}u' + p_x y, \quad -1 \leq y \leq 1, \tag{20}$$

where p_x is a negative constant.

Substituting Equation (20) into Equation (17), one can obtain an implicit expression for u':

$$u' = \frac{2\alpha_G Re(\frac{-\beta}{Re}u' + p_x y)[(1-\beta) + b\sqrt{(1-\beta)^2 - 4\alpha_G^2 Re^2 Wi^2 (\frac{-\beta}{Re}u' + p_x y)^2}]}{[b(1-\beta) + \sqrt{(1-\beta)^2 - 4\alpha_G^2 Re^2 Wi^2 (\frac{-\beta}{Re}u' + p_x y)^2}]^2}. \tag{21}$$

In order to obtain the analytical solution, it is necessary to follow the next steps:

1. Solve Equation (21) to obtain u' for a given p_x;
2. Solve Equation (20) to obtain $T_{xy}(y)$;
3. Solve Equation (11) to obtain $T_{yy}(y)$;
4. Solve Equation (15) to obtain $T_{xx}(y)$.

Again, we should note that Equations (21), (20), (11) and (15) are similar to Equations (5.2) to (5.5), given by Schleiniger and Weinacht [10], respectively, with some differences in the dimensional system. Since sequence 2, 3, and 4 is followed, it is possible to see that all of the components of the extra-stress tensor can be obtained explicitly, just using some algebraic calculations. However, in step 1, it is not easy to solve u' analytically. Therefore, in the next section, we will discuss the assumptions and the numerical strategies adopted to choose p_x properly, to calculate u', and to calculate the component of the velocity u for complementing the present section, and for providing applicability of the mathematical work by Schleiniger and Weinacht [10] in the engineering field.

3.2. Independent Variable Change

The formulation proposed here is based on the change of the independent variable in the equation system. The equation system is rewritten in terms of the component tensor T_{xy}. This change in the equation system allows us to find a solution for y as a function of T_{xy} analytically. From the Equations (6) and (8), two solutions for each equation can be obtained:

$$T_{xx} = \frac{(-1+\beta)\left(1 \pm \sqrt{1 - \frac{4ReT_{xy}Wi^2\alpha_G(ReT_{xy}\alpha_G + 2(-1+\beta)u')}{(-1+\beta)^2}}\right)}{2ReWi\alpha_G}, \tag{22}$$

and

$$T_{yy} = \frac{(-1+\beta)\left(1 \pm \sqrt{1 - \frac{4Re^2T_{xy}^2Wi^2\alpha_G^2}{(-1+\beta)^2}}\right)}{2ReWi\alpha_G}. \tag{23}$$

In the channel center, the extra-stress tensor should be zero. Therefore, the solution adopted is the one with the signal of the minus ($-$) before the square root. From the Equation (4), by integrating in the y direction, one can obtain an equation for $u'(y)$:

$$u' = \frac{Re}{\beta}(p_x y - T_{xy}).$$

Substituting the last equation and the Equations (22) and (23) into Equation (7), the resulting equation is a function of the variables p_x, y and the extra-stress tensor component T_{xy}:

$$(T_{xy} - p_x y)\left[-2 + \frac{2}{\beta} + \frac{1}{\alpha_G}\left(1 - \sqrt{1 - \frac{4Re^2T_{xy}^2Wi^2\alpha_G^2}{(-1+\beta)^2}}\right) + \right.$$
$$\left. + \frac{1}{\alpha_G\beta}\left(\sqrt{1 - \frac{4Re^2T_{xy}^2Wi^2\alpha_G^2}{(-1+\beta)^2}} - 1\right)\right] + T_{xy}\sqrt{1 - \frac{4Re^2T_{xy}^2Wi^2\alpha_G^2}{(-1+\beta)^2}} +$$
$$+ T_{xy}\sqrt{1 - \frac{4Re^2T_{xy}^2Wi^2\alpha_G^2}{(-1+\beta)^2}} + \frac{8Re^2T_{xy}Wi^2(T_{xy} - p_x y)\alpha_G}{(-1+\beta)\beta} = 0. \tag{24}$$

The aim of this formulation is to solve the Equation (24) for y and then obtain a solution y as a function of T_{xy}. After some algebraic manipulations and simplifications, two equations can be obtained, one is given by:

$$y = \frac{T_{xy}}{p_x}.$$

This equation satisfies the hypothesis in the channel center, but from the equation of the model, the relation between y and T_{xy} is non-linear, so this equation should not be adopted. The interesting equation is the second one, which expresses the non-linear relation between the tensorial forces and the width of the channel, as one can see in the equation:

$$y = \left[T_{xy} \left(-2\alpha_G^2 \left((\beta - 1) \left(\beta \left(\sqrt{1 - \frac{4\alpha_G^2 Re^2 T_{xy}^2 Wi^2}{(\beta-1)^2}} - 1 \right) + 1 \right) + Re^2 T_{xy}^2 Wi^2 \right) + \alpha_G(\beta-1)(3\beta-2)\left(\sqrt{1 - \frac{4\alpha_G^2 Re^2 T_{xy}^2 Wi^2}{(\beta-1)^2}} - 1 \right) + (\beta-1)^2 \left(1 - \sqrt{1 - \frac{4\alpha_G^2 Re^2 T_{xy}^2 Wi^2}{(\beta-1)^2}} \right) \right) \right] \Big/ \left[p_x \left(2\alpha_G(\beta-1)^2 \left(\sqrt{1 - \frac{4\alpha_G^2 Re^2 T_{xy}^2 Wi^2}{(\beta-1)^2}} - 1 \right) + (\beta-1)^2 \left(1 - \sqrt{1 - \frac{4\alpha_G^2 Re^2 T_{xy}^2 Wi^2}{(\beta-1)^2}} \right) - 2\alpha_G^2(-\beta + ReT_{xy}Wi + 1)(\beta + ReT_{xy}Wi - 1) \right) \right]. \quad (25)$$

The Equation (25) allows to obtain a distribution of y from a given T_{xy}. Starting from the channel center, where T_{xy} is zero, the values for T_{xy} are increased to obtain, respectively, y, until the channel boundaries, where $y = \pm 1$.

For this procedure, it is necessary to find a step size that is able to capture the extra-stress tensor component T_{xy} distribution until $y = \pm 1$. For that, an assumption that gives us a 6th degree function is accomplished, in which coefficients are all the variables involved in the flow. To obtain this function, it is assumed that, at the wall, the extra-stress tensor component is $T_{xy} = hn$, where h is the size of increment and n is the number of increment needed to arrive in $y = -1$, starting from the channel center (notice that, if it is considered the increment until $y = 1$, the assumption for T_{xy} should be $T_{xy} = -hn$).

Equation (25) is solved at the wall $y = -1$, assuming $T_{xy} = hn$. After some algebraic manipulations, we obtain a function in h, which gives the increment size as the lower real root. The solution of this function has six roots (four complex and two real roots or four real and two complex roots). The required solution is the lower real root. This function is important because it allows one to estimate of the number of points that are needed to obtain the distribution of the extra-stress tensor component in the channel (it is possible because the function is also dependent on n). The function is given by:

$$\begin{aligned}
P(h) = {} & \alpha_G^2 h^6 n^6 Re^4 Wi^4 + 2\alpha_G^2 h^5 n^5 Px Re^4 Wi^4 + h^4(\alpha_G^2 n^4 Px^2 Re^4 Wi^4 + \\
& + 4\alpha_G^4 \beta^2 n^4 Re^2 Wi^2 - 12\alpha_G^3 \beta^2 n^4 Re^2 Wi^2 + 8\alpha_G^3 \beta n^4 Re^2 Wi^2 + \\
& + 11\alpha_G^2 \beta^2 n^4 Re^2 Wi^2 - 12\alpha_G^2 \beta n^4 Re^2 Wi^2 + 2\alpha_G^2 n^4 Re^2 Wi^2 - \\
& - 3\alpha_G \beta^2 n^4 Re^2 Wi^2 + 5\alpha_G \beta n^4 Re^2 Wi^2 - 2\alpha_G n^4 Re^2 Wi^2) + \\
& + h^3(-8\alpha_G^3 \beta^2 n^3 Px Re^2 Wi^2 + 8\alpha_G^3 \beta n^3 Px Re^2 Wi^2 + \\
& + 12\alpha_G^2 \beta^2 n^3 Px Re^2 Wi^2 - 16\alpha_G^2 \beta n^3 Px Re^2 Wi^2 + 4\alpha_G^2 n^3 Px Re^2 Wi^2 - \\
& - 5\alpha_G \beta^2 n^3 Px Re^2 Wi^2 + 9\alpha_G \beta n^3 Px Re^2 Wi^2 - 4\alpha_G n^3 Px Re^2 Wi^2) + \\
& + h^2(-2\alpha_G^2 \beta^3 n^2 + 5\alpha_G^2 \beta^2 n^2 - 4\alpha_G^2 \beta n^2 + \alpha_G^2 n^2 + 3\alpha_G \beta^3 n^2 - \\
& - 8\alpha_G \beta^2 n^2 + 7\alpha_G \beta n^2 - 2\alpha_G n^2 + \beta^3(-n^2) + 3\beta^2 n^2 - 3\beta n^2 + \\
& + 2\alpha_G^2 \beta^2 n^2 Px^2 Re^2 Wi^2 - 4\alpha_G^2 \beta n^2 Px^2 Re^2 Wi^2 + 2\alpha_G^2 n^2 Px^2 Re^2 Wi^2 - \\
& - 2\alpha_G \beta^2 n^2 Px^2 Re^2 Wi^2 + 4\alpha_G \beta n^2 Px^2 Re^2 Wi^2 - 2\alpha_G n^2 Px^2 Re^2 Wi^2 + \\
& + n^2) + h(2\alpha_G^2 \beta^4 n Px - 8\alpha_G^2 \beta^3 n Px + 12\alpha_G^2 \beta^2 n Px - 8\alpha_G^2 \beta n Px + \\
& + 2\alpha_G^2 n Px - 3\alpha_G \beta^4 n Px + 13\alpha_G \beta^3 n Px - 21\alpha_G \beta^2 n Px + 15\alpha_G \beta n Px - \\
& - 4\alpha_G n Px + \beta^4 n Px - 5\beta^3 n Px + 9\beta^2 n Px - 7\beta n Px + 2n Px) + \\
& + \alpha_G^2 \beta^4 Px^2 - 4\alpha_G^2 \beta^3 Px^2 + 6\alpha_G^2 \beta^2 Px^2 - 4\alpha_G^2 \beta Px^2 + \alpha_G^2 Px^2 - \\
& - 2\alpha_G \beta^4 Px^2 + 8\alpha_G \beta^3 Px^2 - 12\alpha_G \beta^2 Px^2 + 8\alpha_G \beta Px^2 - 2\alpha_G Px^2 + \\
& \mid \beta^4 Px^2 - 4\beta^3 Px^2 + 6\beta^2 Px^2 - 4\beta Px^2 + Px^2
\end{aligned} \quad (26)$$

The next step of the present method is to calculate the distribution between the tensor component T_{xy} and y, and use it to find the gradient pressure p_x. In order to find the pressure gradient, we adopted the existing condition of the solution in the real plane given by Schleiniger and Weinacht [10]:

$$T_{xy}^2 \leq \frac{(1-\beta)^2}{(2ReWi\alpha_G)^2}. \quad (27)$$

Using this condition in the Equation (25), at the wall $y = 1$, and solving the inequation for p_x, we obtained an existence condition for the flow in terms of the pressure gradient:

$$p_x \geq \frac{-1 + \beta + 2\alpha_G(2 + 2\alpha_G(-1+\beta) - 3\beta)}{2ReWi(1-2\alpha_G)^2 \alpha_G}. \quad (28)$$

Note that if Re, Wi or α_G is zero (UCM and Oldroyd-B), the pressure gradient does not have a limiting value, the same happens if $\alpha_G = \frac{1}{2}$. From the inequation (28), it is possible to see that the hypothesis of the pressure gradient component p_x lower than zero is satisfied. After these calculations, it is possible to obtain all of the flow components using the equations written in terms of the component tensor T_{xy}.

The last step is the calculation of the velocity profile $u(y)$ as a function of T_{xy}. Equation (29) shows the relation between the component tensor T_{xy} and the derivative of the velocity profile $u(y)$,

$$u' = \frac{Re}{\beta}(p_x y - T_{xy}). \quad (29)$$

For the calculation of the velocity profile, it is necessary to rewrite these equations in terms of the tensor component T_{xy}. For that, an expression for $\frac{du}{dT_{xy}}$ is needed. Using the chain rule, the Equation (29) can be rewritten as:

$$\frac{du}{dT_{xy}} = \frac{du}{dy}\frac{dy}{dT_{xy}} = \frac{Re}{\beta}(p_x y - T_{xy})\frac{dy}{dT_{xy}}. \tag{30}$$

Integrating Equation (30) the following expression can be obtained:

$$u(T_{xy}) = \frac{Re}{\beta}\left[p_x \frac{y^2}{2} - T_{xy} y + \int y\, dT_{xy}\right]. \tag{31}$$

Solving $\int y\, dT_{xy}$, it can be obtained:

$$u(T_{xy}) = \Bigg((\beta-1)^3\left(2\alpha_G + \sqrt{1 - \frac{4\alpha_G^2 Re^2 T_{xy}^2 Wi^2}{(\beta-1)^2}} - 1\right) \times$$

$$\times \Bigg((8(\alpha_G-1)\alpha_G + 1)\left(2\alpha_G + \sqrt{1 - \frac{4\alpha_G^2 Re^2 T_{xy}^2 Wi^2}{(\beta-1)^2}} - 1\right) \times$$

$$\times \log\left(-2\alpha_G - \sqrt{1 - \frac{4\alpha_G^2 Re^2 T_{xy}^2 Wi^2}{(\beta-1)^2}} + 1\right) +$$

$$+ \frac{4\alpha_G(2\alpha_G-1)\left(\alpha_G(2(\beta-1)^2 + Re^2 T_{xy}^2 Wi^2) - 2(\beta-1)^2\right)}{(\beta-1)^2}\Bigg)\Bigg) \Bigg/ \tag{32}$$

$$\Bigg/ \Bigg(4\alpha_G Re Wi^2\left(-2\alpha_G(\beta-1)^2\left(\sqrt{1 - \frac{4\alpha_G^2 Re^2 T_{xy}^2 Wi^2}{(\beta-1)^2}} - 1\right) +$$

$$+ (\beta-1)^2\left(\sqrt{1 - \frac{4\alpha_G^2 Re^2 T_{xy}^2 Wi^2}{(\beta-1)^2}} - 1\right) +$$

$$+ 2\alpha_G^2(-\beta + Re T_{xy} Wi + 1)(\beta + Re T_{xy} Wi - 1)\Bigg)\Bigg),$$

where the logarithm term does not have a solution in the real plane for any value of α_G, Re, Wi, T_{xy}, and β. This information, about the equation for the velocity u, shows that it is not possible, using this formulation, to obtain an analytical solution for velocity. Therefore, the calculation sequence to obtain the main flow components is:

1. From Equation (28), it is possible to obtain the p_x max, used to start the simulation and the recursive process;
2. Equation (26) allows one to obtain the step size required to find the point distribution to increase T_{xy}, to obtain the coordinate y, respectively;
3. Find u' by the Equation (29);
4. Integrate numerically u' to obtain the velocity u. It is important to emphasize that, to calculate the integral numerically, an interpolation for new values of y equally spaced is adopted, since the analytical y obtained from T_{xy} is not equally spaced. A high-order finite difference approximation was adopted for this calculation;
5. After the integral calculation, it is verified if the value of this integration is 4/3. This is the value obtained for Newtonian fluid in a Poiseuille flow with a maximum streamwise velocity equal to 1. If the value of the integral is different from 4/3, the Newton–Raphson method is used to obtain a pressure gradient where the flow resulting from this gradient has numerical integration of the velocity equal 4/3;
6. Using the expressions given in Equations (22) and (23), it is possible to obtain the extra-stress tensor components distribution.

3.3. High-Order Simulation (HOS)

For the numerical simulation of the Giesekus fluid flow, a high-order method is adopted. In order to eliminate the pressure term in the Navier–Stokes Equation (2), the vorticity–velocity formulation is adopted. Thus, the vorticity component in the z direction, ω_z, can be written as:

$$\omega_z = \frac{\partial u}{\partial y} - \frac{\partial v}{\partial x}. \tag{33}$$

The equation system to be solved is given by:

$$\frac{\partial u}{\partial y} + \frac{\partial v}{\partial x} = 0, \tag{34}$$

$$\frac{\partial^2 v}{\partial x^2} + \frac{\partial^2 v}{\partial y^2} = -\frac{\partial \omega_z}{\partial x}, \tag{35}$$

$$\frac{\partial \omega_z}{\partial t} + \frac{\partial (u\omega_z)}{\partial x} + \frac{\partial (v\omega_z)}{\partial y} = \frac{\beta}{Re}\left(\frac{\partial^2 \omega_z}{\partial x^2} + \frac{\partial^2 \omega_z}{\partial y^2}\right) + \frac{\partial^2 T_{xx}}{\partial x \partial y} + \frac{\partial^2 T_{xy}}{\partial y^2} - \frac{\partial^2 T_{xy}}{\partial x^2} - \frac{\partial^2 T_{yy}}{\partial x \partial y}, \tag{36}$$

and the Giesekus constitutive equation:

$$\mathbf{T} + Wi\overset{\nabla}{\mathbf{T}} + \frac{\alpha_G WiRe}{(1-\beta)}(\mathbf{T}\cdot\mathbf{T}) = \frac{2(1-\beta)}{Re}\mathbf{D}. \tag{37}$$

In this equation, the log-conformation method [17,18] is adopted to overcome the high Weissenberg number problem—HWNP. Using this technique, a conformation tensor $\mathbf{A}$ is adopted. The relation between $\mathbf{T}$ and the conformation tensor $\mathbf{A}$ is given by

$$\mathbf{T} = \frac{(1-\beta)}{ReWi}(\mathbf{A}-\mathbf{I}), \tag{38}$$

and

$$\mathbf{\Psi} = \log_a(\mathbf{A}). \tag{39}$$

The equation to be solved using this technique is given as follows:

$$\frac{\partial \mathbf{\Psi}}{\partial t} + \nabla \cdot (\mathbf{u}\mathbf{\Psi}) = (\mathbf{\Omega}\mathbf{\Psi} - \mathbf{\Psi}\mathbf{\Omega}) + 2\mathbf{B} + \frac{1}{Wi}a^{-\mathbf{\Psi}}(\mathbf{I}-a^{\mathbf{\Psi}})[\mathbf{I}+\alpha_G(a^{\mathbf{\Psi}}-\mathbf{I})], \tag{40}$$

where $a = e$ is the Euler's number.

All of the spatial derivatives are approximated by fifth- and sixth-order compact finite differences [19]. Time derivatives are discretized using a classical fourth-order Runge–Kutta scheme [20]. The Poisson equation is solved using a multigrid full approximation scheme (FAS) [21]. The calculation of the vorticity on the wall is performed according to [22] using compact high-order finite difference approximations.

The boundary conditions adopted are Newtonian Poiseuille profile at the channel entrance (left boundary), non-permeability and no-slip conditions at the walls (upper and lower boundaries), and Neumann boundary conditions for the velocity at the channel exit (right boundary). The problem is solved as an unsteady problem and a very long channel is adopted to avoid the influence of the left boundary. The simulation is carried out until the maximum difference between the vorticity in two consecutive time steps is lower than 10^{-9}.

The calculation sequence to obtain the main flow components by solving Equations (34)–(36) and (40) is given by:

1. Apply a time integration for the vorticity and the extra-stress tensors $\mathbf{\Psi}$ (Runge–Kutta method);
2. Calculate the extra-stress tensor components through the log-conformation method;
3. Calculate the right-hand side of the Poisson equation given by Equation (35);

4. Calculate the velocity v by solving the Poisson equation—Equation (35);
5. Calculate the velocity u using the continuity equation—Equation (34);
6. Update the vorticity ω_z at the walls;
7. Apply a filter after the last step of the time integrator.

The filtering strategy adopted in the last step is a 6th order compact filter given by [23]. The filter is applied at the end of each time step integration. It consists of recalculating vorticity distribution through a tridiagonal system to eliminate the spurious oscillations that can appear in the numerical solution.

For the results shown here, the number of points in the streamwise (x) and wall-normal (y) directions were 9049 and 249, respectively. The distance between two consecutive points were $dx = \frac{2\pi}{16}$ and $dy = \frac{2}{248}$, and were constant in all domains.

4. Results

In this section, the results are presented for each formulation described. A comparison and the advantages and disadvantages of each formulation is presented. The relation to the pressure gradient obtained from the second formulation will be explored.

In order to explore the results, advantages, and disadvantages of each formulation, different types of fluid and flows were simulated, varying the dimensionless parameters.

4.1. Agreement Region

In the present section, we present a comparison of the results obtained with the three techniques in a range of parameters, where all are in agreement. In the range of dimensionless parameters adopted in the present section, all of the formulations showed good agreement. The range of parameters adopted here is given by:

- Reynolds number—$0 < Re < 10{,}000$;
- Weissenberg number—$0 < Wi < 10$;
- β parameter—$0.01 < \beta < 1$;
- α_G parameter—$0 < \alpha_G < 0.5$.

Some results were chosen to show the range mentioned above. These results are presented showing the variation of the streamwise velocity $U(y)$ and the three components of the extra-stress tensor T_{xx}, T_{xy} and T_{yy} in the wall-normal direction y.

Figures 1–3 show the comparison among the formulations presented for different values of the dimensionless numbers Re, β, and α_G in the range of the agreement region.

It is possible to observe very good agreement between the results obtained by different formulations, both for β close to zero (close to polymeric fluid) and for β close to one (close to Newtonian fluid).

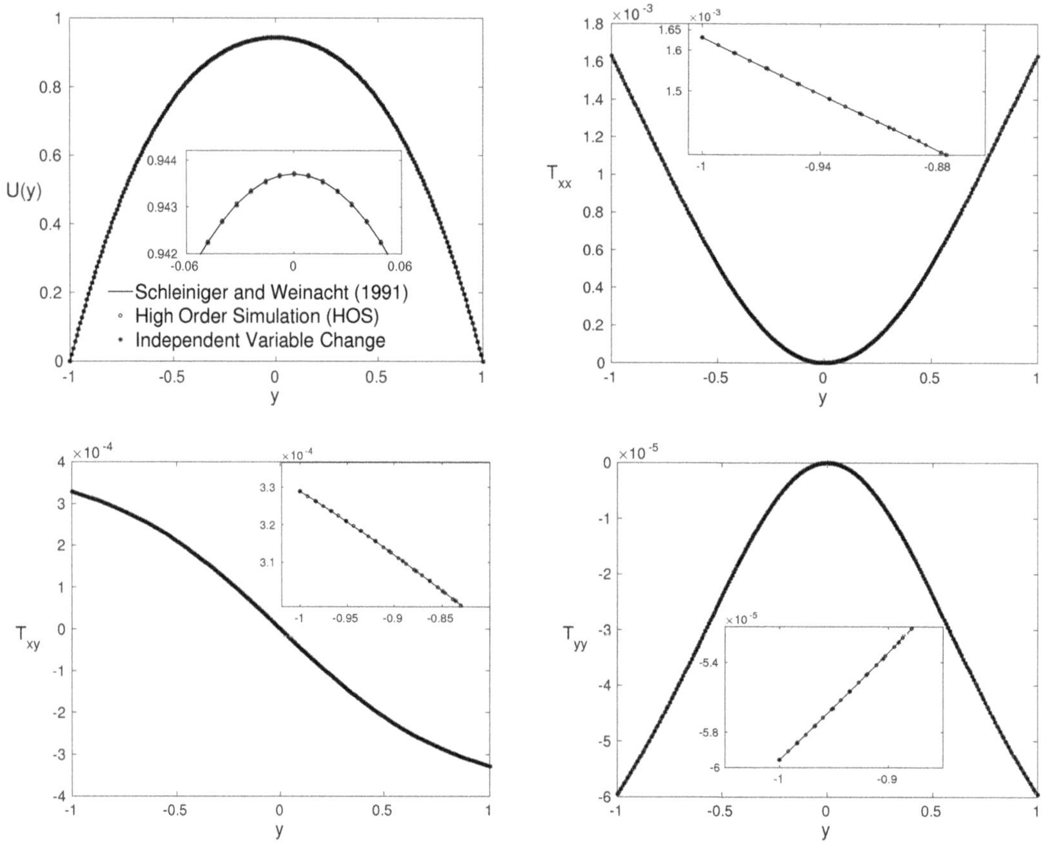

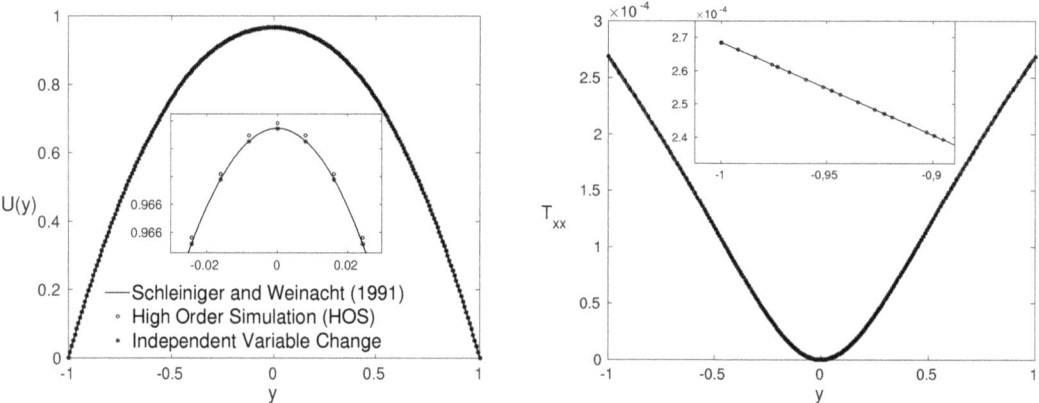

Figure 1. Streamwise velocity $U(y)$ and the components of the extra-stress tensor T_{xx}, T_{xy} and T_{yy} variation in the wall-normal direction y. Dimensionless numbers: $Re = 2000$, $\beta = 0.25$, $\alpha_G = 0.1$ and $Wi = 2$.

Figure 2. *Cont.*

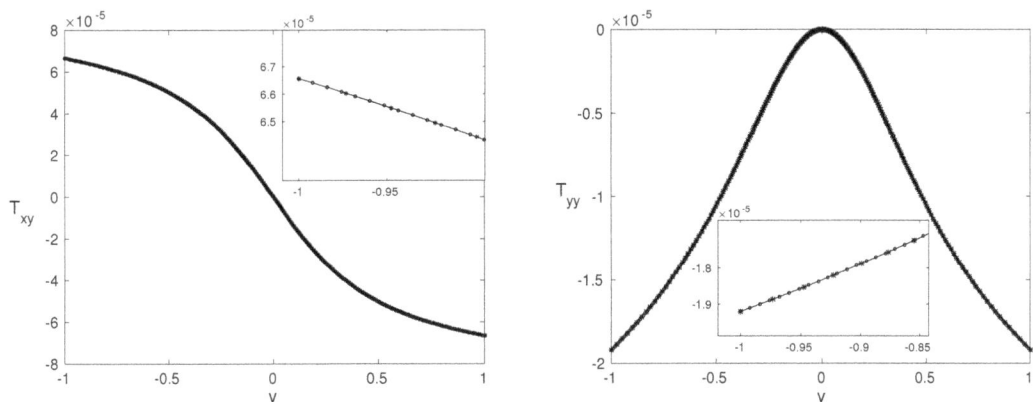

Figure 2. Same as in Figure 1 for dimensionless numbers: $Re = 5000$, $\beta = 0.5$, $\alpha_G = 0.2$ and $Wi = 2$.

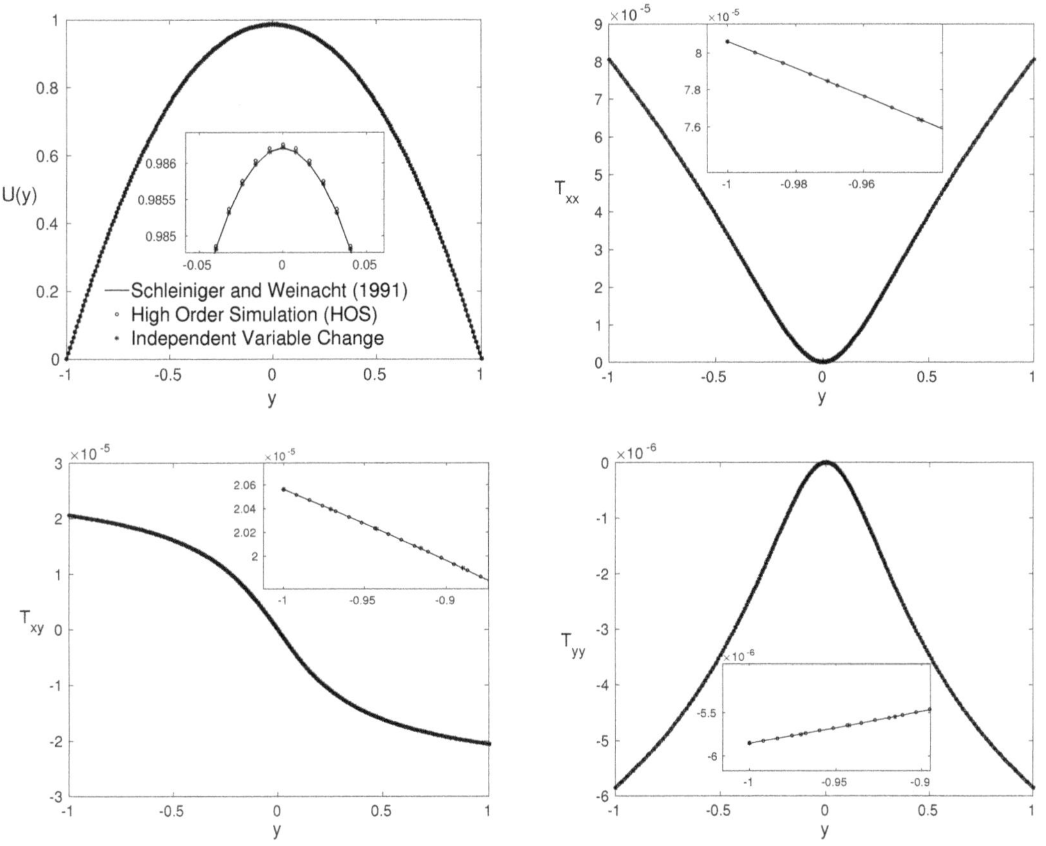

Figure 3. Same as in Figure 1 for dimensionless numbers: $Re = 8000$, $\beta = 0.75$, $\alpha_G = 0.2$ and $Wi = 2$.

For these simulations, different values for the dimensionless parameters of the flow were considered, where all the formulations converge to a solution. It is noteworthy that the formulations have certain limitations for some values for the dimensionless numbers. Out of the agreement region, the results using the first and the second formulation diverges or appears with oscillations on their field. The exploration of these bounds are shown below.

4.2. Purely Polymeric Flows

Considering the dimensionless number $\beta = 0$, the flow of a viscoelastic fluid is known as a purely polymeric fluid flow, since there is no Newtonian solvent contribution in the fluid composition.

The independent variable change formulation does not converge for a purely polymeric fluid. Therefore, for flows with $\beta = 0$, the Schleiniger and Weinacht [10] and HOS formulations were used, and their results were compared.

Figures 4–6 show the variation of the streamwise velocity $U(y)$ and the three components of the extra-stress tensor T_{xx}, T_{xy} and T_{yy} in the wall-normal direction y.

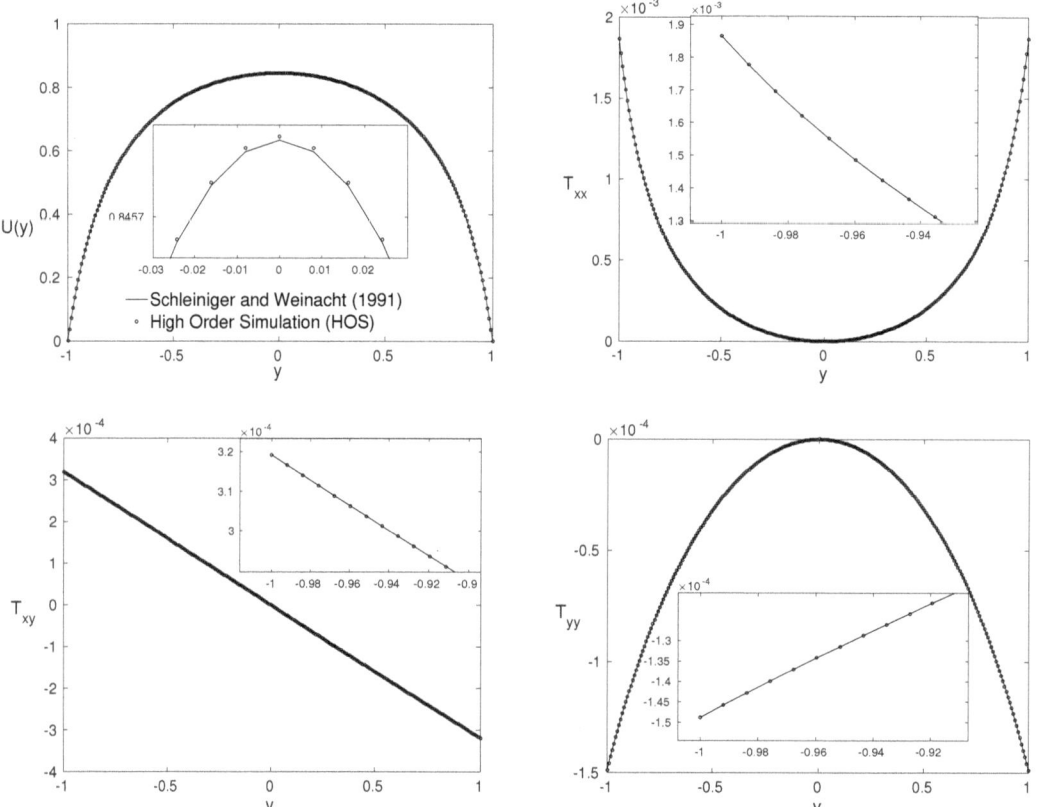

Figure 4. Same as in Figure 1 for dimensionless numbers: $Re = 2000$, $\beta = 0$, $\alpha_G = 0.3$ and $Wi = 2$.

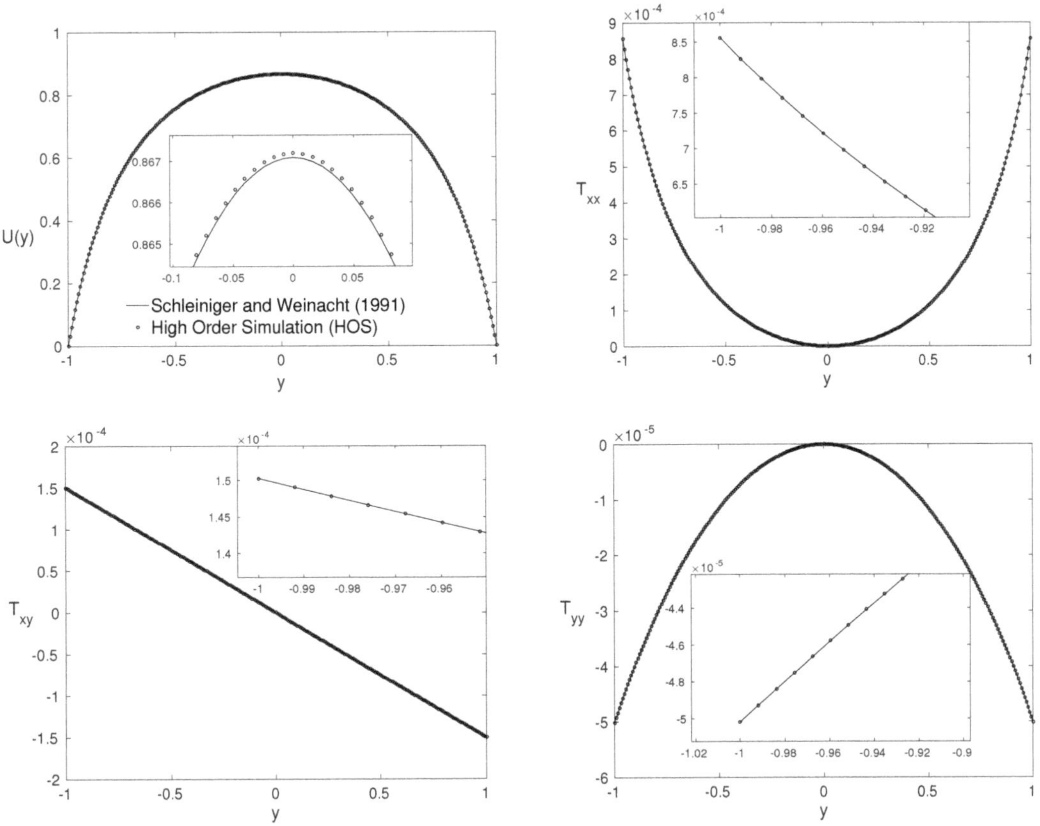

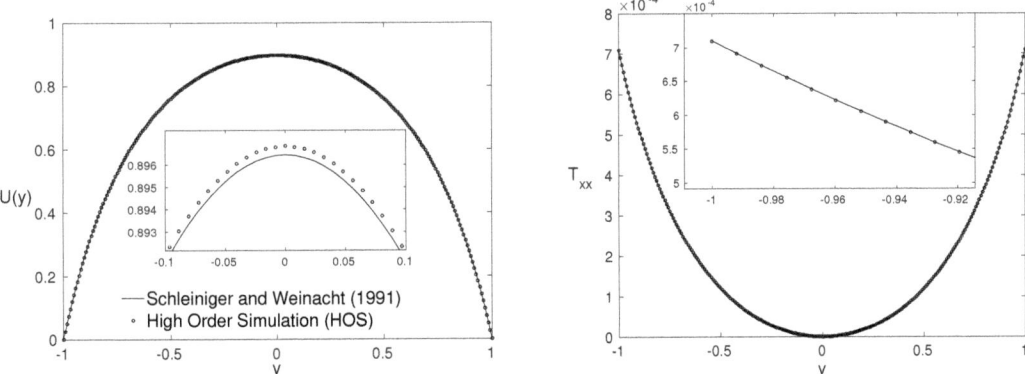

Figure 5. Same as in Figure 1 for dimensionless numbers: $Re = 5000$, $\beta = 0$, $\alpha_G = 0.2$ and $Wi = 2$.

Figure 6. *Cont.*

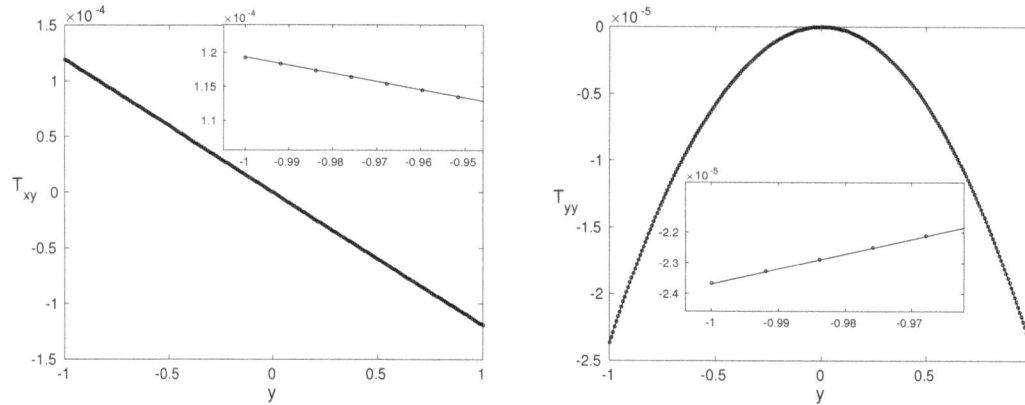

Figure 6. Same as in Figure 1 for dimensionless numbers: $Re = 8000$, $\beta = 0$, $\alpha_G = 0.1$ and $Wi = 2$.

The extra-stress tensor results obtained by the formulations of Schleiniger and Weinacht [10] and HOS, considering the flow for a purely polymeric fluid, showed a good agreement among each other. However, a small difference between the velocity profiles can be observed at the channel center ($u(y = 0)$).

4.3. Low β Number—Close to Purely Polymeric Flows

As mentioned earlier, the formulation using the independent variable change does not work for $\beta = 0$. However, an analysis of the results for this formulation was performed, considering β close to zero. The results obtained by the formulation Schleiniger and Weinacht [10] and by the HOS formulation are compared. The comparisons were carried out for the streamwise velocity $U(y)$ and the three components of the extra-stress tensor T_{xx}, T_{xy} and T_{yy}, as can be seen in Figures 7–9, for $\beta = 0.1, 0.05$ and 0.01, respectively.

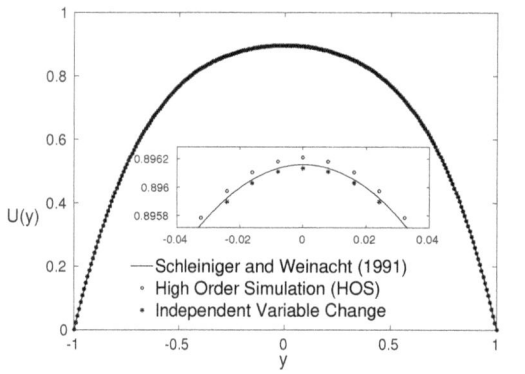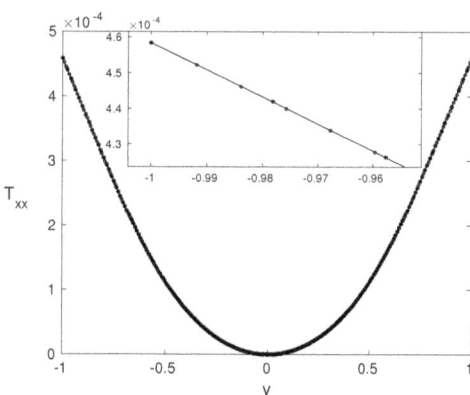

Figure 7. *Cont.*

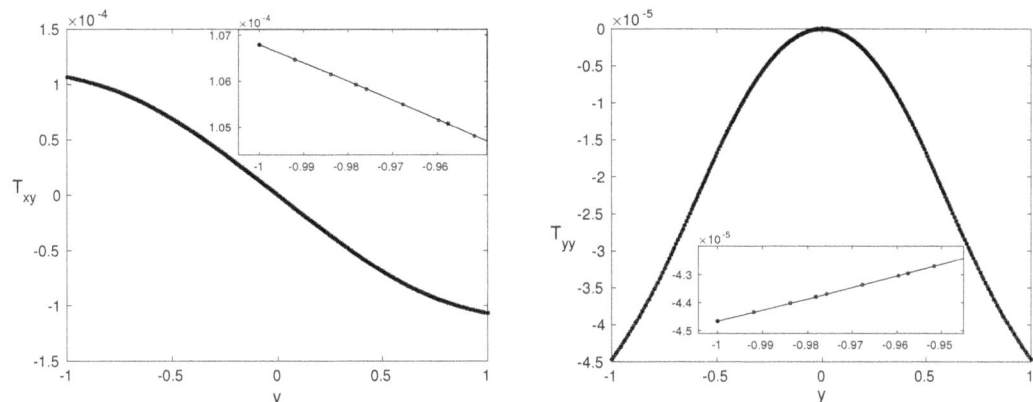

Figure 7. Same as in Figure 1 for dimensionless numbers: $Re = 5000$, $\beta = 0.1$, $\alpha_G = 0.3$ and $Wi = 2$.

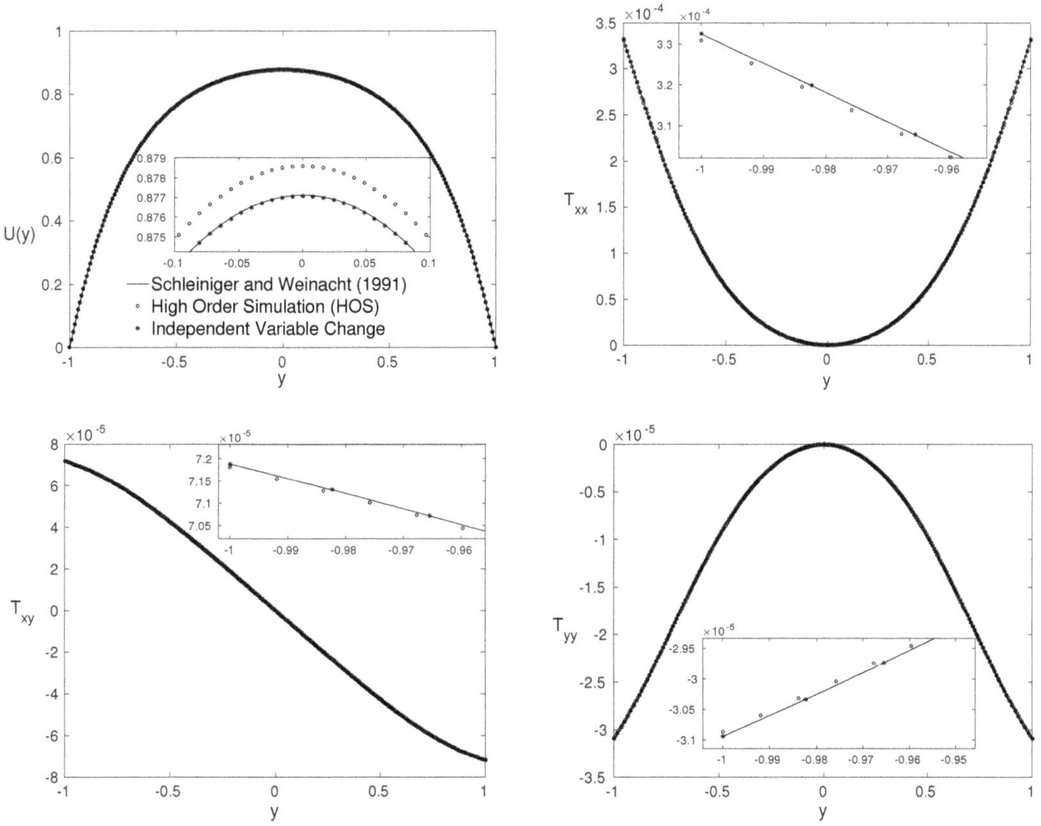

Figure 8. Same as in Figure 1 for dimensionless numbers: $Re = 8000$, $\beta = 0.05$, $\alpha_G = 0.3$ and $Wi = 2$.

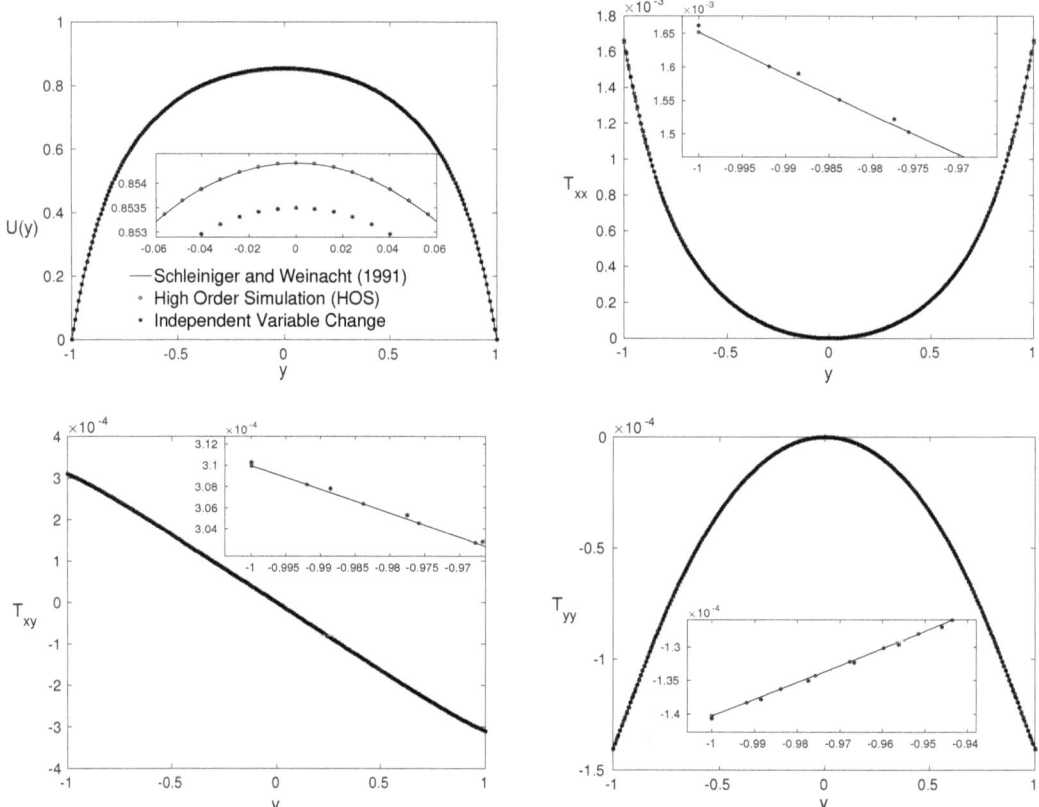

Figure 9. Same as in Figure 1 for dimensionless numbers: $Re = 2000$, $\beta = 0.01$, $\alpha_G = 0.3$ and $Wi = 2$.

The results obtained are in agreement, however, as the value of the dimensionless number β decreases, the difference between the formulations increases. This shows precisely the restriction commented above that, for the independent variable change formulation, the lowest value for β has to be $\beta = 0.01$ in order to obtain acceptable results with this formulation. It is worth mentioning that, as the solution obtained is analytical for all components, except for the streamwise velocity, the increase in the number of points adopted in the solution did not show an influence for lower values of β simulations.

4.4. High Weissenberg Number

The HOS formulation was implemented with the log-conformation technique for flow simulation, considering high values of the Weissenberg number. The SW formulation [10] was not able to converge to the solution when considering Re and Wi higher than 8000 and 10, respectively.

However, the formulation based on the independent variable change was able to obtain solutions for any values of $Re > 0$, $Wi > 0$ and $0 < \alpha_G < 0.5$, with the only restriction to use the dimensionless number $\beta > 0.01$.

Figures 10–12 show the variation of the streamwise velocity $U(y)$ and the three components of the extra-stress tensor T_{xx}, T_{xy} and T_{yy} in the wall-normal direction y for different dimensionless numbers of Re, β, α_G, and with Weissenberg number $Wi = 150, 300$, and 500.

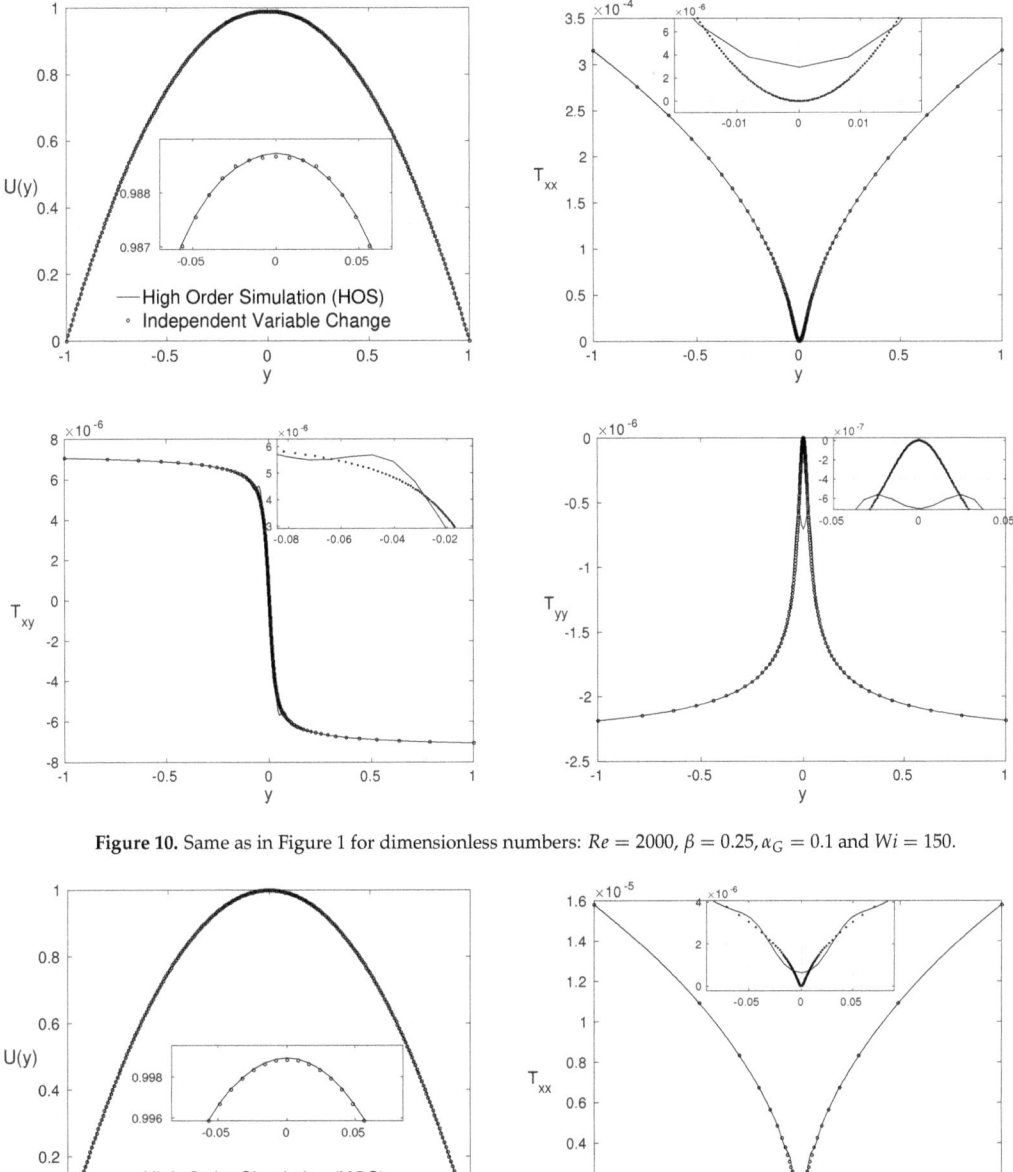

Figure 10. Same as in Figure 1 for dimensionless numbers: $Re = 2000$, $\beta = 0.25$, $\alpha_G = 0.1$ and $Wi = 150$.

Figure 11. *Cont.*

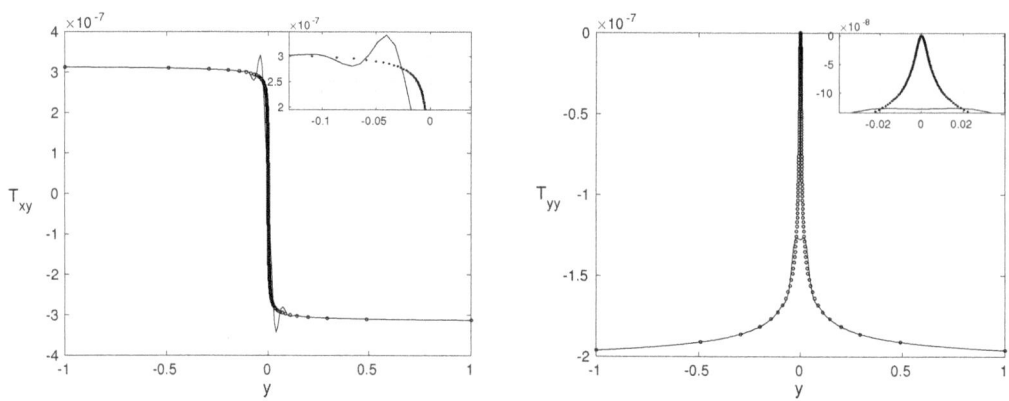

Figure 11. Same as in Figure 1 for dimensionless numbers: $Re = 8000$, $\beta = 0.5$, $\alpha_G = 0.3$ and $Wi = 300$.

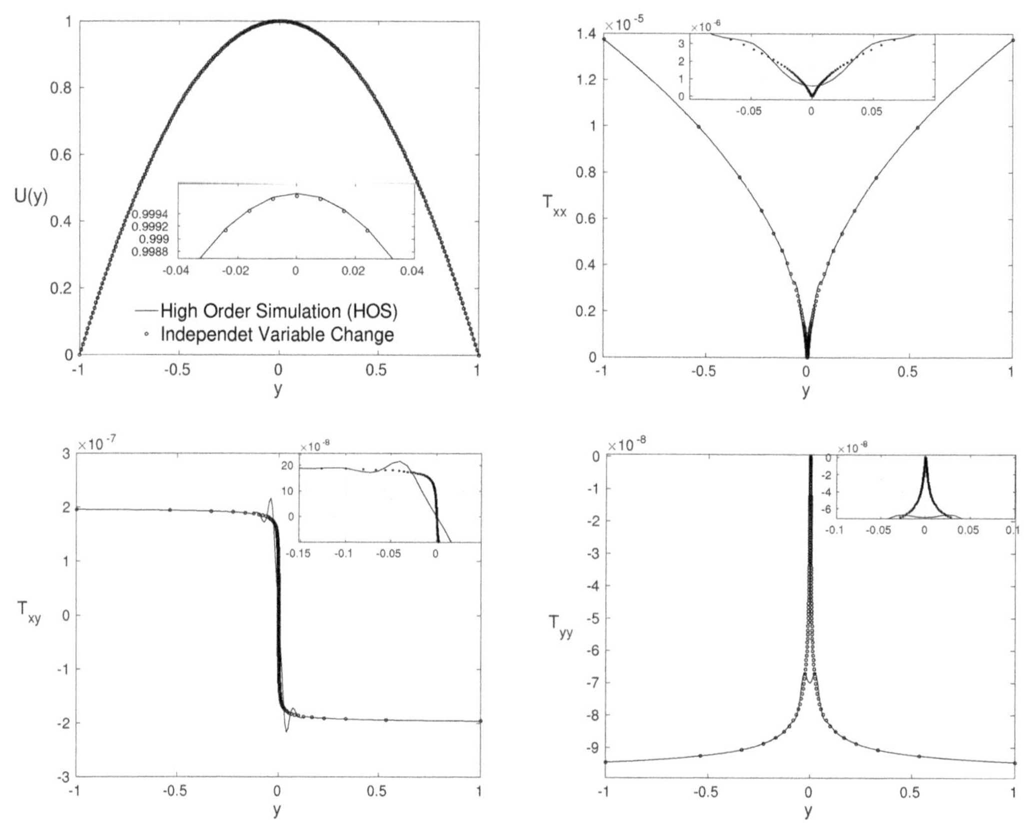

Figure 12. Same as in Figure 1 for dimensionless numbers: $Re = 5000$, $\beta = 0.75$, $\alpha_G = 0.2$ and $Wi = 500$.

One may observe that the results obtained by the different formulations are in agreement. An interesting behavior presented by the HOS formulation can be observed. The results obtained by this formulation, for high Weissenberg numbers, show oscillations for the extra-stress tensor components in the channel center, and as the Weissenberg number increases, the oscillations are more pronounced.

Oscillations usually appear in simulations of flow between parallel plates close to the wall and not in the channel center. However, for high values of the Weissenberg number, a discontinuity is observed in the extra-stress tensor components in the channel center. With the HOS formulation, even using the log-conformation technique to solve the problem of the high Weissenberg number, oscillations were found in the extra-stress tensor components in this region. The use of mesh refinement in the wall-normal direction smooths these oscillations. It is also possible to adopt a stretched mesh, refining only in the region of the channel center to reduce these oscillations and the computational cost.

4.5. Advantages and Disadvantages for Each Formulation

The formulation based on the Schleiniger and Weinacht formulation [10] is advantageous in terms of the velocity cost in obtaining the flow components, because the system of equations is solved in a semi-analytical way. Another advantage of this formulation is the possibility to obtain the flow variables without the contribution of the solvent; that is, flow for the purely polymeric fluid ($\beta = 0$). The disadvantage appears in the direction of convergence towards the adequate pressure gradient for the ideal flow. This disadvantage is related to the term inside the square root of Equation (21), it can becomes negative, and the solution is no longer real. This behavior appears for high values of Weissenberg Wi, Reynolds number Re, and the mobility parameter α_G.

The formulation based on the independent variable change is advantageous for analytically obtaining the main components of the desired flow. Another advantage: there is no restriction for the values of the flow variables, except those already found by Schleiniger and Weinacht [10] (such as $\alpha_G = 0.5$). The disadvantage is the computational cost, since it was not possible to obtain an analytical expression for this. It is necessary to interpolate a new distribution for the tensor T_{xy} to find an equally spaced domain of y, then it is numerically integrated to the expression (29), and finally the streamwise velocity component can be obtained. Another disadvantage of this formulation is the limitation of the flow simulation for purely polymeric fluids ($\beta = 0$). The limitation arises from the equation for the derivative of velocity (29), where the right-hand side terms are divided by β.

The solution obtained by high-order simulation is advantageous because simplifications are not adopted in the equation system that models the flow, so the problem is solved, considering all of the terms in the equation system. The disadvantage of this formulation is the computational cost, since many CPU hours are necessary to obtain the solution for each case. The adopted code uses domain decomposition parallelization, high-order methods for approximating the spatial derivatives, and a classical fourth-order Runge–Kutta method for the temporal derivative.

The simulations were carried out in a computer Intel Xeon E5-2680v2 2.8 GHz. The wall time required for the Schleiniger and Weinacht [10] and the independent variable change formulations were less than 5 s. The high-order simulations were performed using 16 cores and the wall time was about 20 h.

5. Conclusions

This paper presents three different formulations to obtain the solution of a two-dimensional viscoelastic fluid flow between two parallel plates, modeled by the Giesekus constitutive equation.

The first formulation presented is based on the work by Schleiniger and Weinacht [10]. The second formulation presented is based on the idea of the independent variable change. This change allows the flow components to be obtained analytically, except for the velocity profile, obtained using a high-order numerical integration method. The third formulation presented was called high-order simulation (HOS), i.e., the numerical simulation of the flow modeled by the Navier–Stokes equations and the constitutive equation, using high-order methods to obtain the solution.

The Schleiniger and Weinacht [10] formulation is efficient to obtain the flow components accurately and quickly. However, using many numerical methods for the solution makes convergence difficult for specific parameter values. Each step is verified if the value inside the square root is negative, making the solution complex and no longer real. However, this formulation proved efficient in obtaining the flow components for purely polymeric fluids ($\beta = 0$), converging to the solution for almost all values proposed. It did not work for high values of Reynolds Re and Weissenberg Wi numbers.

The formulation based on the independent variable change proved to be very efficient, as it solves all of the flow components analytically, but the velocity profile. The flow components are obtained quickly and accurately, for any values of dimensionless numbers of Reynolds Re, Weissenberg Wi, α_G, and $\beta > 0$. The only limitation of this formulation is when the fluid is composed of purely polymeric fluid flows, or near it $\beta < 0.01$.

The HOS formulation is based on the complete solution of the Navier–Stokes equations and the considered constitutive equation. This formulation has a high computational cost since the simulations take a long time to be solved. However, the numerical methods used proved to obtain good results for the simulations carried out for all of the proposed dimensionless parameters, with a high Weissenberg number, $\beta = 0$, α_G, and Reynolds number Re.

It could be observed that the formulations presented proved to be efficient at obtaining the components of the desired flow. The results of the formulations were explored and analyzed, and their respective limitations and efficiencies were commented on. These results could be used to clarify and help researchers with which formulation is most suitable, depending on the fluid and flow parameters adopted.

Author Contributions: Methodology, L.J.d.S.F., M.T.d.A., A.C.B., D.O.d.A.C. and L.F.d.S.; investigation, L.J.d.S.F., M.T.d.A., A.C.B., D.O.d.A.C. and L.F.d.S.; writing—original draft preparation, L.J.d.S.F., M.T.d.A., A.C.B., D.O.d.A.C. and L.F.d.S.; writing—review and editing, L.J.d.S.F., M.T.d.A., A.C.B., D.O.d.A.C. and L.F.d.S. All authors have read and agreed to the published version of the manuscript.

Funding: Research was carried out using the computational resources of the Center for Mathematical Sciences Applied to Industry (CeMEAI), funded by FAPESP grant 2013/07375-0.

Institutional Review Board Statement: Not applicable.

Informed Consent Statement: Not applicable.

Data Availability Statement: Not applicable.

Acknowledgments: L.J.d.S.F., M.T.d.A. and L.F.d.S. acknowledge the Department of Applied Mathematics and Statistics—University of Sao Paulo, Sao Carlos. A.C.B. acknowledges the Department of Mathematics and Computer Science—Sao Paulo State University, Presidente Prudente and D.O.d.A.C. acknowledges the Department of Mechanical Engineering—Federal University of Rio de Janeiro, Rio de Janeiro.

Conflicts of Interest: The authors declare no conflict of interest.

References

1. Hayat, T.; Khan, M.; Ayub, M. Some simple flows of an Oldroyd-B fluid. *Int. J. Eng. Sci.* **2001**, *39*, 135–147. [CrossRef]
2. Hayat, T.; Khan, M.; Ayub, M. Exact solutions of flow problems of an Oldroyd-B fluid. *Appl. Math. Comput.* **2004**, *151*, 105–119. [CrossRef]
3. Oliveira, P.J. An exact solution for tube and slit flow of a FENE-P fluid. *Acta Mech.* **2002**, *158*, 157–167. [CrossRef]
4. Cruz, D.; Pinho, F.; Oliveira, P. Analytical solutions for fully developed laminar flow of some viscoelastic liquids with a Newtonian solvent contribution. *J. Non-Newton. Fluid Mech.* **2005**, *132*, 28–35. [CrossRef]
5. Duarte, A.; Miranda, A.; Oliveira, P. Numerical and analytical modeling of unsteady viscoelastic flows: The start-up and pulsating test case problems. *J. Non-Newton. Fluid Mech.* **2008**, *154*, 153–169. [CrossRef]
6. Paulo, G.; Oishi, C.; Tomé, M.; Alves, M.; Pinho, F. Numerical solution of the FENE-CR model in complex flows. *J. Non-Newton. Fluid Mech.* **2014**, *204*, 50–61. [CrossRef]
7. Pinho, F.T.; Oliveira, P.J. Axial annular flow of a nonlinear viscoelastic fluid—An analytical solution. *J. Non-Newton. Fluid Mech.* **2000**, *93*, 325–337. [CrossRef]

8. Ferrás, L.L.; Nóbrega, J.M.; Pinho, F.T. Analytical solutions for channel flows of Phan-Thien-Tanner and Giesekus fluids under slip. *J. Non-Newton. Fluid Mech.* **2012**, *171–172*, 97–105. [CrossRef]
9. Yoo, J.Y.; Choi, H.C. On the steady simple shear flows of the one-mode Giesekus fluid. *Rheol. Acta* **1989**, *28*, 13–24. [CrossRef]
10. Schleiniger, G.; Weinacht, R.J. Steady Poiseuille flows for a Giesekus fluid. *J. Non-Newton. Fluid Mech.* **1991**, *40*, 79–102. [CrossRef]
11. James, D.F. Boger Fluids. *Annu. Rev. Fluid Mech.* **2009**, *41*, 129–142. [CrossRef]
12. Giesekus, H. Elasto-viskose Flüssigkeiten, für die in stationären Schichtströmungen sämtliche Normalspannungskomponenten verschieden gross sind. *Rheol. Acta* **1962**, *2*, 50–62. [CrossRef]
13. Raisi, A.; Mirzazadeh, M.; Dehnavi, A.; Rashidi, F. An approximate solution for the Couette–Poiseuille flow of the Giesekus model between parallel plates. *Rheol. Acta* **2008**, *47*, 75–80. [CrossRef]
14. Tomé, M.; Araujo, M.; Evans, J.; McKee, S. Numerical solution of the Giesekus model for incompressible free surface flows without solvent viscosity. *J. Non-Newton. Fluid Mech.* **2019**, *263*, 104–119. [CrossRef]
15. Giesekus, H. A simple constitutive equation for polymer fluids based on the concept of deformation-dependent tensorial mobility. *J. Non-Newton. Fluid Mech.* **1982**, *11*, 69–109. [CrossRef]
16. Giesekus, H. Constitutive equations for polymer fluids based on the concept of configuration-dependent molecular mobility: A generalized mean-configuration model. *J. Non-Newton. Fluid Mech.* **1985**, *17*, 349–372. [CrossRef]
17. Fattal, R.; Kupferman, R. Constitutive laws for the matrix-logarithm of the conformation tensor. *J. Non-Newton. Fluid Mech.* **2004**, *123*, 281–285. [CrossRef]
18. Fattal, R.; Kupferman, R. Time-dependent simulation of viscoelastic flows at high Weissenberg number using the log-conformation representation. *J. Non-Newton. Fluid Mech.* **2005**, *126*, 23–37. [CrossRef]
19. Souza, L.F.; Mendonça, M.T.; Medeiros, M.A.F. The advantages of using high-order finite differences schemes in laminar-turbulent transition studies. *Int. J. Numer. Methods Fluids* **2005**, *48*, 565–592. [CrossRef]
20. Ferziger, J.H.; Peric, M. *Computational Methods for Fluid Dynamics*; Springer: Berlin/Heidelberg, Germany; New York, NY, USA, 1997.
21. Stüben, K.; Trottenberg, U. *Nonlinear Multigrid Methods, The Full Approximation Scheme*; Springer: Koln-Porz, Germany, 1981; Chapter 5, pp. 58–71.
22. Souza, L.F. Instabilidade Centrífuga e Transição Para Turbulência em Escoamentos Laminares Sobre Superfícies Côncavas. Ph.D. Thesis, Instituto Tecnológico de Aeronáutica, Sao Jose dos Campos, Brazil, 2003.
23. Lele, S. Compact Finite Difference Schemes with Spectral-Like Resolution. *J. Comput. Phys.* **1992**, *103*, 16–42. [CrossRef]

Article

Development Length of Fluids Modelled by the gPTT Constitutive Differential Equation

Juliana Bertoco [1,*,†], Rosalía T. Leiva [2,†], Luís L. Ferrás [3,†], Alexandre M. Afonso [4,†] and Antonio Castelo [2,†]

1. Faculdade de Ciências Exatas e Tecnológicas, Universidade do Estado de Mato Grosso, Cáceres 78217-900, MT, Brazil
2. Instituto de Ciências Matemáticas e de Computação, Universidade de São Paulo-USP, São Carlos 13566-590, SP, Brazil; rosalia.taboada@usp.br (R.T.L.); castelo@icmc.usp.br (A.C.)
3. Center of Mathematics—CMAT, University of Minho, Campus de Azurém, 4800-058 Guimarães, Portugal; luislimafr@gmail.com
4. Centro de Estudos de Fenómenos de Transporte, Departamento de Engenharia Mecânica, Faculdade de Engenharia da Universidade do Porto, 4200-465 Porto, Portugal; aafonso@fe.up.pt
* Correspondence: jubertoco@alumni.usp.br
† These authors contributed equally to this work.

Abstract: In this work, we present a numerical study on the development length (the length from the channel inlet required for the velocity to reach 99% of its fully-developed value) of a pressure-driven viscoelastic fluid flow (between parallel plates) modelled by the generalised Phan–Thien and Tanner (gPTT) constitutive equation. The governing equations are solved using the finite-difference method, and, a thorough analysis on the effect of the model parameters α and β is presented. The numerical results showed that in the creeping flow limit ($Re = 0$), the development length for the velocity exhibits a non-monotonic behaviour. The development length increases with Wi. For low values of Wi, the highest value of the development length is obtained for $\alpha = \beta = 0.5$; for high values of Wi, the highest value of the development length is obtained for $\alpha = \beta = 1.5$. This work also considers the influence of the elasticity number.

Keywords: viscoelastic fluids; generalised PTT model; finite-differences; development length

1. Introduction

A variety of functional applications are based on the premise that the flow is fully developed. It is assumed that after a certain time the fluid has travelled a certain length (development length—L) along the channel, after which the flow no longer changes in the direction of flow. This is used, for example, in extrusion dies, lab-on-a-ship, etc. [1–3].

The development length of Newtonian flows in channels and pipes (see Figure 1) is well understood [4].

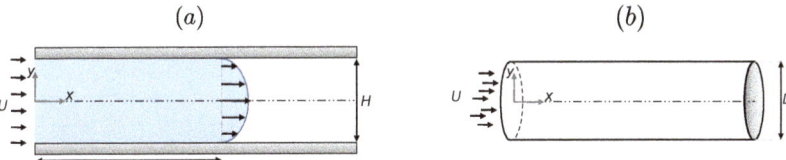

Figure 1. Schematic of the channel and pipe geometries used for the study of the development length. (**a**) Channel flow. (**b**) Pipe flow.

Durst et al. [4] developed two correlations between L (the distance the fluid travels to become fully developed) and the Reynolds number, $Re = \frac{\rho U H}{\eta}$, where U is the imposed mean inlet velocity, ρ is the fluid density, H is the width of the channel (for pipes, one should replace H by D-diameter), and η is the Newtonian viscosity. These correlations are given by,

$$\frac{L}{H} = [(0.631)^{1.6} + (0.0442Re)^{1.6}]^{1/1.6}, \tag{1}$$

$$\frac{L}{D} = [(0.619)^{1.6} + (0.0567Re)^{1.6}]^{1/1.6}, \tag{2}$$

and predict well the development length for channel and pipe flows, respectively. For other works on the development length of Newtonian fluids please consult the following references [5–10].

For generalised Newtonian fluids (with varying viscosity), several works have been proposed in the literature [11–19]. We would like to highlight the works of Fernandes et al. [19] and Poole and Ridley [18], in which they presented two correlations for the development length in channel and pipe flows of power-law fluids (the viscosity is a function of the second invariant of the deformation tensor, $\dot{\gamma}$ (for simple flows, $\dot{\gamma}$ is simply the shear rate). The viscosity is then given by $\eta = k\dot{\gamma}^{n-1}$). The correlations are given by,

$$\frac{L}{H} = [(f(n) - \exp(15.706 - 4.002))^{1.6} + (0.0444Re_{gen})^{-0.209n^2 + 0.645n + 1.225}]^{1/1.6}, \tag{3}$$

$$\frac{L}{D} = [(0.246n^2 - 0.675n + 1.03)^{1.6} + (0.0567Re_{MR})^{1.6}]^{1/1.6}, \tag{4}$$

for channel and pipe flows, respectively. Here, $Re_{gen} = \frac{6\rho U^{2-n} H^n}{k}\left(\frac{n}{4n+2}\right)^n$, $Re_{MR} = \frac{8\rho U^{2-n} D^n}{k}\left(\frac{n}{6n+2}\right)^n$, and $f(n) = \frac{-0.355}{1+2\exp(0.553-4.273n)}$. Note the increasing complexity in the correlations when going from a Newtonian to a power-law fluid.

In the case of viscoelastic fluids, the number of papers on this topic is smaller. This is due to the complexity of viscoelastic flows, such as the presence of singularities at the entrance of the channel, overshoots in the velocity profile, and the high Weissenberg number problem.

We would like to highlight the work of Na and Yoo [20] in which they perform numerical simulations to determine the development length of an Oldroyd-B fluid and conclude that the development length (for a fixed Re) increases slightly with the Weissenberg number, $Wi = \frac{\lambda U}{H}$ (where λ is the relaxation time of the fluid in seconds), but is more strongly affected by Re. Liang [1] proposed a theoretical work for the development length of viscoelastic fluids entering an extrusion die. They presented an expression for estimating the length of the entrance region, which has applications in the extrusion industry. In the work by Philippou et al. [10] the authors present a study on the flow development of a Bingham plastic fluid in tubes and channels considering the Papanastasiou regularisation and the finite element method. They considered the Navier's slip law at the wall and concluded that as slip increases, the development length initially increases exhibiting a global maximum before vanishing rapidly above the critical point corresponding to sliding flow. More recently, Yapici et al. [21] presented a study on the development length of steady flows of Oldroyd-B and Phan–Thien–Tanner (PTT) fluids through a two-dimensional rectangular channel and concluded that the development length determined for the Oldroyd-B fluid varies exponentially with Wi and linearly for the linear PTT model; they also concluded that higher entry lengths are predicted with increasing Wi (at fixed Re).

To remove the unstable numerical effect of the singularity at the entrance corner, a continuous inlet velocity profile is used in both works of Na et al. [20] and Yapici et al. [21]. This regularised profile can affect the true development length, so in the work of Guilherme [22] the log-conformation formulation [23–25] is used, which reduces the rate of increase of the stresses and thus avoids the need to introduce artificial inlet velocity profiles.

It should be noted that the development of a correlation for the prediction of the development length of such complex fluids is still difficult due to the high number of parameters involved and the fact that it is model dependent.

Here we follow the work of Guilherme [22], where a detailed analysis of the development length of the linear PTT model is performed. We extend his work to the exponential and generalised PTT models [22,26,27].

This work is organised as follows. First, we present the differential equations to be solved and their numerical solution. In Section 3, we present the geometry and the meshes. In Section 4, we perform a validation of the numerical method and the meshes, using Newtonian benchmark results. In Section 5, we present and discuss the results for viscoelastic fluids. The paper ends with the main conclusions in Section 6.

2. Governing Equations

The equations governing the flow of an incompressible fluid, under isothermal conditions, are the continuity,

$$\nabla \cdot \mathbf{u} = 0 \tag{5}$$

and the momentum equations,

$$\rho \frac{D\mathbf{u}}{Dt} = -\nabla p + \nabla \cdot \sigma + \mathbf{F}, \tag{6}$$

where $\mathbf{u}$ is the velocity vector, p is the pressure, σ is the stress tensor (to be defined later), ρ is the mass density, and $\mathbf{F}$ represents the external forces. Note that all variables are dimensionless, with: $\mathbf{x} = \mathbf{x}*/H$, $\mathbf{u} = \mathbf{u}*/U$, $t = t * U/H$, $p = p*/(\rho U^2)$, $\sigma = \sigma*/(\rho U^2)$ (the $*$ represents the dimensional variable).

In order to achieve a closed system of equations, a constitutive equation for the extra-stress tensor, σ, is required. Recently, Ferrás et al. [27] proposed a new differential model based on the Phan–Thien–Tanner constitutive equation [26] (see also [28]). This new model considers a more general function for the rate of destruction of junctions, the Mittag–Leffler function, where one or two fitting parameters are included, in order to achieve additional fitting flexibility.

The Mittag–Leffler function is defined as,

$$E_{\alpha,\beta}(z) = \sum_{j=0}^{\infty} \frac{z^j}{\Gamma(\alpha j + \beta)}, \tag{7}$$

with α, β being real and positive. $\Gamma(\cdot)$ is the gamma function, given by:

$$\Gamma(t) = \int_0^\infty x^{t-1} e^{-x} dx. \tag{8}$$

when $\alpha = \beta = 1$, the Mittag–Leffler function reduces to the exponential function, and when $\beta = 1$ the original one-parameter Mittag–Leffler function, E_α is obtained.

The constitutive equation is given by:

$$K(\sigma_{kk})\sigma + Wi \overset{\square}{\sigma} = \frac{2(1-\zeta)}{Re} \mathbf{D}, \tag{9}$$

where σ_{kk} is the trace of the stress tensor, $Wi = \lambda U/H$ is the Weissenberg number, $Re = UH/\nu$ is the Reynolds number ($\nu = \mu_0/\rho$ is the kinematic viscosity), $\mathbf{D} = \frac{1}{2}\left(\nabla \mathbf{u} + (\nabla \mathbf{u})^t\right)$ is the rate of deformation tensor, σ is the elastic stress, and $\zeta = \frac{\mu_S}{\mu_0}$ is the viscosity coefficient, where $\mu_0 = \mu_S + \mu_P$ is the total shear viscosity (μ_S is the solvent/Newtonian viscosity, μ_P is the polymer viscosity) and $\overset{\square}{\sigma}$ represents the Gordon–Schowalter derivative.

The stress function, $K(\sigma_{kk})$, is given by a new formulation that imparts more flexibility and accuracy to the model predictions, as discussed in [27,29,30]. It is given by:

$$K(\sigma_{kk}) = \Gamma(\beta) E_{\alpha,\beta}\left(\frac{\varepsilon Re\, Wi\, \sigma_{kk}}{(1-\zeta)}\right), \tag{10}$$

where ε represents the extensibility parameter, Γ is the Gamma function, and the normalisation $\Gamma(\beta)$ is used to ensure that $K(0) = 1$, for all choices of β.

3. Numerical Method

The numerical method used in this work is based on finite differences. It can deal with tree-like mesh grids (see Figure 2b) and allows fast Cartesian discretizations, flexibility and accuracy, and local mesh refinement. In order to fit the discretization stencil near the interfaces between grid elements of different sizes, a robust method based on a moving least squares meshless interpolation technique is used to compute the weights of the finite difference approximation in a given hierarchical grid, allowing complex mesh configurations and preserving the overall accuracy of the resulting method.

Figure 2a shows a schematic representation of the mesh refinement. Note that some of the points (variables) of the computational cells (red dots) must be approximated because they are not located in the center of the cell (e.g., the center of computational cell 1 is not the same as the location of the red dot used to compute the derivative of the property being evaluated). To solve this problem, we use a special adaptive least square interpolation (MLS). The method is known as HiG-Flow, and a detailed mathematical explanation can be found in [31,32].

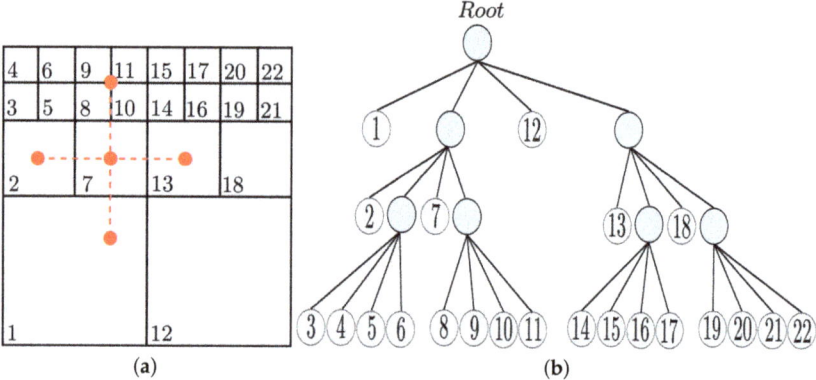

Figure 2. (a) Mesh refinement and the need to use an adaptive least squares method; (b) dependency tree.

For the numerical solution of the Navier–Stokes equations together with the constitutive equation given by the gPTT model, the momentum Equation (6) is rewritten:

$$\frac{\partial \mathbf{u}}{\partial t} + \mathbf{u} \cdot \nabla \mathbf{u} = -\nabla p + \frac{1}{Re} \nabla^2 \mathbf{u} + \nabla \cdot \mathbf{S} + \mathbf{F}, \tag{11}$$

$$\sigma = \frac{2(1-\zeta)}{Re} \mathbf{D} + \mathbf{S}. \tag{12}$$

From Equation (9), the rheological constitutive equation can be written as:

$$\frac{\partial \sigma}{\partial t} + (\mathbf{u} \cdot \nabla)\sigma - \left[(\nabla \mathbf{u})^t \cdot \sigma + \sigma \cdot \nabla \mathbf{u}\right] = \frac{1}{Wi} \mathbf{M}(\sigma), \tag{13}$$

where $\mathbf{M}(\sigma)$ is given by Equation (14),

$$\mathbf{M}(\sigma) = \frac{2(1-\zeta)}{Re} \mathbf{D} - \left[\Gamma(\beta) E_{\alpha,\beta}\left(\frac{\varepsilon Re\, Wi\, \sigma_{kk}}{(1-\zeta)}\right)\right]\sigma - \xi Wi(\sigma \cdot \mathbf{D} + \mathbf{D} \cdot \sigma), \tag{14}$$

and the parameter ξ ($0 \leq \xi \leq 1$) accounts for the slip between the molecular network and the continuous medium. For $\xi = 0$ there is no slip and the motion becomes affine.

The parameter ζ leads to a non-zero second normal-stress difference in shear, resulting in secondary flows in ducts having non-circular cross-sections. Since in this work we only consider 2D flows, and, due to the high number of parameters involved in the numerical simulations and the gPTT model itself, we have only considered the case when $\zeta = 0$.

3.1. Calculation of $\mathbf{u}^{(n+1)}$ and $p^{(n+1)}$

To calculate the velocity $\mathbf{u}^{(n+1)}$ and pressure $p^{(n+1)}$ fields, we use the incremental projection method by Chorin [33], that uncouples the mass conservation and momentum equations, given by Equations (5) and (6), respectively. This method allows one to obtain an intermediate velocity field $\tilde{\mathbf{u}}^{n+1}$ from Equation (11). In the HiG-Flow methodology, this Equation (11) can be approximated using an explicit Euler method, Runge–Kutta RK-2 or RK-4, or, the semi-implicit Euler methods, Cranck–Nicolson, and BDF2. One can also choose a spatial discretization orders of 2 or 4. One can use the the convective central schemes or Upwind (order 1), or, schemes of order 2 like the Cubista [34] and Quick [35].

In this work an Implicit Euler scheme together with a second order spatial approximation and an Upwind Cubista scheme for the convective terms, is used:

$$\frac{\tilde{\mathbf{u}}^{(n+1)} - \mathbf{u}^n}{\delta t} + \mathbf{u}^n \cdot \nabla \mathbf{u}^n = -\nabla p^n + \frac{1}{Re}\nabla^2 \tilde{\mathbf{u}}^{(n+1)} + \nabla \cdot S^n + F^n \quad (15)$$

here, δt is the time step, n represents the known values of velocity, stress, and pressure at instant n, and $n+1$ represents the new velocity field values (unknown) to be obtained from the solution of the equation. At the inlet, (see Figure 3) we consider a constant velocity profile, $u(y) = 1$ (the stress components are set to 0) and at the outlet, we assume fully developed boundary conditions (Neumann boundary conditions) for the velocity and stress (the pressure is imposed). Finally, at the walls ($y = 0$ and $y = 1$), we have the empirical no-slip boundary condition ($u = 0$).

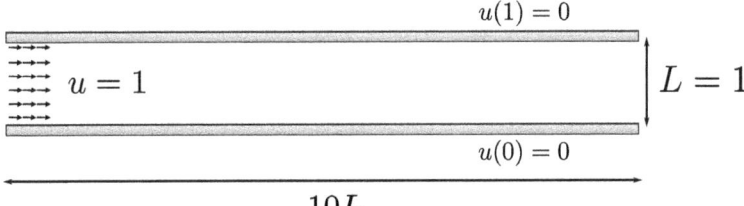

Figure 3. Dimensionless representation of the channel domain.

Using the projection method, it is well known that the velocity field $\tilde{\mathbf{u}}^{n+1}$ obtained from Equation (15) may not satisfy the mass conservation equation. Therefore, in order to solve this problem, the equation for the potential $\psi^{(n+1)} = \delta t (p^n - p^{(n+1)})$ is solved,

$$\nabla^2 \psi^{(n+1)} = \nabla \cdot \tilde{\mathbf{u}}^{(n+1)}, \quad (16)$$

and the Helmholtz–Hodge decomposition (see [31,32,36] for more details) is used to correct the previous non-conservative velocity field $\tilde{\mathbf{u}}^{(n+1)}$,

$$\mathbf{u}^{(n+1)} = \tilde{\mathbf{u}}^{(n+1)} - \nabla \psi^{(n+1)}. \quad (17)$$

The new velocity field $\mathbf{u}^{(n+1)}$ satisfies the mass conservation equation. Finally, the pressure is updated $p^{(n+1)} = p^n + \frac{\psi^{(n+1)}}{\delta t}$.

3.2. Calculation of the Extra-Stress Tensor $\sigma^{(n+1)}$

The velocity and pressure fields were obtained in the previous subsection. We now aim to obtain the extra-stress tensor $\sigma^{(n+1)}$ field. Equation (13) is solved using the Explicit Euler method, and, to calculate $\mathbf{M}(\sigma)$ (see Equation (14)), the Mittag–Leffler function and the term $\Gamma(\beta)E_{\alpha,\beta}$ are obtained numerically from Equation (18) and the approximations presented in the work by R. Gorenflo, J. Loutchko, and Y. Luchko [37],

$$\Gamma(\beta)E_{\alpha,\beta} \approx \Gamma(\beta) \sum_{k=1}^{N} \frac{z^{k-1}}{\Gamma(\alpha(k-1)+\beta)} + O(z^N). \tag{18}$$

The numerical implementation of the Mittag–Leffler function follows the work by Davide Verotta and Eduardo Mendes [38] (developed in Fortran), which is adapted in this work to C++. The original fortran code is based on a Matlab function developed by Igor Podlubny and Martin Kacena [39] which, in turn, was based on the reference [37].

4. Geometry and Meshes

4.1. Geometry

Due to the low Re values considered in this work, and based on the few literature results on the development length of viscoelastic fluids, we considered a geometry where the length of the channel is fixed at 10 times its width (Figure 3).

4.2. Meshes

We performed simulations considering more than 8 levels of mesh refinement. After some numerical experiments, the following meshes were considered:

- M_1—uniform mesh with 160×16 computational cells and a minimum $\Delta x/H$ and $\Delta y/H$ mesh spacing of 0.0625;
- M_2—uniform mesh with 320×32 computational cells and a minimum $\Delta x/H$ and $\Delta y/H$ mesh spacing of 0.03125;
- M_3—uniform mesh with 640×64 computational cells and a minimum $\Delta x/H$ and $\Delta y/H$ mesh spacing of 0.015625.

The tree meshes are shown in Figure 4.

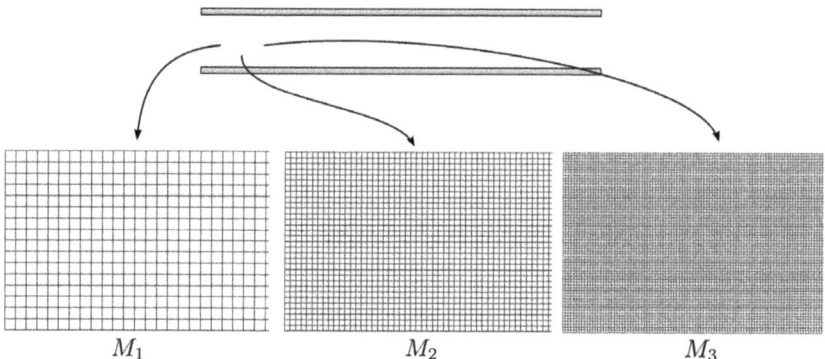

Figure 4. Three consequently refined uniform meshes used in this work.

Numerical simulations were also performed considering a mesh with additional refinement in the centerline of the channel, as shown in Figure 5. A total number of 16,000 cells was used.

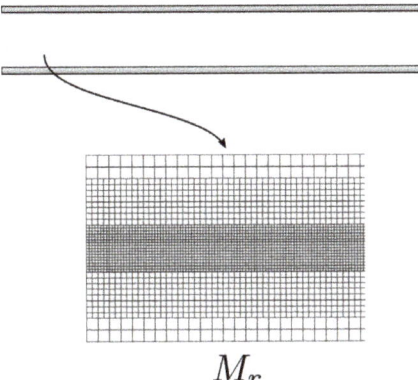

Figure 5. Mesh with refinement in the centerline region.

5. Validation of the Numerical Method

The validation of the numerical method is performed in two steps. First, the numerically-determined fully-developed velocities and stresses are compared with the analytical solution [27]. Then, the development length obtained for the gPTT model with $Wi = 0.001$ (almost Newtonian fluid) is compared with the benchmark results of Durst et al. [4].

5.1. Comparison with the Analytical Solution

Figure 6 shows a comparison between the analytical solution (solid line) and the numerical solution (symbols) for mesh M_3 with $Re = 10^{-3}$, $Wi = 0.1, 0.2, \cdots, 1.0$ and $\varepsilon = 0.25$ (ξ was set to 0). In Figure 6a we have $\alpha = 0.5$ and $\beta = 0.5$ and in Figure 6b $\alpha = 1.5$ and $\beta = 1.5$.

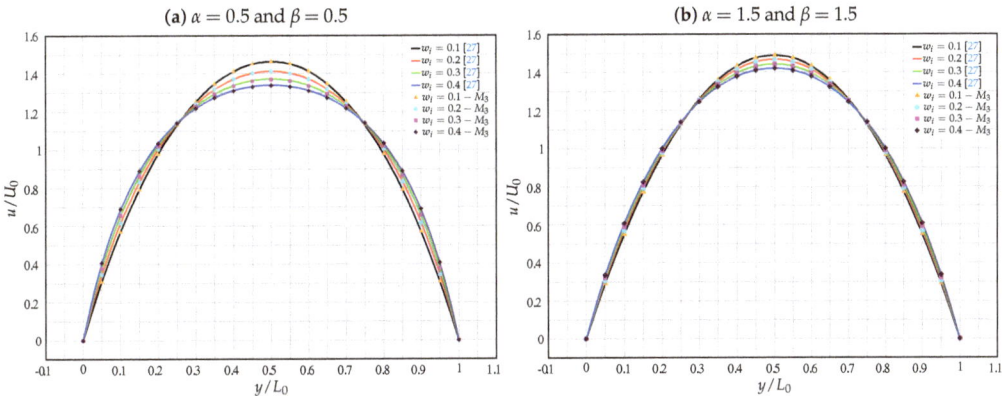

Figure 6. Comparison between the analytical solution (solid line) and the numerical solution (symbols) for mesh M_3 with $Re = 10^{-3}$, $Wi = 0.1, 0.2, \cdots, 1.0$, and $\varepsilon = 0.25$ (ξ was set to 0). (**a**) $\alpha = 0.5$ and $\beta = 0.5$. (**b**) $\alpha = 1.5$ and $\beta = 1.5$.

It can be seen that an excellent agreement is obtained between the analytical and numerical solutions for all the considered values of Wi, which underlines the robustness of the numerical method and the meshes.

For lower values of α and β, we obtain a higher destruction rate of the junctions in the gPTT model. Note that the typical viscoelastic velocity profile is more flattened for lower values of α and β. In this case, the different values of Wi have a stronger impact on the model's behaviour. This result is similar to those found in the literature comparing linear and exponential functions of the trace of the stress tensor.

5.2. Comparison with the Development Length of a Newtonian Fluid

In the limiting case of $Wi \to 0$ we obtain a Newtonian fluid. Therefore, we considered $Wi = 0.001$ and performed simulations for the development length of a gPTT fluid, using the geometry shown in Figure 4. We considered a Reynolds number in the range [0, 100], where the nonlinear variation of the development length with Re is more pronounced. The other parameters of the model were set as follows: $\alpha = 0.1$, $\beta = 0.1$, $\varepsilon = 0.25$, $\zeta = 0$.

Figure 7a shows a comparison between the development length obtained with the gPTT model, a Newtonian fluid, and that obtained by the Durst et al. [4] correlation for the variation of the development length with Re (see Equation (1)). The three results practically overlap, proving once again the robustness of the numerical method.

As Re increases, the results for the gPTT model in the coarse mesh are slightly higher than those obtained for the Newtonian fluid and the correlation. However, in the nonlinear domain the results are quite accurate.

Figure 7b shows the velocity profiles obtained in the fully developed region of the channel (mesh M_3) considering the gPTT and Newtonian models for $Re = 0.001$ and $Re = 100$.

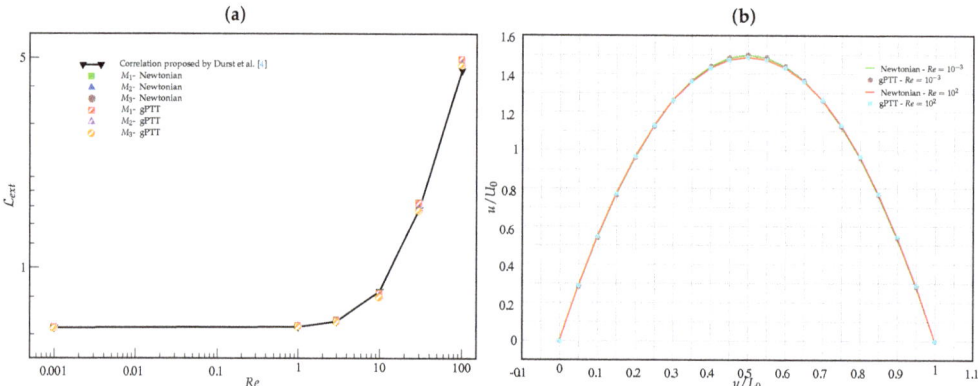

Figure 7. (a) Comparison between the development length obtained with the gPTT model (with $Wi = 0.001$), a Newtonian fluid, and the correlation proposed by Durst et al. [4], for three different meshes M_1, M_2, and M_3. (b) Velocity profiles in the fully developed region for mesh M_3, considering the gPTT and Newtonian models, for $Re = 0.001$ and $Re = 100$.

Again, there is excellent agreement between the two solutions for the two different values of Re. This shows that the value of $Wi = 0.001$ is a good approximation for the Newtonian fluid.

Based on these results, the numerical code is now able to predict the development length of the fluid modelled by the gPTT model considering a wider range of Wi numbers.

6. Development Length of a gPTT Fluid

6.1. Simulations

We performed a large number of simulations by considering Wi = 0.1, 0.2, 0.3, 0.4, 0.5, 0.6, 0.7, 0.8, 0.9, 1, creeping flow ($Re = 0.001$), and the following combination of α and β parameters:

- $(\alpha, \beta) = (0.5; 0.5)$-Meshes M_1, M_2, M_3, M_r-$Wi = 0.1, 0.2, 0.3, 0.4$
- $(\alpha, \beta) = (0.5; 1.5)$-Meshes M_1, M_2, M_3, M_r-$Wi = 0.1, 0.2, 0.3, 0.4$
- $(\alpha, \beta) = (1.5; 0.5)$-Meshes M_1, M_2, M_3, M_r-$Wi = 0.1, 0.2, 0.3, 0.4$
- $(\alpha, \beta) = (1.5; 1.5)$-Meshes M_1, M_2, M_3, M_r-$Wi = 0.1, 0.2, 0.3, 0.4$
- $(\alpha, \beta) = (1.0; 1.0)$-Meshes M_2, M_r-$Wi = 0.1, 0.2, 0.3, 0.4$

- $(\alpha, \beta) = (0.5; 0.5)$-Meshes M_2, M_r-$Wi = 0.5, 0.6, 0.7, 0.8, 0.9, 1$
- $(\alpha, \beta) = (0.5; 1.5)$-Meshes M_2, M_r-$Wi = 0.5, 0.6, 0.7, 0.8, 0.9, 1$
- $(\alpha, \beta) = (1.5; 0.5)$-Meshes M_2, M_r-$Wi = 0.5, 0.6, 0.7, 0.8, 0.9, 1$
- $(\alpha, \beta) = (1.5; 1.5)$-Meshes M_2, M_r-$Wi = 0.5, 0.6, 0.7, 0.8, 0.9, 1$
- $(\alpha, \beta) = (1.0; 1.0)$-Meshes M_2, M_r-$Wi = 0.5, 0.6, 0.7, 0.8, 0.9, 1$

This gives a total number of 132 simulations. The simulations with the finest mesh took about 15 h each.

The first set of 72 simulations allowed conclusions to be drawn about the convergence of the numerical method and the error in calculating the development length using the Richardson extrapolation technique. Based on the results of these simulations, a second set of simulations was performed for higher values of Wi using meshes M_2 (see Figure 4) and M_r (see Figure 5). These meshes were chosen based on a trade-off between accuracy and computational time.

6.2. Creeping Flow

6.2.1. Case of $\mathcal{L}_{99\%}$

The development length, determined as the length from the channel inlet required for the velocity to reach 99% of its fully developed value, and denoted here as $\mathcal{L}_{99\%}$, is shown in Table 1. The results are shown only for Wi up to 0.4, since convergence problems for finer meshes are observed for higher values of Wi. The main problem arises from the singularity at the entrance corner of the channel, which generates an error that propagates along the channel (for more details, see [24,25]).

Table 1. Benchmark development length values for the velocity ($\mathcal{L}_{99\%}$).

Wi	α	β	M_1	M_2	M_3	M_r	$\mathcal{L}_{ext}$	% Error
0.1	0.5	0.5	0.701	0.669	0.660	0.661	0.657	0.48
	0.5	1.5	0.675	0.655	0.649	0.642	0.647	0.29
	1.5	0.5	0.689	0.659	0.652	0.650	0.651	0.24
	1.5	1.5	0.653	0.650	0.648	0.632	0.644	0.68
	1.0	1.0	—	0.658	—	0.641		
0.2	0.5	0.5	0.844	0.737	0.711	0.736	0.703	1.202
	0.5	1.5	0.806	0.731	0.713	0.721	0.707	0.844
	1.5	0.5	0.788	0.701	0.682	0.692	0.677	0.781
	1.5	1.5	0.803	0.736	0.712	0.747	0.699	1.851
	1.0	1.0	—	0.719	—	0.709		
0.3	0.5	0.5	0.984	0.853	0.805	0.865	0.777	3.578
	0.5	1.5	0.980	0.878	0.852	0.822	0.843	1.073
	1.5	0.5	0.883	0.775	0.750	0.784	0.742	1.015
	1.5	1.5	1.033	0.933	0.928	1.012	0.928	0.022
	1.0	1.0	—	0.836	—	0.852		
0.4	0.5	0.5	1.104	1.024	0.949	1.102	-	-
	0.5	1.5	1.175	1.109	1.082	1.165	1.065	1.600
	1.5	0.5	0.984	0.877	0.862	0.931	0.860	0.276
	1.5	1.5	1.307	1.286	1.273	1.365	1.246	2.155
	1.0	1.0	—	1.001	—	1.079		

Note that the error is higher at the lowest and highest values of α and β, being more pronounced when α and β are low. The maximum error was 3.6% and was obtained, as expected, for $Wi = 0.4$ and $\alpha = \beta = 0.5$. It should be noted that the errors are quite low and therefore these solutions can be used as benchmarks.

Figure 8 shows the development lengths for mesh M_2 presented in Table 1. A nonlinear variation of the development length with α, β, and Wi is observed.

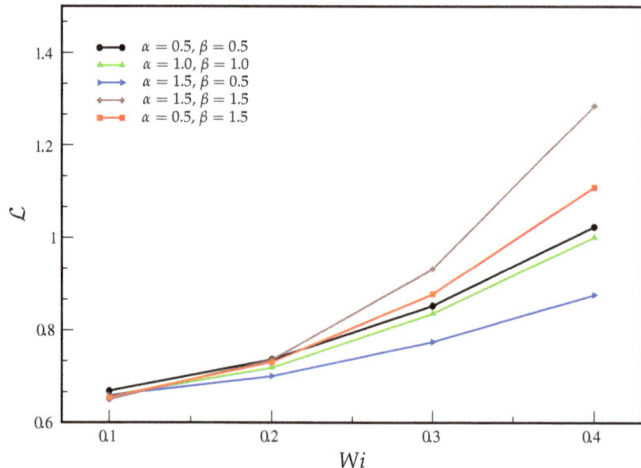

Figure 8. Development length as a function of Wi considering 99% U_{max} ($\mathcal{L}_{99\%}$) and mesh M_2.

The development length increases with Wi, with viscoelastic effects delaying the diffusion and convection of information from the walls to the center of the channel. This diffusion and convection is also strongly influenced by the parameters of the Mittag–Leffler function. For low values of Wi, the highest value of the development length is obtained for $\alpha = \beta = 0.5$; for high values of Wi, the highest value of the development length is obtained for $\alpha = \beta = 1.5$. A molecular continuum explanation of this phenomenon is not an easy task. At high α and β values, the rate of destruction of the junctions is lower than at low α and β values. This means that when Wi values are low and the rate of junction destruction is high, information travels slowly from the wall to the center of the channel (compared to when the rate of junction destruction is low). The opposite was expected. Note that in this case the development lengths are very similar for all tested values of α and β, and therefore the influence of these parameters on the development length is small. These results can be justified by the low value of Wi.

As Wi increases, the highest value of development length is reached with a low rate of destruction of junctions. This result can be justified by the fact that as the rate of destruction of the junctions decreases, the information is transmitted more slowly due to the small number of new contacts between the strands representing the molecules.

6.2.2. Case of $\mathcal{L}_{98\%}$

In addition to the error arising at the entrance corner due to a singularity, we also have the problem of the development length, which takes into account 99% of the fully developed maximum velocity and may not work so well in an intermediate mesh as M_2, for higher Wi. This leads to an increased difficulty for the numerical method to capture $\mathcal{L}_{99\%}$ for the mesh M_2.

Therefore, to capture the essence of the development length for higher Wi values, we considered another development length, $\mathcal{L}_{98\%}$ (length from channel entry required for the velocity to reach 98% of its fully-developed value), which is less restrictive.

The results obtained for $\mathcal{L}_{98\%}$ are shown in Figure 9 for mesh M_2 and Table 2 for the three meshes (along with the extrapolated development length value).

Table 2. Benchmark development length values for the velocity ($\mathcal{L}_{98\%}$).

Wi	α	β	M_1	M_2	M_3	M_r	$\mathcal{L}_{ext}$	% Error
0.1	0.5	0.5	0.578	0.573	0.571	0.560	0.570	0.234
	0.5	1.5	0.564	0.565	0.565	0.552	0.565	0.002
	1.5	0.5	0.571	0.566	0.565	0.554	0.565	0.044
	1.5	1.5	0.547	0.564	0.566	0.545	0.566	0.047
	1.0	1.0	—	0.566	—	0.551		
0.2	0.5	0.5	0.656	0.622	0.612	0.603	0.608	0.685
	0.5	1.5	0.652	0.625	0.619	0.604	0.617	0.278
	1.5	0.5	0.629	0.601	0.592	0.578	0.588	0.725
	1.5	1.5	0.658	0.632	0.621	0.622	0.613	1.316
	1.0	1.0	—	0.616	—	0.594		
0.3	0.5	0.5	0.725	0.696	0.681	0.659	0.665	2.417
	0.5	1.5	0.765	0.736	0.732	0.697	0.731	0.088
	1.5	0.5	0.694	0.656	0.643	0.628	0.636	1.062
	1.5	1.5	0.840	0.789	0.789	0.812	0.789	0.000
	1.0	1.0	—	0.705	—	0.680		
0.4	0.5	0.5	0.782	0.790	0.776	0.730	-	-
	0.5	1.5	0.875	0.885	0.897	0.836	-	-
	1.5	0.5	0.770	0.729	0.720	0.713	0.717	0.353
	1.5	1.5	1.052	1.010	1.068	1.080	-	-
	1.0	1.0	—	0.820	—	0.809		
0.5	0.5	0.5	—	0.893	—	0.812		
	0.5	1.5	—	1.062	—	1.010		
	1.5	0.5	—	0.813	—	0.819		
	1.5	1.5	—	1.265	—	1.367		
	1.0	1.0	—	0.958	—	0.975		
0.6	0.5	0.5	—	0.998	—	0.907		
	0.5	1.5	—	1.258	—	1.216		
	1.5	0.5	—	0.909	—	0.946		
	1.5	1.5	—	1.536	—	1.663		
	1.0	1.0	—	1.106	—	1.163		
0.7	0.5	0.5	—	1.106	—	1.006		
	0.5	1.5	—	1.473	—	1.443		
	1.5	0.5	—	1.009	—	1.083		
	1.5	1.5	—	1.817	—	1.960		
	1.0	1.0	—	1.258	—	1.360		
0.8	0.5	0.5	—	1.207	—	1.110		
	0.5	1.5	—	1.702	—	1.672		
	1.5	0.5	—	1.112	—	1.225		
	1.5	1.5	—	2.103	—	2.254		
	1.0	1.0	—	1.412	—	1.573		
0.9	0.5	0.5	—	1.298	—	1.206		
	0.5	1.5	—	1.947	—	1.870		
	1.5	0.5	—	1.214	—	1.365		
	1.5	1.5	—	2.392	—	2.544		
	1.0	1.0	—	1.562	—	1.752		
1.0	0.5	0.5	—	1.368	—	1.279		
	0.5	1.5	—	2.236	—	2.029		
	1.5	0.5	—	1.313	—	1.499		
	1.5	1.5	—	2.679	—	2.827		
	1.0	1.0	—	1.709	—	1.933		

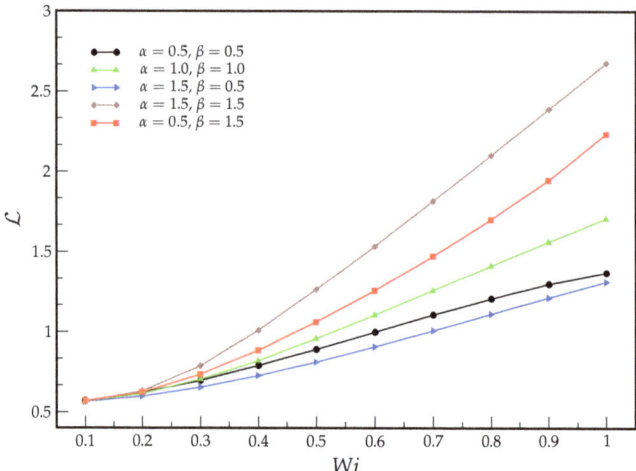

Figure 9. Development length as a function of Wi considering 98% U_{max} ($\mathcal{L}_{98\%}$) and mesh M_2.

The results obtained follow the same trend observed in the case $\mathcal{L}_{99\%}$, with some minor differences. The development length is initially higher for the case $\alpha = \beta = 0.5$ than for the case $\alpha = \beta = 1$, until $Wi = 0.2$, where the growth rate of $\mathcal{L}_{98\%}$ becomes larger with Wi for $\alpha = \beta = 1$. For $\mathcal{L}_{99\%}$, the two development lengths are quite similar.

For $Wi = 1$, we obtain a development length of 2.679 for $\alpha = \beta = 1.5$ and a development length of 1.313 for $\alpha = 1.5$, $\beta = 0.5$ (and 1.368 for $\alpha = \beta = 0.5$). Again, these results are consistent with the idea that the higher the rate of destruction of junctions, the smaller the development length (information travels faster across the channel).

Figure 10–12 show the different velocity profiles obtained at 10 different sections of the channel. The first numerical velocity profile is taken at $x/H = 0.1$ ($x/L = 0.01$) and the last profile is taken at the middle of the channel ($x/H = 5$ or $x/L = 0.5$).

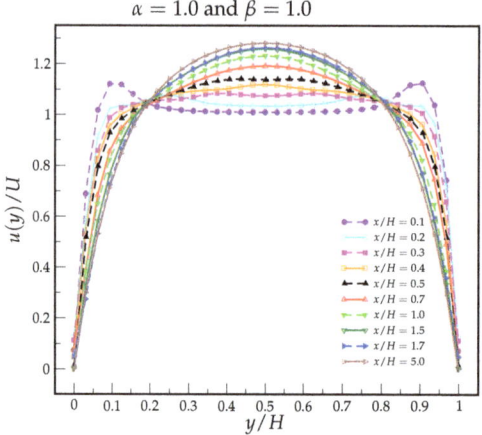

Figure 10. Velocity profiles obtained at 10 different sections of the channel for $Wi = 1.0$, $\alpha = 1$, $\beta = 1$.

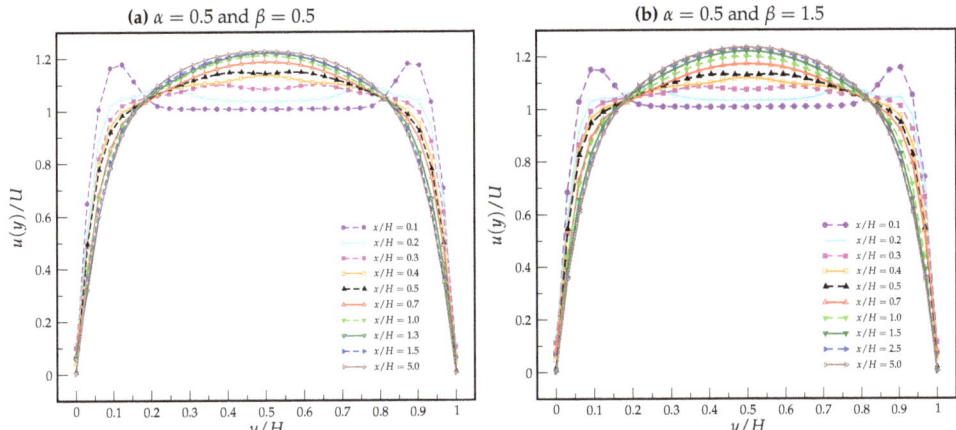

Figure 11. Velocity profiles obtained at 10 different sections of the channel for $Wi = 1.0$. (a) $\alpha = 0.5, \beta = 0.5$; (b) $\alpha = 0.5, \beta = 1.5$.

Figure 10 shows the reference velocity profiles obtained for the classical case of an exponential PTT model ($\alpha = \beta = 1$). The velocity profile evolves from a plug profile in the center of the channel (for profiles near the inlet) to the typical parabolic profile (in the fully developed region). Note the overshoots near the walls that occur when the fluid is still developing. This is due to the different characteristic times of the fluid and the diffusion of information moving from the walls ($y/H = 0$ and $y/H = 1$) to the center of the channel ($y/H = 0.5$).

Figure 11 shows the velocity profiles obtained for $\alpha = 0.5, \beta = 0.5$ and $\alpha = 0.5, \beta = 1.5$. It can be seen that the overshoots are stronger near the inlet (compared to the exponential PTT model) and that a lower maximum velocity is obtained when the fluid is fully developed. The influence of the parameter β is residual.

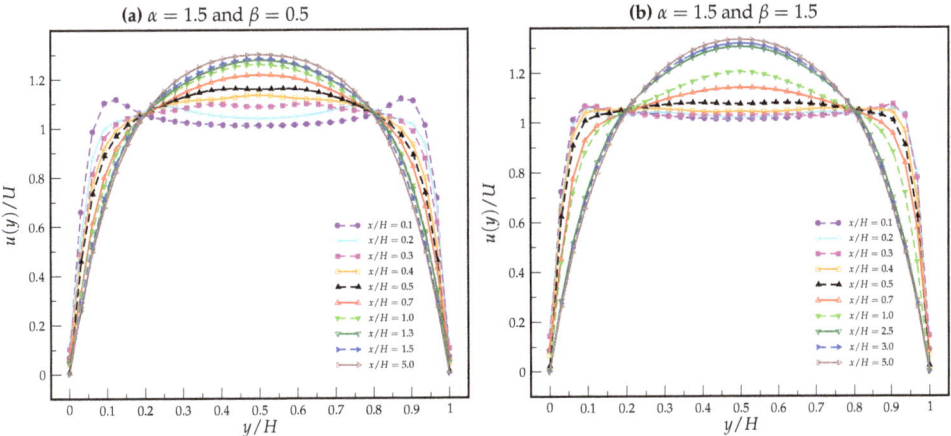

Figure 12. Velocity profiles obtained at 10 different sections of the channel for $Wi = 1.0$. (a) $\alpha = 1.5, \beta = 0.5$; (b) $\alpha = 1.5, \beta = 1.5$.

Figure 12 shows the velocity profiles obtained for $\alpha = 1.5, \beta = 0.5$ and $\alpha = 1.5, \beta = 1.5$. In Figure 12a it can be seen that the overshoots near the inlet resemble the case of the exponential PTT model, and that the maximum velocity increases again. The main difference is that the plug profile is now less pronounced. When β increases (Figure 12b), one can observe a dramatic change in the evolution of the velocity profiles. The velocity overshoots are almost suppressed and the maximum velocity increases. This means that the rate of destruction of junctions improves the diffusion of information.

6.3. The Influence of the Elasticity Number

In this subsection we study the influence of the elasticity number, $El = \frac{Wi}{Re}$, on the development length of the gPTT model. We consider three different values for El (0.1, 1, 10), $\alpha = 0.5$, and two different values for β, 0.5 and 1.5. The mesh used for the simulation is M_2.

The results are shown in Tables 3 and 4 and Figure 13 for $\mathcal{L}_{98\%}$ and $\mathcal{L}_{99\%}$.

As expected, the results for low values of Re are consistent with those obtained earlier in this work (see previous sections for more details). The results obtained for the different definitions of the development length are qualitatively similar. However, higher values of development length are obtained for $\mathcal{L}_{99\%}$, as expected.

For high values of β (1.5), the influence of the elasticity number seems to be neglected by the fluid, since we obtain the same development length for $El = 1$ and $El = 10$. For $El = 0.1$ the results are quite different, which is due to the low values of Wi compared to the $El = 1$ and $El = 10$ cases.

For $\beta = 0.5$, the rate of destruction of junctions increases and the development length decreases by about half. Again, the $El = 1$ and $El = 10$ cases show similar development lengths, which is due to the similar values of Wi and the almost creeping flow conditions.

The results show that for the tested ranges of El, Wi, and Re, no critical value is found for El. This is due to the fact that Mach's Elastic number is less than 1.

Table 3. Influence of the Elasticity number, $El = \frac{Wi}{Re}$, on the development length of the gPTT model. $El = 0.1$, $El = 1.0$, and $El = 10.0$ for $\mathcal{L}_{98\%}$, $\alpha = 1.5$.

		$El = 0.1$				$El = 1.0$				$El = 10.0$	
Re	Wi	$\beta = 1.5$	$\beta = 0.5$	Re	Wi	$\beta = 1.5$	$\beta = 0.5$	Re	Wi	$\beta = 1.5$	$\beta = 0.5$
0.01	0.001	0.5509	0.5509	0.01	0.01	0.5483	0.5486	0.01	0.1	0.5647	0.5671
0.1	0.01	0.5489	0.5493	0.1	0.1	0.5646	0.5671	0.05	0.5	1.2646	0.6751
0.5	0.05	0.5511	0.5559	0.5	0.5	1.2606	0.8005	0.075	0.75	1.9597	1.0564
1.0	0.1	0.5626	0.5673	0.75	0.75	1.9621	1.0310	0.1	1.0	2.6779	1.3049
2.0	0.2	0.6132	0.5801	1.0	1.0	2.6929	1.2440	0.125	1.25	3.3340	1.5250
3.0	0.3	0.7377	0.6052					0.15	1.5	3.8610	1.7201
								0.175	1.75	4.2769	1.9524

Table 4. Influence of the Elasticity number, $El = \frac{Wi}{Re}$, on the development length of the gPTT model. $El = 0.1$, $El = 1.0$, and $El = 10.0$ for $\mathcal{L}_{99\%}$, $\alpha = 1.5$.

		$El = 0.1$				$El = 1.0$				$El = 10.0$	
Re	Wi	$\beta = 1.5$	$\beta = 0.5$	Re	Wi	$\beta = 1.5$	$\beta = 0.5$	Re	Wi	$\beta = 1.5$	$\beta = 0.5$
0.01	0.001	0.6309	0.6310	0.01	0.01	0.6278	0.6285	0.01	0.1	0.6560	0.6628
0.1	0.01	0.6286	0.6293	0.1	0.1	0.6557	0.6626	0.05	0.5	1.5350	0.7590
0.5	0.05	0.6341	0.6433	0.5	0.5	1.5288	0.9888	0.075	0.75	2.4140	1.4070
1.0	0.1	0.6524	0.6611	0.75	0.75	2.4121	1.3760	0.1	1.0	3.3136	1.8657
2.0	0.2	0.7105	0.6756	1.0	1.0	3.3295	1.8284	0.125	1.25	4.0917	2.2971
3.0	0.3	0.8870	0.6810					0.15	1.5	4.7350	2.6578
								0.175	1.75	5.2590	2.9676

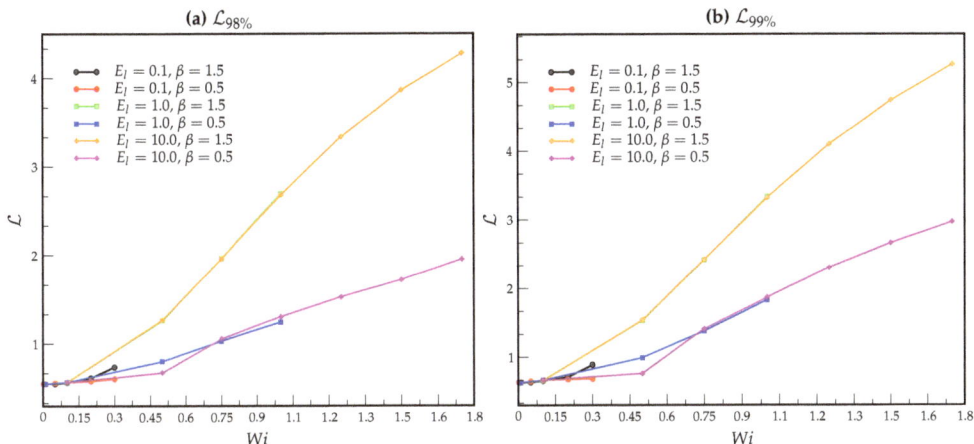

Figure 13. Influence of the elasticity number, $El = \frac{Wi}{Re}$, on the development length of the gPTT model. We consider three different values of El (0.1, 1, 10), $\alpha = 0.5$ and two different values of β, 0.5 and 1.5.

7. Conclusions

In this work, we present a numerical study on the development length of a pressure-driven viscoelastic fluid flow (between parallel plates) modelled by the generalised Phan–Thien and Tanner (gPTT) constitutive equation. The governing equations are solved using the finite-difference method, and, a thorough analysis on the effect of the model parameters α and β is presented. We consider two different definition of the development length: The length from the channel inlet required for the velocity to reach 99% (and 98%) of its fully-developed value.

The numerical results showed that the in the creeping flow limit (i.e., $Re = 0$), the development length for the velocity exhibits a non-monotonic behaviour. The development length increases with Wi, with viscoelastic effects delaying the diffusion and convection of information from the walls to the center of the channel. For low values of Wi, the highest value of the development length is obtained for $\alpha = \beta = 0.5$; for high values of Wi, the highest value of the development length is obtained for $\alpha = \beta = 1.5$. At high α and β values, the rate of destruction of the junctions is lower than at low α and β values. This means that when Wi values are low and the rate of junction destruction is high, information travels slowly from the wall to the center of the channel (compared to when the rate of junction destruction is low). The opposite was expected. Note that in this case, the development lengths are very similar for all tested values of α and β, and therefore the influence of these parameters on the development length is small. These results can be justified by the low value of Wi.

As Wi increases, the highest value of development length is reached with a low rate of destruction of junctions. This result can be justified by the fact that as the rate of destruction of the junctions decreases, the information is transmitted more slowly due to the small number of new contacts between the strands representing the molecules.

We also studied the influence of the elasticity number, El, on the development length of the gPTT model. As expected, the results for low values of Re are consistent with those obtained for creeping flow. For high values of β (1.5), the influence of the elasticity number seems to be neglected by the fluid, since we obtain the same development length for $El = 1$ and $El = 10$. For $El = 0.1$, the results are quite different, which is due to the low values of Wi compared to the $El = 1$ and $El = 10$ cases. For $\beta = 0.5$, the rate of destruction of junctions increases and the development length decreases by about half. Again, the $El = 1$ and $El = 10$ cases show similar development lengths, which is due to the similar values of Wi and the almost creeping flow conditions.

Author Contributions: Conceptualization, J.B., L.L.F., A.M.A., R.T.L. and A.C.; methodology, J.B., L.L.F., A.M.A., R.T.L. and A.C.; software, J.B., R.T.L. and A.C.; validation, J.B., L.L.F., A.M.A., R.T.L. and A.C.; formal analysis, J.B., L.L.F., A.M.A., R.T.L. and A.C.; investigation, J.B., L.L.F., A.M.A., R.T.L. and A.C.; writing—original draft preparation, J.B., L.L.F., A.M.A., R.T.L. and A.C.; writing—review and editing, J.B., L.L.F., A.M.A., R.T.L. and A.C.; funding acquisition, J.B., L.L.F., A.M.A., R.T.L. and A.C. All authors have read and agreed to the published version of the manuscript.

Funding: This research was funded by FCT—Fundação para a Ciência e a Tecnologia, through projects UIDB/00532/2020 and UIDP/00532/2020 of CEFT (Centro de Estudos de Fenómenos de Transporte), and through PTDC/EMS-ENE/3362/2014 and POCI-01-0145-FEDER-016665—funded by FEDER funds through COMPETE2020—Programa Operacional Competitividade e Internacionalização (POCI) and by national funds through FCT. It was also funded by FCT through the CMAT projects UIDB/00013/2020 and UIDP/00013/2020 (CMAT—Centro de Matemática e Aplicações); and was funded by Fapesp-Fundação de Amparo à Pesquisa do Estado de São Paulo, Fapesp-2017/21105-6 and Petrobrás (Project 0050.0075367.12.9). Research was carried out using the computational resources of the Center for Mathematical Sciences Applied to Industry (CeMEAI), funded by FAPESP grant 2013/07375-0.

Institutional Review Board Statement: Not applicable.

Informed Consent Statement: Not applicable.

Data Availability Statement: Not applicable.

Acknowledgments: J. Bertoco acknowledges the support by Faculdade de Ciências Exatas e Tecnológicas—FACET—UNEMAT—MT. J. Bertoco, R.T. Leiva, and A. Castelo acknowledge the support by ICMC—Instituto de Ciências Matemáticas e de Computação, Departamento de Matemática e Estatística, USP, São Carlos, SP. A.M. Afonso acknowledges the support by CEFT (Centro de Estudos de Fenómenos de Transporte). L.L. Ferrás acknowledges the support by CMAT—Centre of Mathematics from the University of Minho.

Conflicts of Interest: The authors declare no conflict of interest.

Abbreviations

The following abbreviations are used in this manuscript:

PTT	Phan–Thien and Tanner
gPTT	generalised Phan–Thien and Tanner
Re	Reynolds
Wi	Weissenberg

References

1. Liang, J.Z. Determination of the Entry Region Length of Viscoelastic Fluid Flow in Channel. *Chem. Eng. Sci.* **1998**, *53*, 3185–3187. [CrossRef]
2. Li, Z.; Haward, S.J. Viscoelastic flow development in planar microchannels. *Microfluid. Nanofluid.* **2015**, *5*, 1123–1137. [CrossRef]
3. Lobo, O.J.; Chatterjee, D. Development of flow in a square mini-channel: Effect of flow oscillation. *Phys. Fluids* **2018**, *30*, 042003. [CrossRef]
4. Durst, F.; Ray, S.; Unsal, B.; Bayoumi, O.A. The Development Lengths of Laminar Pipe and Channel Flows. *ASME J. Fluids Eng.* **2005**, *127*, 1154–1160. [CrossRef]
5. Boussinesq, J. Sur la maniere don't les vitesses, dans un tube cylindrique de section circulaire, evase a son entrée, se distribuent depuis entrée jusqu'aux endroits ou se trouve etabli un regime uniforme. *Compt. Rend.* **1891**, *113*, 49–51.
6. Schiller, L. Die Entwicklung der laminar en Geschwindigkeitsverteilung und ihre Bedeutung Ähnlichkeitsmessungen. *Z. Angew. Math. Mech.* **1922**, *2*, 96–106. [CrossRef]
7. Vrentas, J.S.; Duda, J.L.; e Bargeron, K.G. Effect of Axial Diffusion of Vorticity on Flow Development in Circular Conduits. *AIChE J.* **1966**, *12*, 837–844. [CrossRef]
8. Atkinson, B.; Brocklebank, M.P.; Card, C.C.; e Smith, J.M. Low Reynolds Number Developing Flows. *AIChE J.* **1969**, *154*, 548–553. [CrossRef]
9. Ferrás, L.L.; Afonso, A.M.; Nóbrega, J.M.; Alves, M.A.; Pinho, F.T. Development length in planar channel flows of Newtonian fluids under the influence of wall slip. *J. Fluids Eng.* **2012**, *134*, 104503–104508. [CrossRef]
10. Kountouriotis, Z.; Philippou, M.; Georgiou, G.C. Development lengths in Newtonian Poiseuille flows with wall slip. *Appl. Math. Comput.* **2016**, *291*, 98–114. [CrossRef]

11. Collins, M.; Schowalter, W.R. Behaviour of Non-Newtonian Fluids in the Inlet Region of a Channel. *AIChE J.* **1963**, *9*, 98–102. [CrossRef]
12. Mashelkar, R.A. Hydrodynamic Entrance-Region Flow of Pseudo-plastic Fluids. *Proc. Inst. Mech. Eng.* **1975**, *9*, 683–689.
13. Soto, R.J.; Shah, V.L. Entrance Flow of a Yield-Power Law Fluid. *Appl. Sci. Res.* **1976**, *32*, 73–85. [CrossRef]
14. Mehrota, A.K.; Patience, G.S. Unified Entry Length for Newtonian and Power Law Fluids in Laminar Pipe Flow. *Can. J. Chem. Eng.* **1990**, *68*, 529–533. [CrossRef]
15. Ookawara, S.; Ogawa, K.; Dombrowski, N.; Amooie-Foumeny, E.; Riza, A. Unified Entry Length Correlation for Newtonian, Power Law and Bingham Fluids in Laminar Pipe Flow at Low Reynolds Number. *J. Chem. Eng. Jpn.* **2000**, *33*, 675–678. [CrossRef]
16. Gupta, R.C. On Developing Laminar Non-Newtonian Flow in Pipes and Channels. *Nonlinear Anal. Real World Appl.* **2001**, *2*, 171–193. [CrossRef]
17. Chebbi, R. Laminar Flow of Power-Law Fluids in the Entrance Region of a Pipe. *Chem. Eng. Sci.* **2002**, *57*, 4435–4463. [CrossRef]
18. Poole, R.; Ridley, S.B. Development length requirements for fully-developed laminar pipe flow of inelastic non-Newtonian liquids. *ASME J. Fluids Eng.* **2007**, *129*, 1281–1287. [CrossRef]
19. Fernandes, C.; Ferrás, L.L.; Araujo, M.S.; Nóbrega, J.M. Development length in planar channel flows of inelastic non-Newtonian fluids. *J. Non-Newton. Fluid Mech.* **2018**, *255*, 13–18. [CrossRef]
20. Na, Y.; Yoo, J.Y. A finite volume technique to simulate the flow of a viscoelastic fluid. *Comput. Mech.* **1991**, *8*, 43–55. [CrossRef]
21. Yapici, K.; Karasozen, Y.; Uludag, Y. Numerical Analysis of Viscoelastic Fluids in Steady Pressure-Driven Channel Flow. *ASME J. Fluids Eng.* **2012**, *134*, 051206. [CrossRef]
22. Guilherme, L. Numerical Study of the Development Length in Viscoelastic Fluid Flows. Master's Thesis, University of Porto, Porto, Portugal, 2016.
23. Fattal, R.; Kupferman, R. Constitutive laws of the matrix-logarithm of the conformation tensor. *J. Non-Newton. Fluid Mech.* **2004**, *123*, 281–285. [CrossRef]
24. Afonso, A.; Oliveira, P.J.; Pinho, F.T.D.; Alves, M.A. The log-conformation tensor approach in the finite-volume method framework. *J. Non-Newton. Fluid Mech.* **2009**, *157*, 55–65. [CrossRef]
25. Su, J.; Ouyang, J.; Wang, X.;Yang, B.; Zhou, W. Lattice Boltzmann method for the simulation of viscoelastic fluid flows over a large range of Weissenberg numbers. *J. Non-Newton. Fluid Mech.* **2013**, *194*, 42–59. [CrossRef]
26. Phan-Thien, N. A nonlinear network viscoelastic model. *J. Rheol.* **1978**, *22*, 259–283. [CrossRef]
27. Ferrás, L.L.; Morgado, M.L.; Rebelo, M.; McKinley, G.H.; Afonso, A.M. A generalised Phan–Thien—Tanner model. *J. Non-Newton. Fluid Mech.* **2019**, *269*, 88–99. [CrossRef]
28. Ferrás, L.L.; Ford, N.; Morgado, M.L.; Rebelo, M.; McKinley, G.H.; Nóbrega, J.M. Theoretical and numerical analysis of unsteady fractional viscoelastic flows in simple geometries. *Comput. Fluids* **2018**, *174*, 14–33. [CrossRef]
29. Ribau, A.M.; Ferrás, L.L.; Morgado, M.L.; Rebelo, M.; Afonso, A.M. Analytical and numerical studies for slip flows of a generalised Phan-Thien-Tanner fluid. *Z. Angew. Math. Mech.* **2020**, *100*, e201900183. [CrossRef]
30. Ribau, A.M.; Ferrás, L.L.; Morgado, M.L.; Rebelo, M.; Afonso, A.M. Semi-analytical solutions for the Poiseuille-Couette flow of a generalised Phan-Thien-Tanner fluid. *Fluids* **2019**, *4*, 129. [CrossRef]
31. Sousa, F.S.; Lages, C.F.; Ansoni, J.L.; Castelo, A.; Simao, A. A finite difference method with meshless interpolation for incompressible flows in non-graded tree-based grids. *J. Comput. Phys.* **2019**, *396*, 848–866. [CrossRef]
32. Castelo, A.; Afonso, A.M.; De Souza Bezerra, W. A Hierarchical Grid Solver for Simulation of Flows of Complex Fluids. *Polymers* **2021**, *13*, 3168. [CrossRef] [PubMed]
33. Chorin, A.J. Numerical solution of the Navier-Stokes equations. *Math. Comput.* **1968**, *22*, 745–762. [CrossRef]
34. Alves, M.; Oliveira, P.; Pinho, F.T. A convergent and universally bounded inperpolation schemes for the treatment of advection. *Int. J. Numer. Methods Fluids* **2003**, *41*, 47–75. [CrossRef]
35. Leonard, B.P. A stable and accurate convective modelling procedure based on quadratic upstream interpolation. *Comput. Methods Appl. Mech. Eng.* **1979**, *19*, 59–98.
36. Guermond, J.L.; Quartapelle, L. On stability and convergence of projection methods based on pressure Poisson equation. *Int. J. Numer. Methods Fluids* **1998**, *26*, 1039–1053. [CrossRef]
37. Gorenflo, R.; Loutchko, J.; Luchko, Y. Computation of the Mittag–Leffler function $E_{\alpha,\beta}(z)$ and its derivative. *Fract. Calc. Appl. Anal.* **2002**, *5*, 491–518. Erratum: *Fract. Calc. Appl. Anal.* **2003**, *6*, 111–112. .
38. Available online: https://github.com/emammendes/Fortran-Code---Mittag--Leffler-Function/blob/master/test_mlfv.f90 (accessed on 1 October 2021).
39. Igor Podlubny. Mittag–Leffler Function. MATLAB Central File Exchange. Retrieved 7 October 2021. Available online: https://www.mathworks.com/matlabcentral/fileexchange/8738-mittag-leffler-function (accessed on 1 October 2021).

Article

A FENE-P $k - \varepsilon$ Viscoelastic Turbulence Model Valid up to High Drag Reduction without Friction Velocity Dependence

Michael McDermott [1,*], Pedro Resende [2], Thibaut Charpentier [3], Mark Wilson [1], Alexandre Afonso [4], David Harbottle [5] and Gregory de Boer [1]

1. School of Mechanical Engineering, University of Leeds, Woodhouse, Leeds LS2 9JT, UK; M.Wilson@leeds.ac.uk (M.W.); G.N.deBoer@leeds.ac.uk (G.d.B.)
2. ProMetheus, Escola Superior de Tecnologia e Gestão, Instituto Politécnico de Viana do Castelo, 4900-347 Viana do Castelo, Portugal; pedroresende@estg.ipvc.pt
3. Baker Hughes, Kirkby Bank Road, Knowsley Industrial Park, Liverpool L33 7EU, UK; Thibaut.Charpentier@bakerhughes.com
4. Transport Phenomena Research Center, Faculty of Engineering, University of Porto, Rua Roberto Frais s/n, 4200-465 Porto, Portugal; aafonso@fe.up.pt
5. School of Chemical and Process Engineering, University of Leeds, Woodhouse, Leeds LS2 9JT, UK; D.Harbottle@leeds.ac.uk
* Correspondence: ed11m22m@leeds.ac.uk

Received: 8 October 2020; Accepted: 3 November 2020; Published: 17 November 2020

Abstract: A viscoelastic turbulence model in a fully-developed drag reducing channel flow is improved, with turbulent eddies modelled under a $k - \varepsilon$ representation, along with polymeric solutions described by the finitely extensible nonlinear elastic-Peterlin (FENE-P) constitutive model. The model performance is evaluated against a wide variety of direct numerical simulation data, described by different combinations of rheological parameters, which is able to predict all drag reduction (low, intermediate and high) regimes with good accuracy. Three main contributions are proposed: one with a simplified viscoelastic closure for the NLT_{ij} term (which accounts for the interactions between the fluctuating components of the conformation tensor and the velocity gradient tensor), by removing additional damping functions and reducing complexity compared with previous models; second through a reformulation for the closure of the viscoelastic destruction term, E_{τ_p}, which removes all friction velocity dependence; lastly by an improved modified damping function capable of predicting the reduction in the eddy viscosity and thus accurately capturing the turbulent kinetic energy throughout the channel. The main advantage is the capacity to predict all flow fields for low, intermediate and high friction Reynolds numbers, up to high drag reduction without friction velocity dependence.

Keywords: drag reduction; FENE-P fluid; viscoelastic RANS model; OpenFoam CFD

1. Introduction

Since the pioneering experiment by Toms [1], it is known that the additions of small (parts per million) amounts of long-chain flexible polymers to a turbulent flow can drastically reduce the transport energy by decreasing the turbulent drag. The effects are most evident in turbulent shear flow, in which dissolving the polymers in solution can reduce friction losses by as much as 80% compared to the solvent alone [2]. After the discovery of the drag reduction (DR) phenomena, several comprehensive studies were carried out to understand the physical mechanisms of the interactions between the turbulent structures and polymer chains. Early comprehensive studies in this area come from Lumley [3,4], Hoyt [5] and Virk [6]. Lumley suggests that the DR phenomenon is the result of an increase in effective viscosity in an area outside the viscous sub-layer and in the buffer layer, caused by polymer chains stretching in a turbulent flow.

More recent studies have been proposed for the theory of the mechanisms of drag-reducing polymer additives [7,8]. Several Direct Numerical Simulation (DNS) studies were conducted to understand further the energy exchanges between the polymer chains and turbulence structures [9–14]. It is now known to at least low to moderate levels of DR, that the mechanism is the suppression of the near-wall streamwise vortices by polymers that stretch in the extensional flow, and then relax as they are rolled into other vortices, generating forces that tend to weaken these vortices. Quantitatively, this can be expressed as a polymer body force [9], which is positive in the streamwise direction, with an opposite sign (anti-correlation) in the wall direction.

DNS is a great resource to explore the underlying mechanics of drag-reducing viscoelastic turbulent flows. However, for the majority of engineering motivations, DNS is not practical because of the high number of variables which requires a substantial expense of memory and CPU-time. This cost is more prevalent in high DR (HDR) schemes in which the near-wall velocity streaks become more elongated, requiring an increased demand on computational resources.

An alternative approach in capturing flow features at much less computational demands is the application of Reynolds-averaged Navier–Stokes (RANS) models, whose interest has increased in recent years. One of the original implementations of elastic effects within turbulence models was achieved by Pinho [15] and Cruz et al. [16]. Their work focused on low-Reynolds number $k - \varepsilon$ turbulence models, applying a Generalised Newtonian Fluid (GNF) constitutive equation involving dependency of the fluid strain hardening on the third invariant of the rate of deformation tensor. Following these studies, an anisotropic version was also developed which included an increased Reynolds stress anisotropy [17], along with a Reynolds stress turbulence model [18], both able to satisfactorily predict drag-reducing behavior. Nevertheless, the models are constrained because of the inelastic formulation of the GNF constitutive equations.

Further developments in viscoelastic RANS models became possible owing to the emergence of DNS data regarding turbulent viscoelastic fluids. The first elastic model was developed by Leighton et al. [19], which was based on the finitely extensible nonlinear elastic with Peterlin closure (FENE-P) dumbbell constitutive equation model. Their study involved the development of a polymer strain–stress coupling based on the tensor expansion, which incorporated the conformation tensor and Reynolds stress. From this work, more attention arose to the FENE-P model given the molecular roots of the equations. Later, based on a-priori analysis of DNS data, Pinho et al. [20] developed a low-Reynolds number $k - \varepsilon$ model for FENE-P fluids which could predict flow features up to the low drag reduction regime (LDR < 20%). Turbulent viscoelastic closures were proposed, including the non-linear term involving the conformation tensor and the strain rate fluctuations within the conformation tensor equation (denoted NLT_{ij} following the nomenclature of Housiadas et al. [21] and Li et al. [22]); along with the viscoelastic turbulent transport term of the turbulent kinetic energy. One of the key difficulties that arose in this initial study was the decrease in the magnitude of turbulent kinetic energy as viscoelasticity increased, opposite to that found in the DNS literature [23]. Subsequently, the model closures were improved by Resende et al. [24] and the capacity of the model predictions were extended to the intermediate drag reduction regime (20% < IDR < 40%). However, the model closures involved complex damping functions and model constants which gave spurious results for the high drag reduction regime (HDR > 40%). Resende et al. [25] applied the same viscoelastic closures to a low-Reynolds number $k - \omega$ model with only a mathematical transformation of the governing terms involving ω. The closures had identical limitations as the $k - \varepsilon$ model for predicting DR behaviour but demonstrated great versatility and robustness given its application to alternative two-equation models.

During this time, a $k - \varepsilon - \overline{v^2} - f$ model for FENE-P fluids in fully developed channel flow was proposed by Iaccarino et al. [26], following the initial studies of Dubief et al. [27]. They introduced the idea of a turbulent polymer viscosity which accounts for the effects of viscoelasticity and turbulence on the polymer stress within the momentum equation. The reduction in the Reynolds shear stress is assumed by a-priori DNS data analysis from the decreasing $\overline{v^2}$ shown within the DNS studies [9]. The model closure for the NLT_{ij} is much simpler than the one developed by Resende et al. [24], but contains only the trace and not the individual components. The model was later improved by Masoudian et al. [28] and can predict flow features up to maximum DR (MDR). The key advancement of the closures were an NLT_{ij} closure based on DNS analysis and comparisons to the local eddy viscosity peaks; the viscoelastic stress work in the turbulent kinetic energy equations; viscoelastic stress in the momentum equation; and a viscoelastic destruction term in the dissipation transport equation. The viscoelastic turbulent closures within the $\overline{v^2}$ equation (transverse viscoelastic stress work, ε_{yy}^V) should be strictly a function of NLT_{yy}, which is a key component in the formulation of an effective polymer viscosity. However, because only the trace of the NLT_{ij} term is present within the model, the closure had to be formulated using DNS analysis of alternative parameters.

Subsequently, after this study, a second-order Reynolds stress model for FENE-P fluids was proposed by Masoudian et al. [29], extending on the idea of a correlation between the Reynolds stresses and the NLT_{ij} components, similar to Leighton et al. [19]. The model can predict all DR regimes but is generally unattractive due to the higher number of Newtonian terms resulting from higher-order modeling. Masoudian et al. [30] then further improved the $k - \varepsilon - \overline{v^2} - f$ model capabilities via the NLT_{ij} term by introducing a simple extension to include heat transfer, along with removing wall dependence via the friction velocity. There are concerning features when one examines the Bousinesq-type NLT_{ij} term, which has a zero NLT_{yy} component, along with an opposite sign for NLT_{xy}, both terms being crucial for the polymer shear stress in the momentum balance (see Appendix 1 in Pinho et al. [20]). Further, the increase of k in the buffer layer is small, meaning the decoupling of the $\overline{v^2}$ component may not be enough to decrease the eddy viscosity. This is compensated by an opposite trend in the dissipation rate, ε, for increasing DR, which subsequently balances the momentum equations and causes the necessary increase in the velocity profiles.

An alternative approach in predicting DR flow features other than the use of higher-order models such as the $k - \varepsilon - \overline{v^2} - f$ and Reynolds stress models mentioned previously, is that of a modified damping function or polymer eddy viscosity, accounting for the effect the polymer has on reducing the Reynolds shear stress. Tsukahara and Kawaguchi [31] proposed a modified damping function for a low-Reynolds $k - \varepsilon$ model for fluids described by the Giesekus constitutive equation, following the same ideas as Pinho [15] and Cruz. The closure was developed based on the energy-dissipative range and the dynamic characterization of the viscoelastic fluid. The model successfully captures the increase in the magnitude of the turbulent kinetic energy, along with the shift through the buffer layer. Although the magnitudes of k are largely over-predicted in many cases, which is counterbalanced by a lack of closure for the viscoelastic destruction term. In some instances, the model predicts a DR of 1% with a DNS result of 23%. Resende et al. [32] proposed a modified damping function for a low-Reynolds number $k - \varepsilon$ model for FENE-P fluids which can capture the increase of turbulent kinetic energy as flow viscoelasticity is increased, improving on model predictions made previously by Pinho et al. [20] and Resende et al. [24]. The study also improved largely on the NLT_{ij} closure accuracy and simplicity formulated in the previous work. The model is able to predict flow features for a large range of rheological parameters but is limited to a friction Reynolds number of $Re_{\tau_0} = 395$, along with friction velocity still present in the model. For model applicability in flows with reattachment, the friction velocity dependence poses a problem as the values become null at these points and lead to spurious results or floating point errors within computational solvers.

In the present study, an improved $k - \varepsilon$ model for FENE-P fluids is proposed, validated for all drag reduction regimes (low, intermediate, high) and up to the largest friction Reynolds number ($Re_{\tau_0} = 1000$) available in the DNS data. The important contribution to the current model is improved and simplified NLT_{ij} term that removes complexity from the most recent model developed by Resende et al. [32]; along with a modified damping function which accurately predicts the viscoelastic contributions near and away from the wall, effectively reducing the eddy viscosity and thickening the buffer-layer as DR increases. Further, a reformulation of the viscoelastic destruction term, E_{τ_p}, which removes all friction velocity present in the previous $k - \varepsilon$ models. The model is assessed against DNS data covering a wide range of flow conditions in terms of the friction Weissenberg number, Wi_{τ_0}, maximum polymer extension, L^2, viscosity ratio, β, and friction Reynolds number, Re_{τ_0}; along with comparisons against other turbulent FENE-P models within the literature.

The paper is organized as follows: Section 2 introduces the instantaneous and time-averaged governing equations and identifies the viscoelastic terms that will require modeling; Section 3 explains in detail the development of the viscoelastic turbulent closures; Section 4 summarises the model; Section 5 gives the numerical procedure applied; Section 6 presents the results of the flow fields in fully developed channel flow, covering all range of DR and flow conditions; and finally in Section 7, the main conclusions are presented.

2. Governing Equations

The governing equations for incompressible turbulent flow of dilute polymer solutions are the continuity and momentum equations respectively:

$$\frac{\partial \hat{u}_k}{\partial x_k} = 0, \tag{1}$$

$$\rho \frac{D\hat{u}_i}{Dt} \equiv \rho \left(\frac{\partial \hat{u}_i}{\partial t} + \hat{u}_k \frac{\partial \hat{u}_i}{\partial x_k} \right) = -\frac{\partial \hat{p}}{\partial x_i} + \frac{\partial \hat{\tau}_{ik}}{\partial x_k}, \tag{2}$$

where the hat represents instantaneous quantities of velocity $\hat{u}_i$, pressure $\hat{p}$, stress tensor $\hat{\tau}_{ij}$, and fluid density ρ. The stress tensor is the sum of the Newtonian solvent which obeys Newton's law of viscosity, $\hat{\tau}^s_{ij} = 2\mu_s \hat{s}_{ij}$, with μ_s representing the solvent viscosity coefficient, and polymeric contributions, $\hat{\tau}^p_{ik}$,

$$\hat{\tau}_{ij} = \hat{\tau}^s_{ij} + \hat{\tau}^p_{ij}. \tag{3}$$

The kinematic viscosity is used alternatively throughout this study and is defined as $\nu = \mu/\rho$. The instantaneous rate of strain tensor, $\hat{s}_{ij}$, is defined as

$$\hat{s}_{ij} = \frac{1}{2} \left(\frac{\partial \hat{u}_i}{\partial x_j} + \frac{\partial \hat{u}_j}{\partial x_i} \right). \tag{4}$$

The instantaneous polymer contributions are based on the FENE-P rheological dumbbell model [33], with closure given by

$$\hat{\tau}^p_{ij} = \frac{\mu_p}{\lambda} \left(f(\hat{c}_{kk}) \hat{c}_{ij} - \delta_{ij} \right), \tag{5}$$

with

$$f(\hat{c}_{kk}) = \frac{L^2 - 3}{L^2 - \hat{c}_{kk}}, \tag{6}$$

known as the Peterlin function, and $\hat{c}_{kk}$ is the trace of the instantaneous conformation tensor. The other parameters that are associated with the FENE-P model are: λ, the relaxation time of the polymeric fluid; L^2, the maximum extensibility of the dumbbell model; and μ_p, the polymer viscosity coefficient.

The behaviour of the instantaneous conformation tensor follows a hyperbolic differential equation of the form,

$$\frac{\partial \hat{c}_{ij}}{\partial t} + \hat{u}_k \frac{\partial \hat{c}_{ij}}{\partial x_k} - \left(\hat{c}_{kj} \frac{\partial \hat{u}_i}{\partial x_k} + \hat{c}_{ik} \frac{\partial \hat{u}_j}{\partial x_k} \right) = \overset{\nabla}{\hat{c}}_{ij} = -\frac{\hat{\tau}^p_{ij}}{\mu_p}. \tag{7}$$

The Oldroyd's upper convective derivative of the instantaneous conformation tensor is here denoted with $\overset{\triangledown}{\hat{c}}_{ij}$. The local and advective derivatives are the first and second terms respectively. The bracketed term accounts for the effect of polymer stretching by the instantaneous flow.

The Reynolds averaging process [34] is applied to the governing equations via a Reynolds decomposition of the flow fields such that, $\hat{u}_i = U_i + u_i$; where the use of overbars or upper-case represents the averaged quantity; and primes or lower-case represent the instantaneous quantities. The continuity and momentum equations now take the form:

$$\frac{\partial U_k}{\partial x_k} = 0, \tag{8}$$

$$\rho \frac{\partial U_i}{\partial t} + \rho U_k \frac{\partial U_i}{\partial x_k} = -\frac{\partial \overline{p}}{\partial x_i} + \mu_s \frac{\partial^2 U_i}{\partial x_k \partial x_k} - \frac{\partial}{\partial x_k}(\rho \overline{u_i u_k}) + \frac{\partial \overline{\tau}_{ik}^p}{\partial x_k}, \tag{9}$$

referred to as the Reynolds-averaged Navier–Stokes (RANS). The Reynolds stress tensor is $\overline{u_i u_k}$ and requires a closure model. The Reynolds-averaged polymer stress is $\overline{\tau}_{ik}^p$ and written fully as

$$\overline{\tau}_{ij}^p = \frac{\mu_p}{\lambda}\left[f(C_{kk})C_{ij} - \delta_{ij}\right] + \frac{\mu_p}{\lambda}\overline{f(C_{kk}+c_{kk})c_{ij}}, \tag{10}$$

where the additional term on the right requires a closure. The Peterlin function becomes

$$f(C_{kk}) = \frac{L^2 - 3}{L^2 - C_{kk}}. \tag{11}$$

After Reynolds averaging, the instantaneous conformation tensor equation becomes

$$\frac{DC_{ij}}{Dt} - M_{ij} + CT_{ij} - NLT_{ij} = \frac{\overline{\tau}_{ij}^p}{\mu_p}, \tag{12}$$

$$M_{ij} = C_{jk}\frac{\partial U_i}{\partial x_k} + C_{ik}\frac{\partial U_j}{\partial x_k}, \tag{13}$$

$$CT_{ij} = \overline{u_k \frac{\partial c_{ij}}{\partial x_k}}, \tag{14}$$

$$NLT_{ij} = \overline{c_{jk}\frac{\partial u_i}{\partial x_k}} + \overline{c_{ik}\frac{\partial u_j}{\partial x_k}}, \tag{15}$$

which is referred to as the Reynolds-averaged conformation evolution (RACE). M_{ij} is the mean flow distortion term; it is non-zero, but requires no closure. The remaining two terms are named following the nomenclature of Li et al. [22] and Housiadas et al. [21]. They are labelled with CT_{ij}; representing the contribution to the transport of the conformation tensor due to the fluctuating advective terms; and NLT_{ij}, which accounts for the interactions between the fluctuating components of the conformation tensor and the velocity gradient tensor.

Following the analysis of Pinho et al. [20], the nonlinear fluctuating correlation of the average polymeric stress, $\overline{f(C_{kk}+c_{kk})c_{ij}}$ in Equation (10) was shown to be negligible for LDR and HDR when compared with the linear part. This was later neglected in the models of Resende et al. [24] and Masoudian et al. [28] and is also neglected here. The CT_{ij} term can also be omitted for all DR regimes following a budget analysis of the RACE carried out by Housiadas et al. [21] and Li et al. [22]. The NLT_{ij} term cannot be neglected since it is a significant contributor to the RACE and therefore requires a suitable closure.

2.1. Model for the Reynolds Stress Tensor

The Reynolds stress tensor is computed by adopting the Boussinesq turbulent stress strain relationship,

$$-\rho \overline{u_i u_j} = 2\rho \nu_T S_{ij} - \frac{2}{3}\rho k \delta_{ij}, \qquad (16)$$

where k is the turbulent kinetic energy, S_{ij} is the mean rate of strain tensor and $\mu_T = \rho \nu_T$ is the eddy viscosity. ν_T is modelled by the typical isotropic $k - \varepsilon$ turbulence model for low Reynolds numbers, which includes a damping function f_μ to account for near-wall effects:

$$\nu_T = C_\mu f_\mu \frac{k^2}{\tilde{\varepsilon}^N}, \qquad (17)$$

where $\tilde{\varepsilon}^N = \nu_s \overline{\frac{\partial u_i}{\partial x_j}\frac{\partial u_i}{\partial x_j}}$ is the viscous dissipation of k by the Newtonian solvent,

$$f_\mu = \left[1 - \exp\left(-\frac{y^+}{a_\mu}\right)\right]^2, \qquad (18)$$

and $a_\mu = 26.5$. The dimensionless wall scaling is $y^+ = u_{\tau_0} y / \nu_0$, where u_{τ_0} is the friction velocity, y is the distance to the nearest wall, and ν_0 is the sum of solvent and polymer viscosity coefficients ($\nu_0 = \nu_s + \nu_p$). The damping function requires additional modelling to capture the anisotropy of the drag reducing flow as a result of viscoelastic flow effects, to be discussed further in this study (Section 3.2).

2.2. Transport Equation for the Turbulent Kinetic Energy

The governing transport equation for the turbulent kinetic energy of turbulent flow with FENE-P fluids is given by,

$$\rho \frac{\partial k}{\partial t} + \rho U_i \frac{\partial k}{\partial x_i} = \rho \frac{\partial}{\partial x_i}\left[\left(\nu_s + \frac{f_t \nu_T}{\sigma_k}\right)\frac{\partial k}{\partial x_i}\right] + P_k - \rho(\tilde{\varepsilon}^N + D) \\ + Q^V - \rho \varepsilon^V, \qquad (19)$$

with

$$D = 2\nu_s \left(\frac{d\sqrt{k}}{dx_i}\right)^2. \qquad (20)$$

$P_k = -\rho \overline{u_i u_j}\frac{\partial u_i}{\partial x_j}$ is the rate of production of k.

The Newtonian closures of Equation (19) are those present in the Nagano et al. [35,36] models. To increase numerical stability, a modified Newtonian rate of dissipation of k is applied instead of the true dissipation, which are related by $\varepsilon^N = \tilde{\varepsilon}^N + D$. For better model performance and to correct for the turbulent diffusion near walls, a turbulent variable Prandtl number is added of the form, $f_t / \sigma_k = 1 + 3.5 \exp(-(Re_T/150)^2)$ with $Re_T = k^2/(\nu_s \tilde{\varepsilon})$ and model constant $\sigma_k = 1.1$. More details of the form of Equation (19) can be found in Pinho et al. [20] and Resende et al. [24].

The last two terms on the right side of the Equation (19) are:

$$Q^V = \frac{\overline{\partial \tau'_{ik,p} u_i}}{\partial x_k} \quad \text{and} \quad \varepsilon^V = \frac{1}{\rho}\overline{\tau'_{ik,p}\frac{\partial u_i}{\partial x_k}}, \qquad (21)$$

which are the viscoelastic turbulent transport and the viscoelastic stress work, respectively. They represent the fluctuating viscoelastic turbulent part of the k transport equation and require suitable closure models.

A budget analysis for each term in the k transport equation was performed by Pinho et al. [20] for different regimes of DR. They demonstrated that the magnitude of Q^V has more impact on the overall budget in the IDR, and also developed a closure. In the HDR, the amplitude of Q^V is the same as ε^V but has a different location in the buffer layer, in which the effects of Q^V are overcome by turbulent diffusion, thus, revealing negligible effects to overall flow predictions. Masoudian et al. [28] had chosen to neglect the Q^V contributions in the $k - \varepsilon - v^2 - f$ model and is also not included here as well.

2.3. Transport Equation for the Rate of Dissipation of Turbulent Kinetic Energy

The corresponding governing transport equation for the modified Newtonian rate of dissipation of k is given by,

$$\rho \frac{\partial \tilde{\varepsilon}^N}{\partial t} + \rho U_i \frac{\partial \tilde{\varepsilon}^N}{\partial x_i} = \rho \frac{\partial}{\partial x_i} \left[\left(\nu_s + \frac{f_t \nu_T}{\sigma_\varepsilon} \right) \frac{\partial \tilde{\varepsilon}^N}{\partial x_i} \right] + f_1 C_{\varepsilon_1} \frac{\tilde{\varepsilon}^N}{k} P_k \\ - f_2 C_{\varepsilon_2} \rho \frac{(\tilde{\varepsilon}^N)^2}{k} + \rho E + E_{\tau_p}, \quad (22)$$

with

$$E = \nu_s \nu_T (1 - f_\mu) \left(\frac{\partial^2 U_j}{\partial x_i \partial x_k} \right)^2. \quad (23)$$

As mentioned in the previous sub-section, all terms are modelled in the Newtonian context (excluding E_{τ_p}). The damping functions of Equation (22) are $f_1 = 1$ and $f_2 = 1 - 0.3 \exp(-(Re_T)^2)$; with model coefficients $\sigma_\varepsilon = 1.3$, $C_{\varepsilon_1} = 1.45$ and $C_{\varepsilon_2} = 1.90$.

The last term in Equation (22) is the viscoelastic contribution to the overall $\tilde{\varepsilon}^N$ balance. This term acts as a Newtonian destruction to the dissipation and is given by,

$$E_{\tau_p} = 2\mu_s \frac{\mu_p}{\lambda(L^2 - 3)} \overline{\frac{\partial u_i}{\partial x_m} \frac{\partial}{\partial x_k} \left\{ \frac{\partial}{\partial x_m} [f(C_{nn}) f(\hat{C}_{pp}) c_{qq} C_{ik}] \right\}}. \quad (24)$$

It has non-negligible effects on flow predictions for all DR regimes and thus requires a suitable model.

3. Development of Viscoelastic Closures

In this section, the turbulent viscoelastic cross-correlations that were isolated in the previous section are presented with model closures. The closures are developed on the basis of the DNS data case (19) (Table 1), and then subsequently compared with other DNS data sets for accurate model predictions. The DNS data in Table 1 pertain to all DR regimes with a large variation in rheological parameters and flow viscosity for fully-developed channel flow established by: Li et al. [23]; Thais et al. [37,38]; Masoudian et al. [28,30,39] and Iaccarino et al. [26].

The non-dimensional numbers that define the different DNS data sets are defined as follows: the friction Reynolds number $Re_{\tau_0} = h u_\tau / \nu_0$ is based on the friction velocity (u_τ), the channel half-height (h), the zero shear-rate kinematic viscosity of the solution, which is the sum of the kinematic viscosity of the solvent and polymer ($\nu_0 = \nu_s + \nu_p$); The Weissenberg number $Wi_{\tau_0} = \lambda u_\tau^2 / \nu_0$; and the ratio between the solvent viscosity and the solution viscosity at zero shear rate is $\beta = \nu_s / \nu_0$.

In the following sub-sections, closures are developed for: the NLT_{ij} term of Equation (12) with focus on the dominant NLT_{xx} component; a modification of the damping function f_μ (Equation (18)), named f_V, which accounts for the reduction of the Reynolds shear stress due to viscoelastic effects; the viscoelastic stress work, ε^V of Equation (19); and the viscoelastic destruction, E_{τ_p}, of Equation (22).

Table 1. Independent Direct Numerical Simulation data for turbulent channel flow of finitely extensible nonlinear elastic-Peterlin (FENE-P) fluids at $\beta = 0.9$, with drag reduction (DR) model predictions.

Case	Reference	Rheological Parameters			DNS	Drag Reduction (%)	
		Re_{τ_0}	Wi_{τ_0}	L^2		Current Model	Model [32]
(1)	Li et al. [23]	125	25	900	19	20	-
(2)	Li et al. [23]	125	25	3600	22	23	-
(3)	Li et al. [23]	125	25	14,400	24	25	-
(4)	Li et al. [23]	125	50	900	31	30	35
(5)	Li et al. [23]	125	100	900	37	36	39
(6)	Li et al. [23]	125	100	1800	45	43	-
(7)	Li et al. [23]	125	100	3600	56	51	51
(8)	Masoudian et al. [28]	180	25	900	19	19	-
(9)	Li et al. [23]	180	50	900	31	30	34
(10)	Masoudian et al. [28]	180	100	900	38	38	39
(11)	Masoudian et al. [28]	180	100	3600	54	53	51
(12)	Thais et al. [37]	180	116	10,000	64	60	-
(13)	Iaccarino et al. [26]	300	36	3600	33	32	34
(14)	Iaccarino et al. [26]	300	36	10,000	35	35	32
(15)	Iaccarino et al. [26]	300	120	10,000	59	59	58
(16)	Masoudian et al. [30]	395	25	900	19	22	19
(17)	Masoudian et al. [30]	395	50	900	30	30	-
(18)	Masoudian et al. [30]	395	50	3600	38	38	-
(19)	Masoudian et al. [30]	395	100	900	37	37	38
(20)	Masoudian et al. [30]	395	100	3600	48	47	52
(21)	Masoudian et al. [39]	395	100	10,000	55	55	-
(22)	Masoudian et al. [30]	395	100	14,400	61	60	62
(23)	Thais et al. [37]	395	116	10,000	62	60	-
(24)	Li et al. [23]	395	200	14,400	75	69	67
(25)	Masoudian et al. [30]	590	50	3600	39	40	64
(26)	Thais et al. [37]	590	116	10,000	61	59	74
(27)	Thais et al. [38]	1000	50	900	30	33	60

3.1. Closure for NLT_{ij}

The NLT_{ij} exact transport equation is greatly simplified based on the DNS analysis of Pinho et al. [20]: Following the transport equation of $\overline{f(\hat{c}_{mm})c_{kj}\frac{\partial u_i}{\partial x_k}} + \overline{f(\hat{c}_{mm})c_{ik}\frac{\partial u_j}{\partial x_k}}$, it is assumed that

$$\overline{f(\hat{c}_{mm})c_{kj}\frac{\partial u_i}{\partial x_k}} + \overline{f(\hat{c}_{mm})c_{ik}\frac{\partial u_j}{\partial x_k}} \approx f(C_{mm})\left(\overline{c_{kj}\frac{\partial u_i}{\partial x_k}} + \overline{c_{ik}\frac{\partial u_j}{\partial x_k}}\right) = f(C_{mm})NLT_{ij}. \tag{25}$$

The full details of this approximation and the exact transport equation of NLT_{ij} can be found in Pinho et al. [20] and Resende et al. [24].

The complete closure of NLT_{ij} is presented below and was developed to improve model predictions based on better physical modeling compared with the most recent model developed by Resende et al. [32].

$$NLT_{ij} = \overline{c_{kj}\frac{\partial u_i}{\partial x_k}} + \overline{c_{ik}\frac{\partial u_j}{\partial x_k}}$$

$$\approx \underbrace{f_N C_{N1}\frac{\lambda\sqrt{\tilde{L}}\varepsilon^N}{\nu_0 f(C_{mm})}\delta_{ij}}_{I} - \underbrace{f_N^{1/4} C_{N2} M_{ij}}_{II} + \underbrace{C_{N3}\frac{k}{\nu_0}\sqrt{\tilde{L}M_{nn}\frac{\overline{\frac{\partial u_i}{\partial x_k}\frac{\partial u_j}{\partial x_k}}}{\hat{\gamma}^2}}}_{III}, \tag{26}$$

where $f_N = \nu_T/\nu_0$ is the local eddy viscosity, $\hat{\gamma} = \sqrt{2S_{pq}S_{pq}}$ is the shear rate invariant, $\tilde{L} = \sqrt{L^2/900}$ is the normalised maximum extension with the lowest DR, with model constants $C_{N1} = 0.11$, $C_{N2} = 0.3$ and $C_{N3} = 0.3$.

The closure of Equation (26) is modelled in three parts: parts I and II are modeled in the same fashion as the model of Resende et al. [32], part III is greatly improved and is the main contribution to the NLT_{ij} closure.

Part I is approached by introducing the Taylor's longitudinal micro-scale, λ_f, to the relationship between the double correlation of fluctuating strain rates and the turbulent kinetic energy in homogeneous isotropic turbulence. More details can be found in Resende et al. [32], with adjustments $L^{0.42}$ to $\sqrt{L}$ and $f(C_{mm})^{0.8}$ to $f(C_{mm})$.

Part II is primarily responsible for capturing the shear component, NLT_{xy}. The correlation here is with the exact term, M_{ij} (see Equation (13)), and by the local eddy viscosity, $f_N^{1/4}$. The $L^{0.15}$ variation is removed from the model developed by Resende et al. [32]. The negative part of the NLT_{xx} component is also captured here via the M_{xx} term, which according to Dubief et al. [27], is the region where polymers inject energy into turbulence.

Part III is developed to predict the NLT_{xx} component which is the dominant term in the trace of NLT_{ij}, responsible for the stretch of the polymer chains due to turbulent fluctuations. Following the same assumption as Masoudian et al. [29], one can see that $NLT_{xx} \sim \overline{u'_x u'_x} \sim k$. In physical terms, the turbulent stretching terms represent the ability of the turbulent fluctuations to act on the polymer chains. This stretching is effective if the polymer shear and maximum extensibility are large enough. So, $\sqrt{\tilde{L} M_{nn}/\tilde{\gamma}}$ is included here with k. Note that for fully developed channel flow, this term reduces to $\sqrt{\tilde{L} C_{xy}}$ which increases proportionally to drag reduction. This new term includes the same physical assumption as Masoudian et al. [28,30], and is simplified from the very complex ad-hoc approach of Resende et al. [32], viz

$$NLT_{III}^{Resende} = f_N^{0.9} \exp\left(-\frac{-f_N}{1.05\sqrt{\beta}(10 + 0.3L + \tilde{L} - (\tilde{L}-1)^2)}\right)$$
$$\times \left(\frac{C_{mm}}{(\beta/0.9)^{0.7\beta}}\left(2 - \left[1 - \exp\left(-\frac{2U_b h/\nu_s}{3500}\right)\right]^4\right)\right)^{0.7} \frac{\frac{dU_i}{dx_k}\frac{dU_j}{dx_k}}{\sqrt{\frac{dU_p}{dx_q}\frac{dU_p}{dx_q}}}. \quad (27)$$

The performance of the NLT_{ij} closure can be analysed in Figure 1 by comparing the predictions with DNS data case (19) in Table 1, and with the model of Resende et al. [32]. Figure 1a–c plots each normal component of NLT_{ij}, with the predictions as accurate as the previous model [32]. The new NLT_{xx} component is capable of predicting the maximum value and peak location of the destruction effect away from the wall along with the negative part near the wall, but requires a much simpler closure. The closure performance becomes more noticeable at higher Reynolds numbers, in which the polymer extension is largely overestimated previously. The NLT_{yy} component is the leading order term in the C_{yy} component away from the wall, which is the dominant contributor to an effective polymeric viscosity. This strongly influences the turbulent dynamics according to Thais et al. [40] and Benzi et al. [41] with their DNS and toy model analysis respectively. This term is represented by the first term in Equation (26), along with NLT_{zz}. The NLT_{zz} component was shown by Pinho et al. [20] to have low impact, and thus $NLT_{zz} = NLT_{yy}$ is an appropriate approximation. The shear component, NLT_{xy}, can be viewed in Figure 1e, where the predictions omit similar results compared with the previous model [32], but do not require additional L^2 variation via $L^{0.15}$.

Overall, all main features of NLT_{ij} are well captured such as the peak locations and magnitudes, but with a much simpler closure for the dominant contributor of polymer stretch, NLT_{xx}. Further, the NLT_{xy} and NLT_{yy} terms responsible for the polymer shear stress contribution in the momentum balance are featured, which were previously represented ad-hoc with friction velocity dependence [26,28] or misrepresented [30].

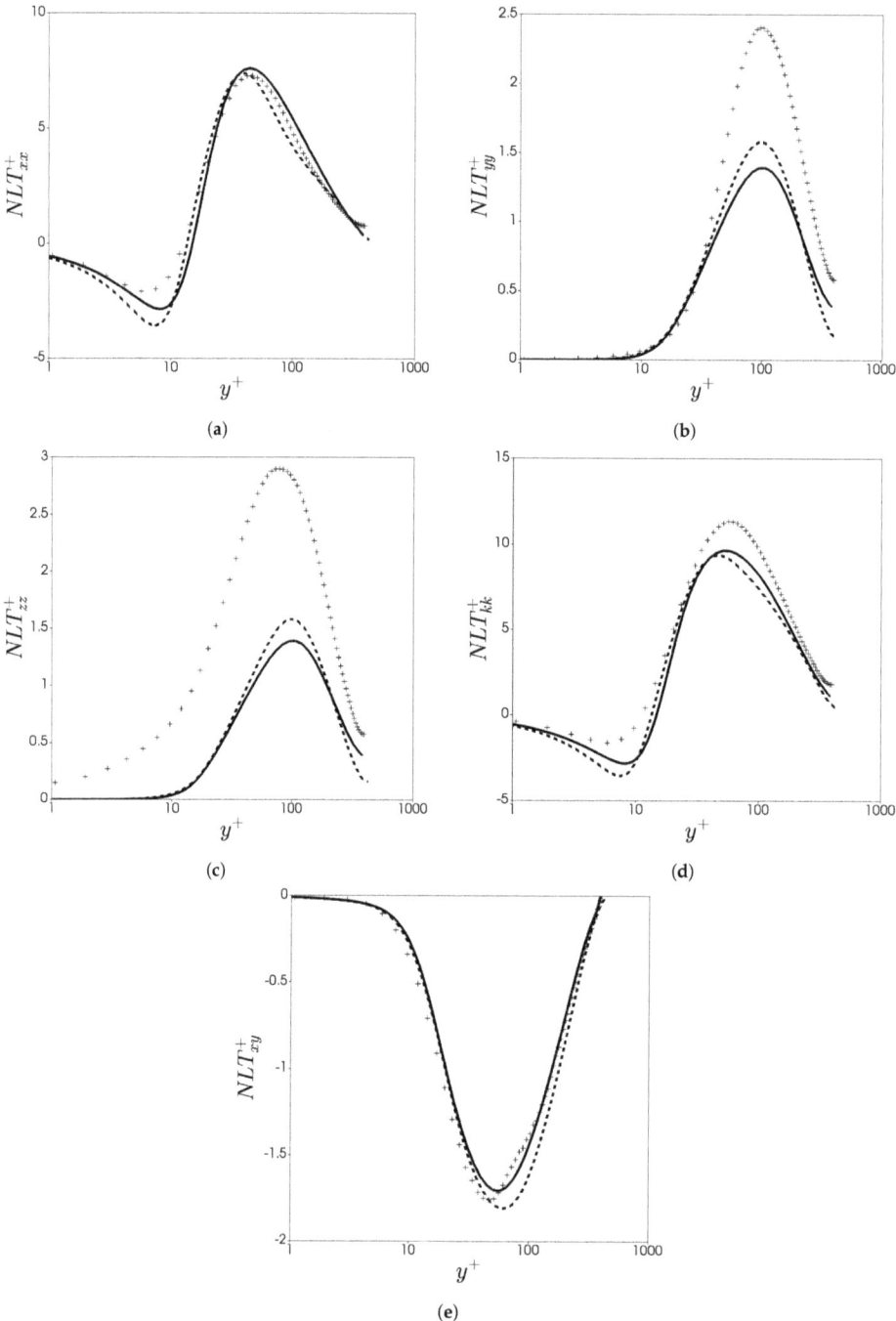

Figure 1. Comparison of the NLT_{ij} model between DNS data (+DR = 37%, case (19)) and predictions with the new model (continuum lines), and previous model (dash lines): (**a**) NLT_{xx}^+; (**b**) NLT_{yy}^+; (**c**) NLT_{zz}^+; (**d**) NLT_{kk}^+ and (**e**) NLT_{xy}^+.

3.2. Model for the Modified Damping Function, f_v

There have been many attempts to predict the eddy viscosity reduction as flow viscoelasticity increases for drag-reducing flows. In the case of low-Reynolds $k - \varepsilon$ models for FENE-P fluids, this was examined firstly by Pinho et al. [20] for the LDR regime; then later by Resende et al. [24] for the IDR regime. In both cases, there was a consistent reduction in the magnitude of k as DR increased, contrary to the DNS findings [23]. Similar attempts to model a modified damping function were made by Pinho [15]; Cruz et al. [16]; Resende et al. [17] and Tsukahara and Kawaguchi [31] to develop viscoelastic turbulence models using different constitutive equations.

Recently, Resende et al. [32] proposed a modified damping function which was able to predict the correct behavior of the eddy viscosity close to the wall, leading to the appropriate increase for the magnitude of k, and the shift away from the wall into the buffer layer as DR increased. This proposal was founded from the a-priori DNS data analysis by Resende et al. [42], demonstrating the necessary increase to the production of k close to the wall. The model derived by Resende et al. [32] is based on the DNS analysis of Li et al. [23], with an approximation of the form $DR \sim C_{kk}/L$, giving rise to the correct damping of near-wall eddies as DR increases. In the $k - \varepsilon - v^2 - f$ models proposed by Iaccarino et al. [26] and Masoudian et al. [28], the near-wall eddy viscosity damping effect is achieved by v^2, as $\nu_T = C_\mu v^2 k / \varepsilon$. However, the reduction in v^2 is not enough to increase k as given by the DNS data.

The approach by Resende et al. [32] works well in increasing k in the buffer layer, but fails to capture the viscoelastic effects away from the wall, due to the fact that $f_\mu^{\text{Previous}} \to 1$ as $y \to h$, which is contrary to the DNS data of Li et al. [22] and the analogous behavior of v^2 away from the wall. Therefore an additional model is required to capture the effect of nonequilibrium away from the wall, similarly to the Newtonian model of Park et al. [43]. Benzi et al. [41] demonstrated that the overall effect of polymer stretching is to introduce an effective viscosity proportional to C_{yy}, which is dominated by the NLT_{yy} component (modeled here with the first term in Equation (26)). An additional term is multiplied to the eddy viscosity to account for the global reduction of eddy structures for increasing DR. This approach is similar to the model of Resende et al. [24] and the study using DNS data of Resende et al. [42] which multiplies the damping function by a factor of $1 - g(\text{VE})$, where $g(\text{VE})$ is a function of the viscoelastic terms, VE.

The final model presented for the modified damping function, f_v, is

$$f_v = (1 - A)\left[1 - \exp\left(-\frac{y^*}{a_\mu(1 + B/a_\mu)}\right)\right]^2, \tag{28}$$

$$A = C_A \left(f_N \frac{\lambda^2 \tilde{L}^{3/2}}{f(C_{kk})^2} \frac{\varepsilon}{\nu_0}\right)^{0.3}, \tag{29}$$

$$B = C_B(C_{kk} - 3)^{1.25}/L, \tag{30}$$

with model constants $C_A = 0.071$ and $C_B = 0.44$. An additional contribution in the present model comes from an alternative representation of the dimensionless wall scaling $y^+ = u_\tau/\nu_w$, where ν_w is the wall viscosity. The presence of the wall friction velocity poses a problem for flows with re-circulation or reattachment were the friction velocity becomes null at these points, causing floating point errors within computational solvers. Possibilities other than y^+ that solve this issue are $Re_y \equiv ky/\nu_0$ or the turbulent Reynolds number, Re_t. Wallin and Johansson [44] formulated an alternative scaling, y^*, in terms of Re_y so that $y^* \approx y^+$ for $y^+ \leq 100$ in channel flows. The form proposed is

$$y^* = C_{y1} Re_y^{1/2} + C_{y2} Re_y^2, \tag{31}$$

where $C_{y1} = 2.4$ and $C_{y2} = 0.003$. The Re_y-term is motivated by the fact that the near-wall asymptotic behaviour for $Re_y^{1/2}$ is $\sim y^2$. The Re_y^2-term is artificially introduced to obtain a near linear relation in the buffer region.

The performance of the f_v closure can be analyzed in Figure 2 by comparing the predictions with DNS data cases (16, 19, 20) for LDR, IDR and HDR respectively in Table 1 and with the model of Resende et al. [32]. The predictions offer significant improvement away from the wall compared to the previous model. The effects can be viewed for the turbulent kinetic energy and the eddy viscosity in the results section, offering improved results for various levels of DR and Reynolds numbers. The f_v closure more accurately represents the anisotropic effect akin to the $v^2 - f$ models of Masoudian et al. [28,30], with the thickening of the buffer layer from the stretched polymers, along with a global reduction with the new closure.

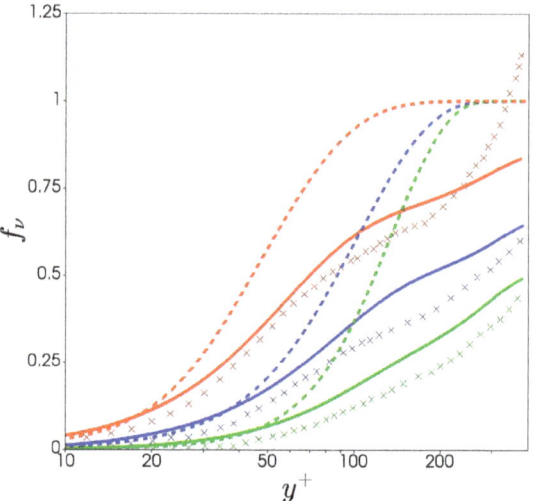

Figure 2. Comparison of the f_v model between DNS data ($\times$ crosses) and predictions with the new model (continuum lines), and previous model (dash lines): each colour represents a different drag reduction regime: red (low drag reduction (LDR), case 16); blue (intermediate drag reduction (IDR), case 19); green (high drag reduction (HDR), case 20).

3.3. Development of Closures for ε^V and E_{τ_p}

The closure model for ε^V is approached following the DNS budget analysis of the governing terms proposed by Pinho et al. [20]. In their work, they verified that the double correlation can be neglected with respect to the triple correlation at LDR. This was later confirmed for IDR and HDR by Resende et al. [24] and Masoudian et al. [28], respectively. Pinho et al. [20] extended this analysis and demonstrated that the triple correlation can be decoupled and modeled as a function of $NLT_{mm}/2$. Following this, Masoudian et al. [28] confirmed the model capabilities within 5% accuracy for all DR regimes via an extensive pdf study, and is the model used here given by

$$\varepsilon^V \approx \frac{\nu_p}{2\lambda} f(C_{mm}) NLT_{mm}. \tag{32}$$

The closure model derived for E_{τ_p} assumes that it depends on the same quantities as the classical Newtonian destruction term of the transport equation of ε, but involving a viscoelastic quantity, typically with the viscoelastic stress work used by Resende et al. [24,32] and Masoudian et al. [28,30]. However, as ε^V contains a negative part close to the wall via the NLT_{mm} contribution, it is not feasible to include ε^V in a suitable model for E_{τ_p}, based on the DNS analysis of ε being strictly decreasing near the wall for increasing DR.

The closure derived by Resende et al. [32] is complex with Wi_{τ_0} dependence to force the correct trend in ε. Here, a much simpler approach is obtained with dependence through k and some viscoelastic quantities which increases proportional with DR. The closure is given by

$$E_{\tau_p} \approx -C_{N4} \frac{\tilde{\varepsilon}^N}{k} \left[\nu_p \sqrt{C_\mu f_\mu} \tilde{L}^{3/4} \left(\frac{k}{\nu_0} \right)^2 \right], \tag{33}$$

with model constant $C_{N4} = 0.083$. The effect of Equation (33) on ε predictions can be viewed in the results section for LDR and HDR.

Overall, it is clear that all the developed viscoelastic closures presented in this study perform well compared with DNS data. Most importantly, this was achieved without the use of friction velocity dependence. The simplicity of the governing closures allows easy implementation into 3D codes and can be extended to flows with reattachment when DNS data becomes available.

4. Summary of the Present Model

The governing equations with complete closure models that were developed in the previous sections are presented here.

Momentum equation:

$$\rho \frac{DU_i}{Dt} = -\frac{\partial \overline{p}}{\partial x_i} + \rho \frac{\partial}{\partial x_k} \left[(\nu_s + \nu_T) \frac{\partial U_i}{\partial x_k} \right] + \rho \frac{\partial}{\partial x_k} \left(\frac{\nu_p}{\lambda} \left[f(C_{nn}) C_{ik} - \delta_{ik} \right] \right), \tag{34}$$

where the eddy viscosity is given by

$$\nu_T = C_\mu f_\nu \frac{k^2}{\tilde{\varepsilon}^N}, \tag{35}$$

with modified damping function

$$f_\nu = (1 - A) \left[1 - \exp\left(-\frac{y^*}{a_\mu (1 + B/a_\mu)} \right) \right]^2, \tag{36}$$

$$A = C_A \left(f_N \frac{\lambda^2 \tilde{L}^{3/2}}{f(C_{kk})^2} \frac{\varepsilon}{\nu_0} \right)^{0.3}, \tag{37}$$

$$B = C_B (C_{kk} - 3)^{1.25} / L, \tag{38}$$

with constants $a_\mu = 26.5$, $C_A = 0.071$ and $C_B = 0.44$. y^* is given by Equation (31).

Conformation tensor equation:

$$\frac{DC_{ij}}{Dt} - M_{ij} - NLT_{ij} = -\frac{1}{\lambda} [f(C_{kk}) C_{ij} - \delta_{ij}], \tag{39}$$

with

$$NLT_{ij} \approx f_N C_{N1} \frac{\lambda \sqrt{\tilde{L}} \varepsilon^N}{\nu_0 f(C_{mm})} \delta_{ij} - f_N^{1/4} C_{N2} M_{ij} + C_{N3} \frac{k}{\nu_0} \sqrt{\tilde{L} M_{nn}} \frac{\frac{\partial U_i}{\partial x_k} \frac{\partial U_j}{\partial x_k}}{\dot{\gamma}^2}, \tag{40}$$

where $f_N = \nu_T/\nu_0$ is the local eddy viscosity, $\dot{\gamma} = \sqrt{2 S_{pq} S_{pq}}$ is the shear rate invariant, $\tilde{L} = \sqrt{L^2/900}$ is the normalised maximum extension with the lowest DR, with model constants $C_{N1} = 0.11$, $C_{N2} = 0.3$ and $C_{N3} = 0.3$.

Transport equation of k:

$$\rho \frac{Dk}{Dt} = \rho \frac{\partial}{\partial x_i} \left[\left(\nu_s + \frac{f_t \nu_T}{\sigma_k} \right) \frac{\partial k}{\partial x_i} \right] + P_k - \rho(\tilde{\varepsilon}^N + D) - \frac{\nu_p}{\lambda} f(C_{mm}) \frac{NLT_{mm}}{2}, \quad (41)$$

where $P_k = -\rho \overline{u_i u_j} \frac{\partial U_i}{\partial x_j}$ is the rate of production of k.

Dissipation transport equation:

$$\rho \frac{D\tilde{\varepsilon}^N}{Dt} = \rho \frac{\partial}{\partial x_i} \left[\left(\nu_s + \frac{f_t \nu_T}{\sigma_\varepsilon} \right) \frac{\partial \tilde{\varepsilon}^N}{\partial x_i} \right] - f_2 C_{\varepsilon_2} \rho \frac{(\tilde{\varepsilon}^N)^2}{k} + \rho E$$
$$+ \left(C_{\varepsilon_1} P_k - C_{N4} \nu_p \sqrt{C_\mu f_\mu} \tilde{L}^{3/4} \left(\frac{k}{\nu_0} \right)^2 \right) \frac{\tilde{\varepsilon}^N}{k}, \quad (42)$$

with model constant $C_{N4} = 0.083$.

The remaining constants are from the Newtonian model and are $C_{\varepsilon_1} = 1.45$, $C_{\varepsilon_2} = 1.90$, $C_\mu = 0.09$, $\sigma_k = 1.1$ and $\sigma_\varepsilon = 1.3$.

5. Numerical Procedure

This section presents the numerical methods applied in order to examine the viscoelastic turbulence model against the available DNS data identified within the literature. A new finite volume C++ computational solver was developed in the OpenFOAM software by modifying the $k - \varepsilon$ sub-class files and introducing the FENE-P viscoelastic quantities such as: the polymer stress to the momentum equation; conformation tensor transport equation; and modified damping function to include elastic effects.

A fully-developed channel flow using half of the channel height, h, is applied given the symmetry of the governing geometry. We assigned 100 cells in the transverse (wall) direction with approximately 10 cells located inside the viscous sublayer. This is to provide mesh independent results, with errors within 0.5% for the mean velocity and the friction factor compared with a very fine mesh, similarly with [30]. The initial state of the simulation is the Newtonian solution until a steady-state solution was reached for each run case, except for HDR where a similar IDR developed case is applied to reduce computational time. Relaxation factors for the additional conformation tensor field are set to 0.2, along with residual control set to 10^{-5}. To improve numerical stability, an artificial diffusion term is added to the RACE of the form, $\kappa \partial_k^2 C_{ij}$, where κ denotes a constant, isotropic, artificial numerical diffusivity. In earlier studies [10], the dimensionless artificial numerical diffusivity is taken to be $\kappa/h u_\tau \sim O(10^{-2})$. Here, $\kappa/h u_\tau \sim O(10^{-3})$ and has negligible effect on mean values.

A pressure gradient is forced in the stream-wise direction to be unity, with periodic boundary conditions for all other flow fields, mimicking the DNS procedure of Li et al. [22]. No-slip boundary conditions were imposed on the solid wall for the velocity field U, along with k and $\tilde{\varepsilon}$ set to zero (or very small, $\sim 10^{-15}$). A Dirichlet boundary condition for C_{ij} is reported in Appendix A (similar to [26], but for all components), which is imposed within OpenFOAM under the swak4Foam library using the groovyBC functionality developed by Gschaider [45].

When normalizing the governing equations and inherently the various physical quantities, the velocity scale is taken to be the friction velocity (leading to the use of superscript +) and the length scale is the viscous length, $x_i = x_i^+ \nu_0 / U_\tau$. The conformation tensor is already in dimensionless form.

6. Results and Discussion

Following the numerical procedure proposed in the previous section, the model performance is assessed against a range of different flow and rheological parameters presented in the DNS data within Table 1.

6.1. Analysis of Conformation Tensor

Figure 3 compares the individual components of the conformation tensor with the present model against the model of Resende et al. [32] and selected DNS data covering L^2, Wi_{τ_0} and Re_{τ_0} variations (cases 16, 19, 20 and 26 in Table 1). as can be viewed in Figure 3a, the C_{xx} predictions are consistent with the DNS data. The new closure for NLT_{xx} (see term III in Equation (26)) is responsible for the improved predictions and can capture the Re_{τ_0}, L^2 and Wi_{τ_0} variations with much greater simplicity, especially for increased Reynolds number ($Re_{\tau_0} = 590$) compared with the model of Resende et al. [32].

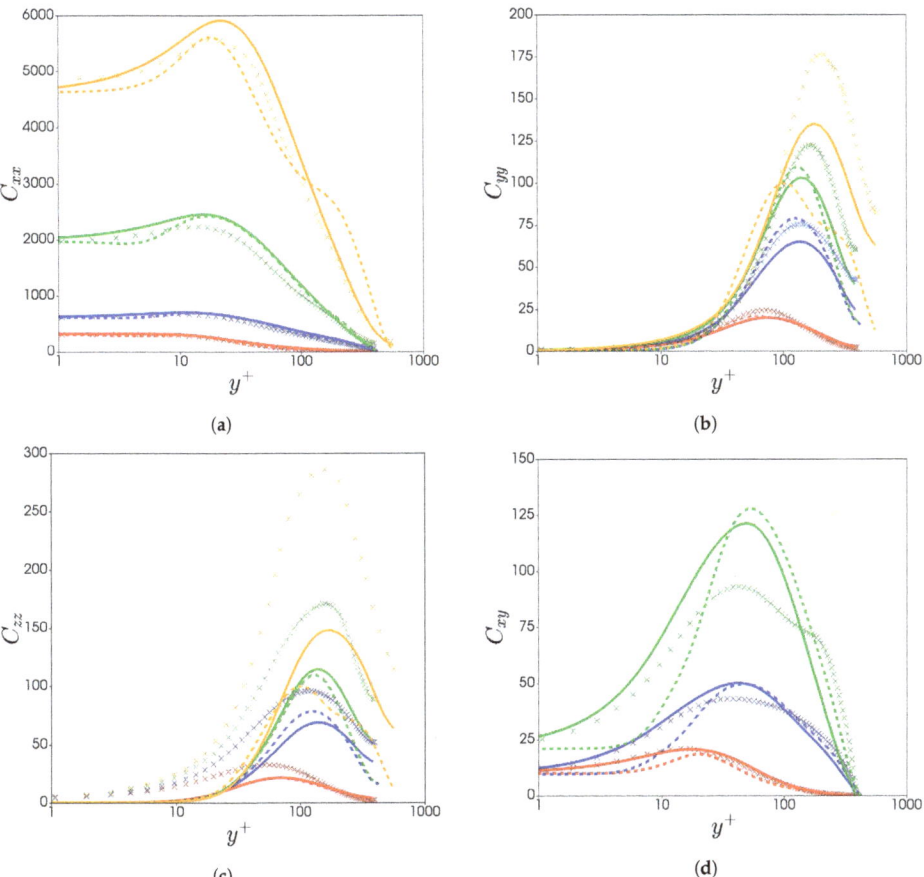

Figure 3. Comparison of the conformation tensor between DNS data (× crosses) and predictions with the new model (continuum lines), and previous model (dash lines): (**a**) C_{xx}; (**b**) C_{yy}; (**c**) C_{zz} and (**d**) C_{xy}. Each colour represents a different drag reduction regime: red (LDR, case 16); blue (IDR, case 19); green (HDR, case 20) and orange (very HDR, case 26. DNS data not available for C_{xy}).

Figure 3b plots the C_{yy} component, showing good agreement with the DNS data and improving upon the most recent model, especially away from the wall. The important feature is the location of the value at the centre-line and the peak location which both show good improvement, especially for higher Reynolds numbers ($Re_{\tau_0} = 590$). The improvements are a result of the new E_{τ_p} closure (see Equation (33)) which directly impacts ε^N in the NLT_{yy} closure (see term I in Equation (26)). Figure 3c plots the C_{zz} component and shows an under-prediction due to the isotropic assumption used in the model of NLT_{ij}, however, its impact is not significant.

The model predictions of the C_{yy} term are important in capturing the features of the C_{xy} component. As can be observed in Figure 3d, the model is able to capture the near-wall region, which, according to the findings of Li et al. [22], is the region of high chain dumbbell extension (limited to $y^+ < 50$) where the effect of C_{xy} acts towards the polymer shear stress.

It is evident that the overall predictions of the individual conformation tensor components are improved compared to the model of Resende et al. [32]. This is a result of the new NLT_{ij} and E_{τ_p} closures developed in the present work, which allows more scope of predictability and increased numerical stability with simpler closures.

6.2. Analysis of k, ε and ν_T

The predicted k profiles are shown in Figure 4a for cases 16 and 19 in Table 1, and Figure 4b for low and high Reynolds number cases (7 and 27). There is reasonable improvement of the profile away from the wall as a result of the new f_ν closure for increasing drag reduction and for various Reynolds numbers.

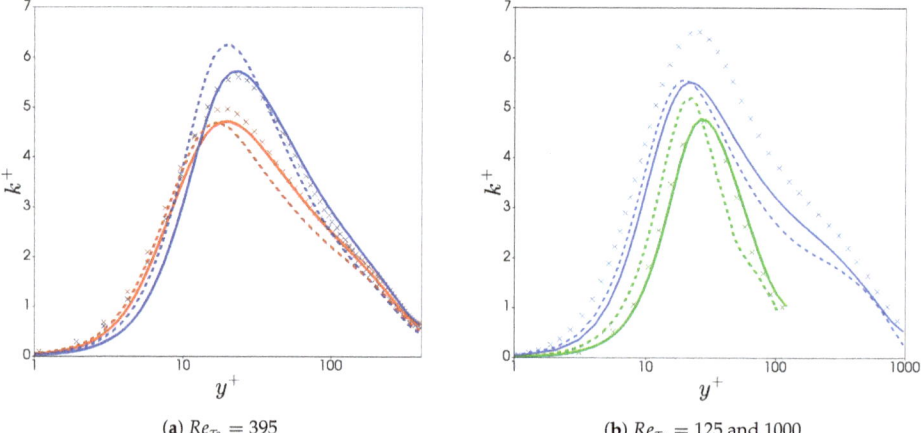

(a) $Re_{\tau_0} = 395$ (b) $Re_{\tau_0} = 125$ and 1000

Figure 4. Comparison of turbulent kinetic energy between DNS data (× crosses) and predictions with the new model (continuum lines), and previous model (dash lines): (**a**) $Re_{\tau_0} = 395$—red (LDR case 16)—and blue (IDR case 19); (**b**) $Re_{\tau_0} = 125$—green (HDR, case 7) and $Re_{\tau_0} = 1000$—blue (IDR, case 27).

In Figure 5, the prediction of the dissipation rate are compared with the DNS data of both LDR (case 16) and very HDR (case 22), along with predictions for the $v^2 - f$ model of Masoudian et al. [28]. The predictions for LDR are captured well with the DNS for both near and far from the wall. For HDR, there is a significant improvement near the wall compared with the $v^2 - f$ model. This is a result of the E_{τ_p} closure formulated (See Equation (33)) which decreases ε as flow viscoelasticity increases. The model of Resende et al. [32] shows similar results to the current model and is not plotted so that the figure is clearer. However, the complexity of the present E_{τ_p} closure model is reduced substantially and removes all friction velocity dependence, but can still predict all the main flow features with good performance.

The local eddy viscosity is plotted in Figure 6a for all ranges of DR. The combined performance of f_ν, k and ε gives rise to the predictions shown. We observe a reduction in the eddy structures within the buffer-layer and log-layer for increasing DR, as the DNS suggests. The damping function predicts well this behavior with the near-wall polymer extension via C_{kk} and the global reduction via $(1 - A)$.

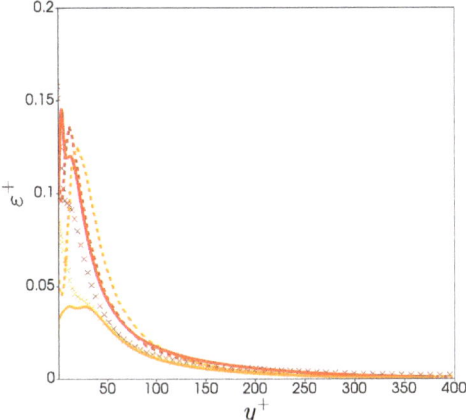

Figure 5. Comparison of the rate of Newtonian dissipation of k between DNS data ($\times$ crosses) and predictions with the new model (continuum lines), and $v^2 - f$ model of Masoudian et al. [28] (dash lines). Each colour represents a different drag reduction regime: red (LDR case 16); orange (HDR case 22).

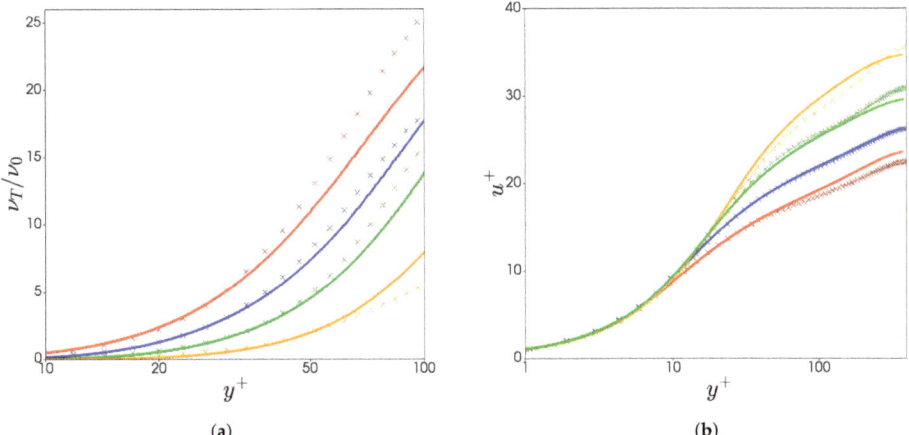

Figure 6. Comparison of the (**a**) local eddy viscosity and (**b**) mean stream-wise velocity profile, between DNS data ($\times$ crosses) and model predictions (continuum lines). Each colour represents a different drag reduction regime: red (LDR case 16); blue (IDR case 19); green (HDR case 20); orange (very HDR case 22).

6.3. Analysis of Velocity Profiles

Figure 6b shows the mean stream-wise velocity profiles for all ranges of DR at $Re_{\tau_0} = 395$. All of the profiles reduce to the linear distribution $u^+ = y^+$ in the viscous sub-layer. Further from the wall, the velocity profiles are well-captured for all ranges of DR.

The model can also predict well a range of Reynolds numbers with varying rheological parameters as can be viewed in Figure 7a. This is extended in Figure 7b for high Reynolds numbers, where there is a significant improvement compared with the model of Resende et al. [32]. This is a result of the new closure model for NLT_{xx} which scales well with Reynolds number and with reduced complexity.

The advantage of the current model is the ability to capture all velocity profiles well within the model limits, with more simplicity with regards to model closures and without friction velocity dependence.

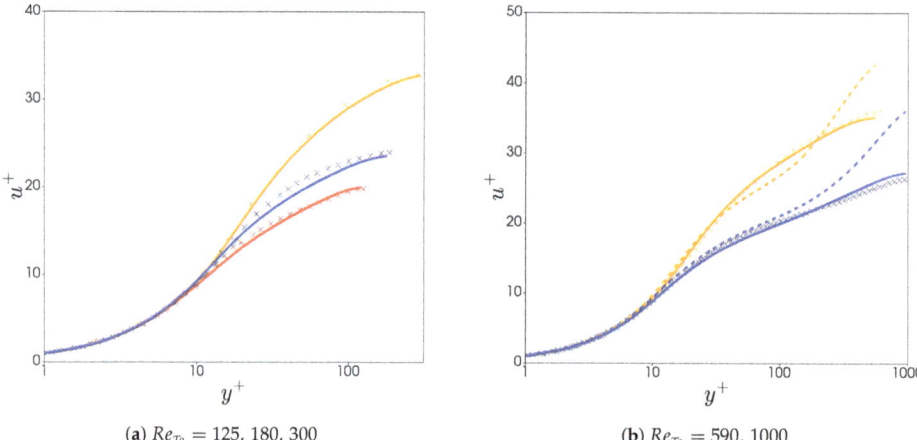

(a) $Re_{\tau_0} = 125, 180, 300$ (b) $Re_{\tau_0} = 590, 1000$

Figure 7. Comparison of the velocity profiles between DNS data (× crosses), current model predictions (continuum lines) and previous model predictions [32] (dashed lines). Each colour represents a different drag reduction regime: (**a**) red (LDR case 1); blue (IDR case 10); orange (very HDR case 15). (**b**) blue (IDR case 27); orange (very HDR case 26).

7. Conclusions

A viscoelastic turbulence model in fully-developed drag-reducing channel flow is improved, with turbulent eddies modeled under a $k - \varepsilon$ representation, along with polymeric solutions described by the finitely extensible nonlinear elastic-Peterlin (FENE-P) constitutive model. A new finite volume C++ computational solver was developed in the OpenFOAM software by modifying the $k - \varepsilon$ sub-class files and introducing the FENE-P viscoelastic quantities such as: the polymer stress to the momentum equation; conformation tensor transport equation; and modified damping function to include elastic effects. The model performance is evaluated against a variety of rheological parameters within the DNS data literature, including: friction Reynolds number $Re_{\tau_0} = 125, 180, 300, 395, 590, 1000$; Wiessenberg number $Wi_{\tau_0} = 25, 36, 50, 100, 116, 200$; and maximum molecular extensibility of the dumbbell chain $L^2 = 900, 1800, 3600, 10,000, 14,400$. The DNS data case (19) in Table 1 ($Re_{\tau_0} = 395$, DR = 37%) is used for the calibration of the closures developed for the turbulent cross-correlations identified in Section 3. The model is capable of predicting all flow features for low and high Reynolds numbers at all regimes of DR and improves significantly on the model of Resende et al. [32], with its ability to capture higher Reynolds numbers with simpler and physical-based closures.

The main feature is the formulation of the NLT_{ij} term which accounts for the interactions between the fluctuating components of the conformation tensor and the velocity gradient tensor. The advantage of the closure is the reduction in the complexity and use of damping functions in the dominant contribution, NLT_{xx}, modeled here to increase with turbulent kinetic energy as the flow viscoelasticity increases, demonstrating significant improvement with a range of rheological parameters and flow conditions.

Further improvements are developed for the viscoelastic destruction term, E_{τ_p}, within the dissipation rate transport equation. Modeled here with dependence on k and viscoelastic quantities, showing the ability to predict ε for low and high drag reduction.

An improved modified damping function, f_v, is also presented, which is able to predict the global reduction of the eddy viscosity and shift away from the wall for increasing viscoelasticity, whilst also improving the profiles of turbulent kinetic energy.

Overall, predictions compare very well with a wide range of DNS data and significantly improves on capturing all flow features with simplicity and performance compared with the most recent $k - \varepsilon$ model developed by Resende et al. [32]. The simplicity of the present model allows easy implementation into 3D codes and increases numerical stability. All friction velocity dependence is removed in the present model which is the first of its kind for damping function $k - \varepsilon$ models, whose main advantage is the realization of simulations in geometries with reattachment. Future work to extend to this study includes the development of an improved $k - \omega$ model based on the present model [25]. This would require the same concept of the modified damping function developed in this paper to be applied, with capabilities to predict flow behavior in industrially represented geometries such as pipes and constrictions.

Author Contributions: Supervision: P.R., T.C., M.W., A.A., D.H. and G.d.B.; software: M.M., G.d.B. and A.A.; formal analysis: M.M., P.R., M.W. and G.d.B.; project administration: T.C., M.W., D.H. and G.d.B. All authors have read and agreed to the published version of the manuscript.

Funding: This research was funded by Engineering and Physical Sciences Research Council (EPSRC) [Grant number: EP/N5059681/1, Award Reference: 1958044].

Conflicts of Interest: The authors declare no conflict of interest.

Appendix A

The FENE-P equations simplify considerably if we consider 1D, laminar, parallel flow: $u_x = u_z = 0$ and $u_x = u_x(y) \equiv u$. The system becomes:

$$C_{yy}\frac{du}{dy} - \frac{1}{\lambda}(f(C_{kk})C_{xy}) = 0 \tag{A1}$$

$$\frac{1}{\lambda}\left(1 - f(C_{kk})C_{yy}\right) = 0 \tag{A2}$$

$$\frac{1}{\lambda}\left(1 - f(C_{kk})C_{zz}\right) = 0 \tag{A3}$$

$$2C_{xy}\frac{du}{dy} - \frac{1}{\lambda}\left(1 - f(C_{kk})C_{xx}\right) = 0, \tag{A4}$$

where $f(C_{kk}) = \frac{L^2}{L^2 - C_{kk}}$. Introducing the Weissenberg number as $Wi = \lambda \frac{du}{dy}$ and solving the system of equations above, one finds the following cubic equation in $f(C_{kk}) \equiv f$:

$$f^3 - f^2 - \frac{2Wi^2}{L^2} = 0, \tag{A5}$$

which omits one real solution, f_R, that satisfies the laminar equations applicable at the wall:

$$C_{xx} = \frac{1}{f_R}\left(\frac{2Wi^2}{f_R^2} + 1\right) \tag{A6}$$

$$C_{yy} = \frac{1}{f_R} \tag{A7}$$

$$C_{zz} = \frac{1}{f_R} \tag{A8}$$

$$C_{xy} = \frac{Wi}{f_R^2} \tag{A9}$$

where

$$f_R = \frac{1}{3}\left(\frac{B}{2^{1/3}} + \frac{2^{1/3}}{B} + 1\right) \tag{A10}$$

with

$$B = (A + [(A+2)^2 - 4]^{1/2} + 2)^{1/3} \quad \text{and} \quad A = 54\left(\frac{Wi}{L}\right)^2. \tag{A11}$$

In our numerical simulations, the explicit definition of the wall value using Equations (A6)–(A9) as a Dirichlet boundary condition considerably improves the stability of the solution procedure and is preferred over a zero-flux Neumann boundary condition for the conformation tensor.

References

1. Toms, B.A. Some observations on the flow of linear polymer solutions through straight tubes at large Reynolds numbers. *Proc. Int. Cong. Rheol.* **1948**, *2*, 135–141.
2. Virk, P.S. Drag reduction fundamentals. *AIChE J.* **1975**, *21*, 625–656. [CrossRef]
3. Lumley, J.L. Drag reduction by additives. *Annu. Rev. Fluid Mech.* **1969**, *1*, 367–384. [CrossRef]
4. Lumley, J.L. Drag reduction in turbulent flow by polymer additives. *J. Polym. Sci. Macromol. Rev.* **1973**, *7*, 263–290, [CrossRef]
5. Hoyt, J. A Freeman scholar lecture: The effect of additives on fluid friction. *J. Fluids Eng.* **1972**, *94*, 258–285. [CrossRef]
6. Virk, P.S.; Merrill, E.; Mickley, H.; Smith, K.; Mollo-Christensen, E. The Toms phenomenon: Turbulent pipe flow of dilute polymer solutions. *J. Fluid Mech.* **1967**, *30*, 305–328. [CrossRef]
7. White, C.M.; Mungal, M.G. Mechanics and prediction of turbulent drag reduction with polymer additives. *Annu. Rev. Fluid Mech.* **2008**, *40*, 235–256. [CrossRef]
8. Procaccia, I.; L'vov, V.S.; Benzi, R. Colloquium: Theory of drag reduction by polymers in wall-bounded turbulence. *Rev. Mod. Phys.* **2008**, *80*, 225. [CrossRef]
9. Dubief, Y.; Terrapon, V.E.; White, C.M.; Shaqfeh, E.S.; Moin, P.; Lele, S.K. New answers on the interaction between polymers and vortices in turbulent flows. *Flow Turbul. Combust.* **2005**, *74*, 311–329. [CrossRef]
10. Sureshkumar, R.; Beris, A.N.; Handler, R.A. Direct numerical simulation of the turbulent channel flow of a polymer solution. *Phys. Fluids* **1997**, *9*, 743–755. [CrossRef]
11. L'vov, V.S.; Pomyalov, A.; Procaccia, I.; Tiberkevich, V. Drag reduction by polymers in wall bounded turbulence. *Phys. Rev. Lett.* **2004**, *92*, 244503. [CrossRef]
12. De Angelis, E.; Casciola, C.M.; L'vov, V.S.; Pomyalov, A.; Procaccia, I.; Tiberkevich, V. Drag reduction by a linear viscosity profile. *Phys. Rev. E* **2004**, *70*, 055301. [CrossRef]
13. Dallas, V.; Vassilicos, J.C.; Hewitt, G.F. Strong polymer-turbulence interactions in viscoelastic turbulent channel flow. *Phys. Rev. E* **2010**, *82*, 066303. [CrossRef]
14. Min, T.; Yoo, J.Y.; Choi, H.; Joseph, D.D. Drag reduction by polymer additives in a turbulent channel flow. *J. Fluid Mech.* **2003**, *486*, 213–238. [CrossRef]
15. Pinho, F. A GNF framework for turbulent flow models of drag reducing fluids and proposal for a k–ε type closure. *J. Non-Newton. Fluid Mech.* **2003**, *114*, 149–184. [CrossRef]
16. Cruz, D.; Pinho, F.; Resende, P. Modelling the new stress for improved drag reduction predictions of viscoelastic pipe flow. *J. Non-Newton. Fluid Mech.* **2004**, *121*, 127–141. [CrossRef]
17. Resende, P.; Escudier, M.; Presti, F.; Pinho, F.; Cruz, D. Numerical predictions and measurements of Reynolds normal stresses in turbulent pipe flow of polymers. *Int. J. Heat Fluid Flow* **2006**, *27*, 204–219. [CrossRef]
18. Resende, P.; Pinho, F.; Cruz, D. A Reynolds stress model for turbulent flows of viscoelastic fluids. *J. Turbul.* **2013**, *14*, 1–36. [CrossRef]
19. Leighton, R.; Walker, D.T.; Stephens, T.; Garwood, G. Reynolds stress modeling for drag reducing viscoelastic flows. In Proceedings of the ASME/JSME 2003 4th Joint Fluids Summer Engineering Conference, Honolulu, HI, USA, 6–10 July 2003; American Society of Mechanical Engineers: New York, NY, USA, 2003; pp. 735–744.
20. Pinho, F.; Li, C.; Younis, B.; Sureshkumar, R. A low Reynolds number turbulence closure for viscoelastic fluids. *J. Non-Newton. Fluid Mech.* **2008**, *154*, 89–108. [CrossRef]
21. Housiadas, K.D.; Beris, A.N.; Handler, R.A. Viscoelastic effects on higher order statistics and on coherent structures in turbulent channel flow. *Phys. Fluids* **2005**, *17*, 035106. [CrossRef]
22. Li, C.; Gupta, V.; Sureshkumar, R.; Khomami, B. Turbulent channel flow of dilute polymeric solutions: Drag reduction scaling and an eddy viscosity model. *J. Non-Newton. Fluid Mech.* **2006**, *139*, 177–189. [CrossRef]
23. Li, C.F.; Sureshkumar, R.; Khomami, B. Influence of rheological parameters on polymer induced turbulent drag reduction. *J. Non-Newton. Fluid Mech.* **2006**, *140*, 23–40. [CrossRef]

24. Resende, P.; Kim, K.; Younis, B.; Sureshkumar, R.; Pinho, F. A FENE-P k–ε turbulence model for low and intermediate regimes of polymer-induced drag reduction. *J. Non-Newton. Fluid Mech.* **2011**, *166*, 639–660. [CrossRef]
25. Resende, P.; Pinho, F.; Younis, B.; Kim, K.; Sureshkumar, R. Development of a Low-Reynolds-number k-ω Model for FENE-P Fluids. *Flow Turbul. Combust.* **2013**, *90*, 69–94. [CrossRef]
26. Iaccarino, G.; Shaqfeh, E.S.; Dubief, Y. Reynolds-averaged modeling of polymer drag reduction in turbulent flows. *J. Non-Newton. Fluid Mech.* **2010**, *165*, 376–384. [CrossRef]
27. Dubief, Y.; Laccarino, G.; Lele, S. *A Turbulence Model for Polymer Flows*; Center for Turbulence Research: Stanford, CA, USA, 2004.
28. Masoudian, M.; Kim, K.; Pinho, F.; Sureshkumar, R. A viscoelastic k-ε-v2-f turbulent flow model valid up to the maximum drag reduction limit. *J. Non-Newton. Fluid Mech.* **2013**, *202*, 99–111. [CrossRef]
29. Masoudian, M.; Kim, K.; Pinho, F.; Sureshkumar, R. A Reynolds stress model for turbulent flow of homogeneous polymer solutions. *Int. J. Heat Fluid Flow* **2015**, *54*, 220–235. [CrossRef]
30. Masoudian, M.; Pinho, F.; Kim, K.; Sureshkumar, R. A RANS model for heat transfer reduction in viscoelastic turbulent flow. *Int. J. Heat Mass Transf.* **2016**, *100*, 332–346. [CrossRef]
31. Tsukahara, T.; Kawaguchi, Y. Proposal of damping function for low-Reynolds-number-model applicable in prediction of turbulent viscoelastic-fluid flow. *J. Appl. Math.* **2013**, *2013*. [CrossRef]
32. Resende, P.; Afonso, A.; Cruz, D. An improved k-ε turbulence model for FENE-P fluids capable to reach high drag reduction regime. *Int. J. Heat Fluid Flow* **2018**, *73*, 30–41. [CrossRef]
33. Bird, R.B.; Armstrong, R.C.; Hassager, O. *Dynamics of Polymeric Liquids, Volume 1: Fluid Mechanics*; Wiley: New York, NY, USA, 1987.
34. Alfonsi, G. Reynolds-averaged Navier–Stokes equations for turbulence modeling. *Appl. Mech. Rev.* **2009**, *62*, 040802. [CrossRef]
35. Nagano, Y.; Hishida, M. Improved form of the k-ε model for wall turbulent shear flows. *J. Fluids Eng.* **1987**, *109*, 156–160. [CrossRef]
36. Nagano, Y. Modeling the dissipation-rate equation for two-equation turbulence model. In Proceedings of the 9th Symposium on Turbulent Shear Flows, Kyoto, Japan, 16–18 August 1993; p. 23.
37. Thais, L.; Gatski, T.B.; Mompean, G. Some dynamical features of the turbulent flow of a viscoelastic fluid for reduced drag. *J. Turbul.* **2012**, *13*, N19. [CrossRef]
38. Thais, L.; Gatski, T.B.; Mompean, G. Analysis of polymer drag reduction mechanisms from energy budgets. *Int. J. Heat Fluid Flow* **2013**, *43*, 52–61. [CrossRef]
39. Masoudian, M.; da Silva, C.; Pinho, F. Grid and subgrid-scale interactions in viscoelastic turbulent flow and implications for modelling. *J. Turbul.* **2016**, *17*, 543–571. [CrossRef]
40. Thais, L.; Tejada-Martinez, A.; Gatski, T.; Mompean, G. Temporal large eddy simulations of turbulent viscoelastic drag reduction flows. *Phys. Fluids* **2010**, *22*, 013103. [CrossRef]
41. Benzi, R. A short review on drag reduction by polymers in wall bounded turbulence. *Phys. D Nonlinear Phenom.* **2010**, *239*, 1338–1345. [CrossRef]
42. Resende, P.; Cavadas, A. New developments in isotropic turbulent models for FENE-P fluids. *Fluid Dyn. Res.* **2018**, *50*, 025508. [CrossRef]
43. Park, T.S.; Sung, H.J. A nonlinear low-Reynolds-number κ-ε model for turbulent separated and reattaching flows—I. Flow field computations. *Int. J. Heat Mass Transf.* **1995**, *38*, 2657–2666. [CrossRef]
44. Wallin, S.; Johansson, A.V. An explicit algebraic Reynolds stress model for incompressible and compressible turbulent flows. *J. Fluid Mech.* **2000**, *403*, 89–132. [CrossRef]
45. Gschaider, B.F. The incomplete swak4Foam reference. *Tech. Rep.* **2013**, *131*, 202.

Publisher's Note: MDPI stays neutral with regard to jurisdictional claims in published maps and institutional affiliations.

© 2020 by the authors. Licensee MDPI, Basel, Switzerland. This article is an open access article distributed under the terms and conditions of the Creative Commons Attribution (CC BY) license (http://creativecommons.org/licenses/by/4.0/).

Article

The Role of Elasticity in the Vortex Formation in Polymeric Flow around a Sharp Bend

Brian Wojcik [1], Jason LaRuez [2], Michael Cromer [2] and Larry A. Villasmil Urdaneta [3],*

[1] Rochester Institute of Technology, School of Physics and Astronomy, New York, NY 14623-5603, USA; brianswojcik@gmail.com
[2] Rochester Institute of Technology, School of Mathematical Sciences, New York, NY 14623-5603, USA; jpl9982@rit.edu (J.L.); mec2sma@rit.edu (M.C.)
[3] College of Engineering Technology, Rochester Institute of Technology, New York, NY 14623-5603, USA
* Correspondence: luvmet@rit.edu; Tel.: +1-585-475-5304

Featured Application: Elasticity is predicted to play a significant role in the formation of vortices in polymeric flow in a sharp bend. The polymer dilution and the flow rate determine if the formation of the vortices occurs upstream or downstream of the bend corner.

Abstract: Fluid dynamic simulations using the FENE-P model of polymer physics are compared to those of an incompressible Newtonian fluid base case in order to understand the role of elasticity in the formation of vortices in a 90° bend narrow channel. The analysis bridges the flow behavior of a purely elastic fluid and that of a Newtonian fluid. We evaluated how four dimensionless numbers—Reynolds number (Re), Weissenberg number (Wi), viscosity ratio (β), and elasticity number (El)—affect the formation of vortices. It is shown that increasing Re and Wi, or lowering β will cause vortices to grow in size. Two phase space diagrams, β vs. El and β vs. Re, were created to show the range of values where inertial and elastic vortices form. Both diagrams have three zones. Depending on the polymer viscosity ratio and the elasticity number, the vortices form either upstream of the bend (elasticity driven) or form downstream of the bend (inertia driven), are suppressed. Our predictions are in good agreement with previous experimental and numerical works.

Keywords: elasticity; polymer; vortex; bend; flow

Citation: Wojcik, B.; LaRuez, J.; Cromer, M.; Villasmil Urdaneta, L.A. The Role of Elasticity in the Vortex Formation in Polymeric Flow around a Sharp Bend. *Appl. Sci.* **2021**, *11*, 6588. https://doi.org/10.3390/app11146588

Academic Editor: Francesca Scargiali

Received: 1 June 2021
Accepted: 15 July 2021
Published: 17 July 2021

Publisher's Note: MDPI stays neutral with regard to jurisdictional claims in published maps and institutional affiliations.

Copyright: © 2021 by the authors. Licensee MDPI, Basel, Switzerland. This article is an open access article distributed under the terms and conditions of the Creative Commons Attribution (CC BY) license (https://creativecommons.org/licenses/by/4.0/).

1. Introduction

Many modern-day products, such as rubber tires and plastic bags, are made from polymers. These items are made by specialized processing machines that handle polymer solutions in pipes, conduits, and accessories such as bends and elbows. These polymeric solutions, as with any other liquid, are subjected to many fluid dynamic effects. One such effect is the formation of recirculation zones, or vortices in abrupt, changes of flow direction as in contractions and bends. These vortices are a feature where some amount of the fluid becomes trapped in a cyclone like structure near or around corners, justifying those authors that refer to them as separation bubbles [1]. Recirculation zones can have detrimental effects on the flow of polymers and polymer solutions, affecting their manufacturing processes. As a result, it is important to know the conditions under which vortices form. The presence of vortices near or around corners is one of the most significant alterations in channel flow for both non-Newtonian and Newtonian fluids. Vortices are well known to form in Newtonian fluids due to the effect of inertia [2]. However, for polymers and polymer solutions not all the physics involved are fully understood [3,4].

Understanding the flow behavior of polymers and polymer solutions is essential to design and optimize fluid flow systems in many practical applications. For example, in microfluidics the polymer concentration has a significant effect on viscoelastic behavior by altering the base flow or result in flow instabilities. In that regard, Gulati et al. [5] studied the flows of dilute and semi-dilute polymer solutions in sharp 90° micro-bends in channels of rectangular cross-section. Their flow visualizations show that a vortex is present in the inner, upstream corner of the bend and grows with increasing Reynolds and Weissenberg numbers for flows of shear-thinning, semi-dilute polymeric solutions. They reported that secondary flows were not present for Newtonian flows under similar conditions and that a vortex is absent for flow of a dilute, non-shear thinning PEO solution. They concluded that shear-thinning appears to be central to the presence of an elastic secondary flow in this geometry. Their experiments were carried at very low Reynolds number (10^{-6} < Re < 0.03), and Weissenberg numbers ranging from 0.42 to 126.

More recently, Kim et al. [6] reported instabilities in viscoelastic flow in a 90° bent channel. They observed that the flow instability in an aqueous PEO solution occurs when the concentration of PEO is as low as 50 ppm. Investigating the effects of the polymer concentration, flow rate, and elasticity number, they found that the flow is stabilized in shear-thinning fluids, whereas the flow instability is amplified when both elastic and inertial effects are pronounced. Their experiments were carried out at Reynolds numbers ranging from 0.3 to 3.0 and Weissenberg numbers of 0 (Newtonian flow) to 40.

Both studies [5,6] coincide in pointing to the shear thinning properties of the solution, a decrease in viscosity under shear strain, as the main reason for the formation of vortices and secondary flows.

At the macroscopic level, it is also well known that polymers, as well as surfactants, are frequently added to Newtonian fluids with the purpose of reducing friction losses in straight pipe turbulent flow. In some cases, the drag-reducing rate is as high as 75% [7] making them attractive for using in complex industrial pipe flow systems. Friction losses in industrial piping systems are mostly due to accessories such as bends, tees, and valves, rather than in the straight pipes. Understanding how polymers affect the Newtonian flow in bends and accessories would help in evaluating their drag reduction potential and application in intricate piping systems.

Munekata et al. [8] studied the bend flow characteristics of two surfactant solutions, experimentally and numerically. They found that the drag, friction coefficient, increases or reduces depending on the average bulk velocity of the solution. As the solution concentration increases a larger bulk velocity is required to observe a reduction in the drag. Although the drag reduction's effects of the surfactants in the solution are lower in bend flow than in a straight pipe, the authors attribute the drag reduction to the suppression of the centrifugal effect and reducing the secondary flow due to the viscoelastic properties of the solutions (normal stress effect). Their analysis was performed at a high Reynolds number for both a Newtonian fluid and a highly viscoelastic fluid. They observed and predicted smaller velocity gradients near the wall for viscoelastic fluid flow than that for Newtonian fluid flow.

Even in purely Newtonian fluid flow, understanding the hydrodynamic behavior in a bend channel or a pipe elbow remains relevant today. Matsumoto et al. [1] investigated the flow dynamics for a Newtonian fluid in a bent channel via two-dimensional direct numerical simulations. They investigated the flow structure along the channel as a function of both the bend angle and the Reynolds number. Their numerical work suggests a scaling relation between the shape of the separation bubble, a downstream vortex after the bend corner, and the flow conductance. Their simulations were carried at high Reynolds number but only for a Newtonian fluid. Nevertheless, they present an integrated phase diagram for the flow dynamics, where depending on bend angle and Reynolds number either the flow is uniform with no recirculation vortex forming, a vortex forms downstream of then bend corner, or vortices shed intermittently from the bend corner.

In this paper we address the role of elasticity in the formation of vortices. With our analysis we are bridging the flow behavior of a polymeric fluid and that of a Newtonian fluid in a 90° bend narrow channel.

In agreement with previous works [5,6], we predict the formation of elastic and inertial vortices for polymer solutions with specific rheological properties and flow conditions. Our primary interest is exploring the possibility of controlling the alteration of the flow, either suppressing or promoting a vortex or separation bubble, by modifying the underlying properties of the polymer solution.

The remainder of this paper is organized as follows. In Section 2, we introduce our hydrodynamic and polymer models. In Section 3, we describe in detail our numerical procedure including simulation parameters, and the computational domain including a few details about the meshing. In Section 4, we present the numerical results, specifically comparing Newtonian and polymer flow and discussing the role of elasticity and Reynolds number on vortex formation. We also construct two flow phase diagrams. In Section 5, we summarize our results and briefly comment on the future perspectives of the present work.

2. Materials and Methods

2.1. Governing Equations

Two types of fluids were studied in this research, a Newtonian fluid and a polymeric fluid. Newtonian flows are characterized by Navier–Stokes equations. Polymer flows require additional equations to characterize the elastic behavior of the polymer chains. In this work, the polymer chains are characterized using the FENE-P model (finitely extensible nonlinear elastic model with Peterlin closure [9]). The FENE-P model was chosen due to its versatility and simplicity in characterizing polymer behavior. In the FENE-P model, polymers are represented as dumbbells, two masses connected by a spring, as in many other polymer models [10]. Nevertheless, in the FENE-P model the spring has a finite stretching limit.

For this work it is assumed that the fluid is incompressible for both Newtonian and polymer-based flows, for which the continuity equation becomes:

$$\nabla \cdot \vec{v} = 0, \tag{1}$$

where $\vec{v}$ is the velocity of the fluid. The general momentum equation of motion is given by:

$$\rho \left(\frac{\partial \vec{v}}{\partial t} + \vec{v} \cdot \nabla \vec{v} \right) = -\nabla \cdot \underline{\underline{\Pi}} + \rho \vec{g}, \tag{2}$$

where $\underline{\underline{\Pi}}$ is the total stress tensor:

$$\underline{\underline{\Pi}} = \underline{\underline{\tau}} + p \underline{\underline{I}}, \tag{3}$$

with $\underline{\underline{\tau}}$ as the extra stress tensor, p the hydrostatic pressure, and $\underline{\underline{I}}$ the identity tensor.

For an incompressible Newtonian fluid, the stress tensor $\underline{\underline{\tau}}$ is given by:

$$\underline{\underline{\tau}} = -\eta_s \underline{\underline{\dot{\gamma}}}, \tag{4}$$

$$\underline{\underline{\dot{\gamma}}} = \left[\nabla \vec{v} + \left(\nabla \vec{v} \right)^T \right], \tag{5}$$

where η_s is the Newtonian shear viscosity, and $\underline{\underline{\dot{\gamma}}}$ is the strain rate tensor.

Combining (3)–(5), substituting into (2), and neglecting the effects of gravity yields the well-known Navier–Stokes equation of fluid dynamics, see below. For a detailed derivation of (6) see ref. [11].

$$\rho \left(\frac{\partial \vec{v}}{\partial t} + \vec{v} \cdot \nabla \vec{v} \right) = -\nabla p + \eta_s \nabla^2 \vec{v}. \tag{6}$$

For complex fluids such as polymers and polymer solutions, Equation (4) is not sufficient to describe the dynamics of the flows. An additional stress term is added to the Newtonian constitutive equation. For complex fluids:

$$\underline{\underline{\tau}} = \underline{\underline{\tau}}_N + \underline{\underline{\tau}}_{P'} \tag{7}$$

where $\underline{\underline{\tau}}_N$ denotes the Newtonian stress component, and $\underline{\underline{\tau}}_p$ denotes the stress component characterizing the polymer behavior.

The equation used to estimate $\underline{\underline{\tau}}_p$ depends on the particular model of polymer physics being employed. For the purposes of this research, the FENE-P constitutive equation is given by [10]:

$$\underline{\underline{\tau}}_p = G_O f(\underline{\underline{I}} - \underline{\underline{A}}), \tag{8}$$

$$\lambda \underline{\underline{A}}_{(1)} + f(\underline{\underline{A}} - \underline{\underline{I}}) = 0, \tag{9}$$

where G_O is the elastic modulus, and λ is the relaxation time of the polymers. The conformation tensor $\underline{\underline{A}}$ measures the stretch and orientation of the polymers. The upper-convective derivative [12], denoted by $(\ )_{(1)}$, is given by:

$$\frac{\partial (\ )}{\partial t} + \vec{v} \cdot \nabla (\ ) - \left(\left(\nabla \vec{v} \right)^T \cdot (\ ) + (\ ) \cdot \nabla \vec{v} \right), \tag{10}$$

and:

$$f = \frac{b}{b - tr\underline{\underline{A}}}, \tag{11}$$

with b representing the square of the maximum extensibility.

The total stress in the polymer solution is then obtained by combining Equations (7) and (8), and substituting that result back into (3):

$$\underline{\underline{\Pi}} = G_O f(\underline{\underline{I}} - \underline{\underline{A}}) - \eta_s \underline{\underline{\dot{\gamma}}} + p\underline{\underline{I}}. \tag{12}$$

2.2. Computational Methodology

2.2.1. Numerical Solver

The numerical solver implemented to solve the governing equations for the present work is the open-source programming package RheoTool. RheoTool is an open-source toolbox based on the OpenFOAM® library to simulate the flow of Generalized Newtonian Fluids (GNF) and viscoelastic fluids [13].

RheoTool is a modification of the viscoelastic solver available in the OpenFOAM® toolbox [14]. The main goal of the modification was to improve its stability for differential-type constitutive equations. The major contributions of RheoTool are using the log-conformation approach to solve Oldroyd-B type constitutive equations, handling high-resolution schemes with a componentwise and deferred correction approach to discretize the convective terms, and introducing a new stress–velocity coupling term together with the well-known SIMPLEC algorithm for pressure–velocity coupling [15].

In the present work, the polymer is characterized by the FENE-P model which is solved using the log-conformation approach [16,17], and selecting the stress–velocity coupling as the stabilization method.

In terms of discretization, gradient terms are discretized using the Gauss scheme with linear interpolation, Laplacian terms are discretized using the Gauss scheme (only choice) with linear interpolation and the corrected scheme for the surface normal gradient, and the convective terms are discretized using the CUBISTA scheme [18].

The solver chosen for all (asymmetric) equations is the Preconditioned (bi-) Conjugate Gradient with the Diagonal incomplete-Cholesky (LU) preconditioner.

The solution is advanced in time using the Euler scheme with adjustable time steps. A minimum time step of 10^{-6} s is set together with a maximum Courant number of 2.0 and a maximum time step of 10^{-2} s. This time step condition allows for speed up of the convergence of transient simulations to steady state.

2.2.2. Simulation Parameters

The main goal of this research is to evaluate the role that elasticity plays in the formation of vortices in flow around a sharp corner. For that purpose, we performed multiple simulations to create two phase diagrams in order to show the overall effects of four dimensionless parameters, Wi, Re, El, and β. These parameters are defined as follows:

- The Weissenberg number Wi = λ·U/H, where λ is the relaxation time of the polymer, U is the characteristic velocity, and H is the characteristic length scale of the geometry (channel height). The Weissenberg number is a ratio of the polymeric timescale to a convective timescale, akin to an elastic to viscous forces ratio in the context of this work.
- The Reynolds number Re = ρ·U·H/$η_0$, where ρ and $η_0$ are, respectively, the density and the zero-shear rate viscosity of the solution. The Reynolds number is a measure of the ratio of inertia to viscous forces.
- The elasticity number, a derived parameter, characterizes the balance of elastic and inertial forces in the fluid and is defined as El = Wi/Re.
- The solvent viscosity ratio β = $η_s$/$η_o$, where $η_s$ is the Newtonian solvent viscosity and $η_o$ is the zero-shear rate viscosity of the solution.

An important note about the elasticity number El and the viscosity ratio β is that they are both material parameters that depend on $η_o$, and are independent of the fluid velocity. The zero-shear rate viscosity, $η_o$, physically corresponds to the concentration of polymers in the solution. A decreasing El indicates a decrease in the polymer concentration. Conversely, an increase in β corresponds also to a decrease in the polymer concentration, which is the parameter typically varied during experimental work.

Two sets of simulations were conducted in order to meet this goal. The first group of simulations conducted during this research were used to create a Weissenberg number versus Reynolds number phase space diagram. The Weissenberg numbers evaluated were 0.5, 1, and 1.5. Simulations were run with viscosity ratios, β values, ranging from 0 to 1 in 0.1 increments. A β value of 0 represents a polymer melt. A β value of 1 represents the viscosity of a Newtonian fluid (solvent). A second group of simulations was conducted in order to create a viscosity ratio, β, versus elasticity number, El, phase diagram. All of these simulations were performed with a constant value for the Weissenberg number (Wi = 1).

2.2.3. Computational Domain

Figure 1 shows a representation of the computational domain used to model the right-angle bend geometry. The fluid flows into the system at the top left, where a uniform inlet velocity (for a given flowrate Q) is imposed, travels around the sharp bend, and then exits out at the bottom, where the outlet pressure is set. At the walls, no-slip and no-penetration conditions are set for the velocities, and zero gradient for the pressure and the stress tensors.

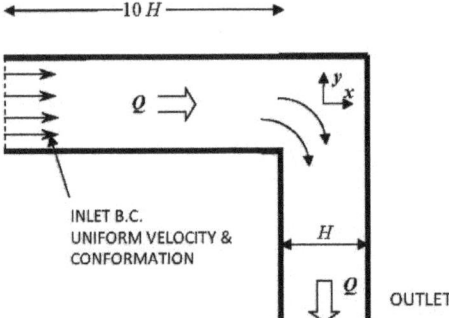

Figure 1. Computational domain, right-angle bend geometry. Q, volumetric flow rate; H, channel height; inlet/outlet branches length is 10 times H.

The bend used in the simulations has a ten to one ratio of length to channel height, H, in both branches. This ratio is more than enough to obtain a fully developed flow well upstream of the corner bend and for the exit, being far enough away to prevent any upstream effect from the outlet boundary conditions, see references [19,20], and also Appendix A. The geometry used follows the experimental work of Gulati et al. [5] and is typical of benchmark cases and similar configurations that have been evaluated both numerically and experimentally, see [1,6,21,22].

2.2.4. Meshing

Figure 2 presents an enlarged portion of the bend corner geometry with the progression in mesh refinement used for the simulations. Preliminary runs were performed in the coarse mesh shown in Figure 2a. The mesh was halved twice for the final run of every simulation, Figure 2c. An even finer fourth mesh was used in a few simulations to estimate grid convergence regarding the size predicted for inertial and elastic vortices including the phase maps.

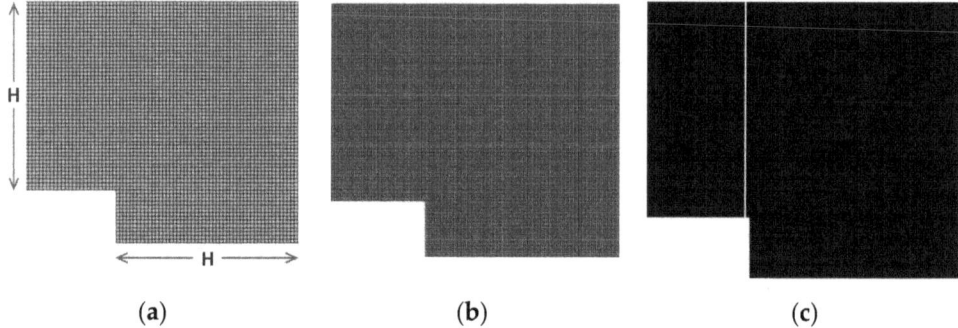

Figure 2. Uniform meshing size progression: (**a**) coarse, $\Delta x = \Delta y = 1/50 \cdot H$; (**b**) medium, $\Delta x = \Delta y = 1/100 \cdot H$; (**c**) fine, $\Delta x = \Delta y = 1/200 \cdot H$.

Figure 3 presents a sample of the grid convergence analysis performed in choosing the grid for the simulations and estimate the numerical errors of the predictions. The profiles in the figure correspond to a case with an elastic vortex located upstream of the bend corner. They were created along the white vertical line highlighted in Figure 2c; the grid chosen for the numerical analysis. The line crosses the elastic vortex roughly in the middle.

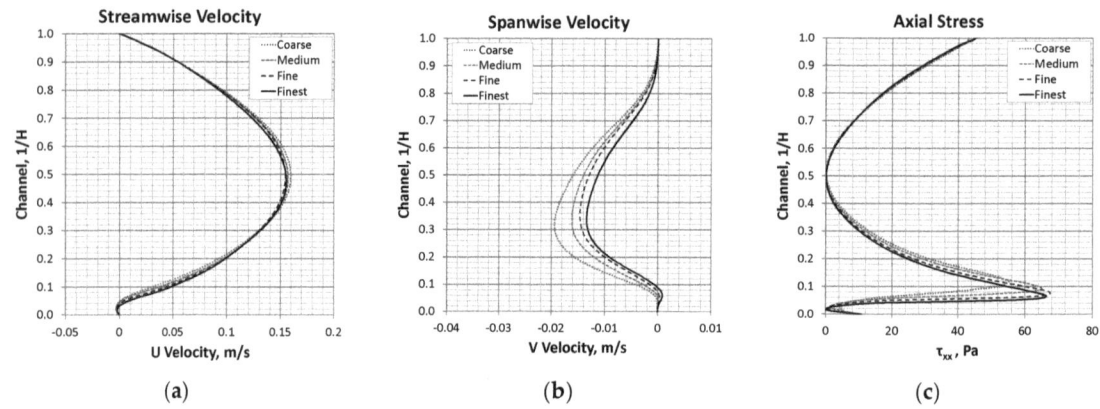

Figure 3. Grid convergence analysis, Wi = 1.0, El = 100, β = 0.1: (**a**) horizontal velocity; (**b**) vertical velocity; (**c**) axial stress.

Figure 3a compares the streamwise velocity component for the four grids evaluated. The weighted average deviation error outside the vortex for the fine grid is 1.3% and 5.5%

inside the vortex. Similarly, Figure 3b compares the spanwise velocity component for the same four grids. The weighted average deviation error outside the vortex is 14% and 26% inside the vortex, respectively (in both cases, note the very low actual value of the predicted velocity). Lastly, Figure 3c compares the axial stress. The weighted average deviation error outside the vortex is 4.0% and 8.3% inside the vortex, respectively.

Similar analyses were conducted for many cases, including those with inertial vortices located downstream of the bend corner. In a typical high Reynolds number case, Re = 100, the weighted average deviation errors of all variables are significantly smaller. For the streamwise velocity component, the weighted average deviation error outside the vortex for the fine grid is 0.02% and 0.4% inside the vortex, respectively. For the spanwise velocity component, the weighted average deviation error outside the vortex is 0.09% and 0.5% inside the vortex, respectively. Additionally, for the strain rate, the weighted average deviation error outside the vortex is 0.04% and 0.28% inside the vortex, respectively.

In general terms, it was found that predicting the location and length of the vortex is the most significant challenge in attaining grid convergence. Outside the vortex and far from it, the grid convergence index (GCI) for the fine grid is excellent, ranging from 0.0 to 1.5% in all variables. Within the vortex, the GCI for the fine grid is acceptable, ranging from 0.1% to 25% in all variables. Again, it should be noted that actual values for the spanwise velocity and the normal stress within the vortex are significantly smaller when compared to the streamwise velocity and the axial stresses, by two to three orders of magnitude. The mesh and grid convergence analyses were conducted following the Procedure for Estimation and Reporting of Uncertainty Due to Discretization in CFD Applications by Roache et al. [23].

The formation of vortices in polymer flow has been studied extensively in planar contractions and the size and strength have been shown to be strongly dependent on the mesh [15,24,25]. For example, the effect of mesh refinement on the size of the vortices was highlighted by Alves et al. [26] when developing benchmark solutions for the flow of Oldroyd-B and PTT fluids in planar contractions. They found that a very high degree of mesh fineness was required to obtain accurate results with the Oldroyd-B fluid, while the PTT fluid in general did not require the finest meshes. These authors relied on Richardson's extrapolation to measure the level of convergence.

3. Results

3.1. Newtonian Flow

To better understand the effects the elasticity of polymers have on the formation of vortices in a right angle bend geometry, simulations of Newtonian fluid flow were conducted first in order to obtain a baseline or reference. It is well known that in Newtonian flow vortices form downstream of a bend corner at sufficiently high Reynolds number [22,27].

To establish the baseline, the Reynolds number value at which the vortices first appear was sought. This value, called Re_{crit} for the duration of this paper, was found to be approximately 30. For Reynolds number values larger than 30, a clear vortex forms downstream of the bend corner. This critical Reynolds number value is in line with other numerical work [1,2].

For Newtonian fluids, the Reynolds number is the single parameter required to describe the flow. It is expected that higher Reynolds numbers should lead to larger vortices downstream of the bend corner. We refer to these vortices forming downstream as inertial vortices from now on. Figure 4 presents two inertial vortices. The vortex in Figure 4a corresponds to a flow with a Reynolds number of 50, while Figure 4b corresponds to Re = 100.

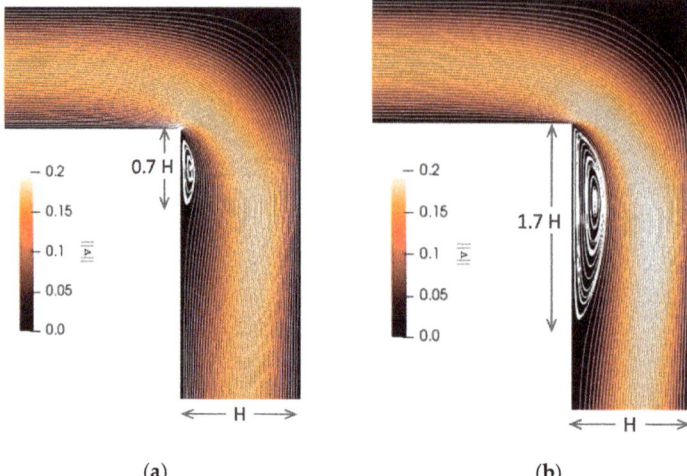

Figure 4. Downstream inertial vortices, Newtonian Flow: (**a**) Reynolds number Re = 50; (**b**) Reynolds number Re = 100.

Comparing both figures, it can be seen that the second vortex at Re = 100 is predicted to be significantly larger than the vortex that forms at Re = 50. This comparison shows that a higher Reynolds number value leads to a larger inertial vortex in Newtonian flow. It should be noted that the size (length) of the vortices in these and all figures that follow was determined numerically by the change in direction of the tangential velocity near the wall, equivalent to a change in sign of the shear stress (skin friction), see Appendix B.

To validate both, the statement that the inertial vortex size increases with Reynolds number and that 'RheoTool' is a valid numerical open source tool for the solution of both Newtonian and non-Newtonian flow, we modeled the bend Newtonian flow using the commercial tool ANSYS Fluent v19.1. Figure 5 compares the non-dimensional vortex length (vortex length to channel height ratio) predicted with both Fluent and Rheotool up to a Reynolds number of 250. The values predicted by RheoTool are within 1% of those predicted by Fluent. The figure clearly shows that the vortex length (l) increases with Reynolds number. Additionally, the comparison also indicates that RheoTool is as good as Fluent in solving the Navier–Stokes equations for Newtonian flow. It should be noted that this Newtonian inertial vortex would become unstable at a sufficiently high Reynolds number value [1].

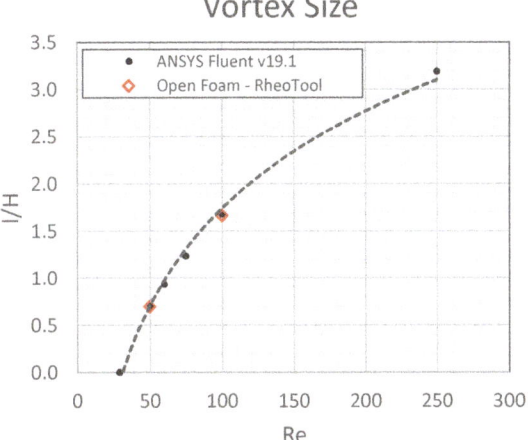

Figure 5. Non-dimensional inertial vortex length for Newtonian flow. The dashed line (logarithmic regression) is added to highlight the trend of the Newtonian vortex length solution.

3.2. Polymer Flow

As previously discussed, there are four dimensionless numbers—Wi, Re, El, and β—typically used to characterize polymer fluid dynamics. Nevertheless, the elasticity number (El) is defined as the ratio of the Weissenberg and Reynolds numbers. Consequently, only three dimensionless parameters are effectively needed to characterize polymer flow. Evaluating and predicting the effect on these parameters on polymer flow and vortex formation is the main goal of this research.

It has already been shown, for the Newtonian case, that a difference in Reynolds number means a difference in vortex size. For polymer fluids, even though elasticity is present, this should be no different. Holding the Weissenberg number constant comparing two simulations with different elasticity numbers should give insight into how the Reynolds number affects the vortices.

3.2.1. Inertial Vortices

The vortices in Figure 6 correspond to flows at different Reynolds numbers for the same Weissenberg number of 1.5 and a viscosity ratio of 0.1. Larger downstream vortices are predicted as the Reynolds number is increased. With both Wi and β held constant, it should be this variation in Reynolds number that is accountable for the increase in vortex size. It should be noted that increasing the Reynolds number while holding both Wi and β constant is equivalent to reducing the elasticity number.

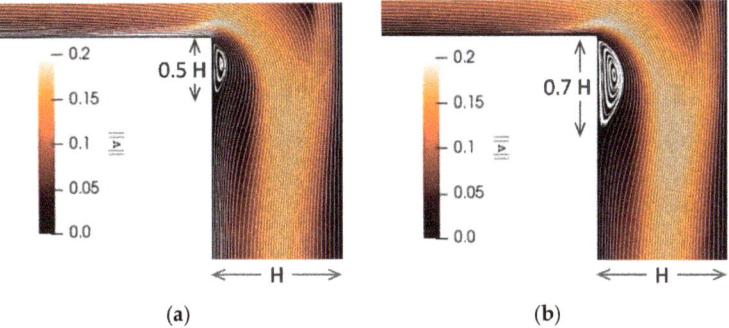

Figure 6. *Cont.*

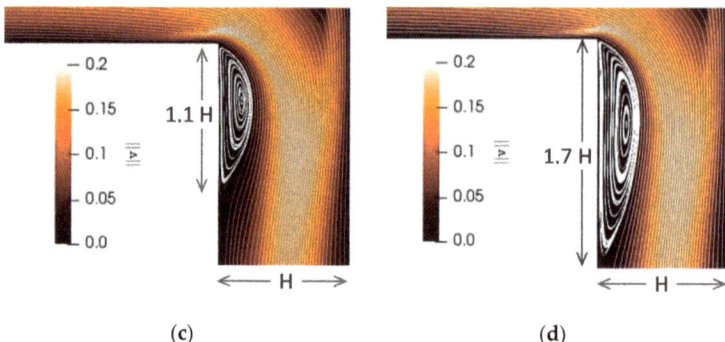

Figure 6. Downstream inertial vortices, Polymer Flow, Wi = 1.5 and β = 0.1: (**a**) Re = 50 (El = 0.03); (**b**) Re = 60 (El = 0.025); (**c**) Re = 75 (El = 0.02); (**d**) Re = 100 (El = 0.015).

It can then be stated that larger Reynolds numbers lead to larger inertial vortices in polymer flow. Conversely, reducing the Reynolds number by increasing the elasticity of the polymer solution should reduce the inertial vortex size.

Like Newtonian flow, current predictions indicate that larger Reynolds numbers in polymer flow should lead to larger inertial vortex sizes. This is consistent with laminar Newtonian flow physics, but what role do the polymers play in inertial vortex sizes? A way to evaluate the effect of the polymers on the flow would be to compare a Newtonian inertial vortex to a polymer vortex at the same Reynolds number.

Figure 7 compares predicted inertial vortices between Newtonian Flow and Polymer flow at the same Reynolds number but different Weissenberg numbers. The vortex in Figure 7a corresponds to Newtonian flow at a Reynolds number equal to 100. The vortex in Figure 7b corresponds to polymer flow with Wi = 0.5. (El = 0.005). The vortex in Figure 7c corresponds to polymer flow with Wi = 1.0 (El = 0.01). Additionally, the vortex in Figure 7d corresponds to a polymer flow with Wi = 1.5 (El = 0.015). All three polymer inertial vortices are predicted to be comparable in size, within 5%; with the one corresponding to a Weissenberg number of 1.0 being the only one significantly larger than its Newtonian counterpart, but only about 10%, just slightly above the average grid convergence error of 8%. This is an indication that adding polymers to the flow at this high Reynolds number is predicted to have a slight to nonexistent effect on the fluid dynamics of the flow around a bend. The Weissenberg number effect on inertial vortices at high Reynolds numbers is predicted to only be moderate.

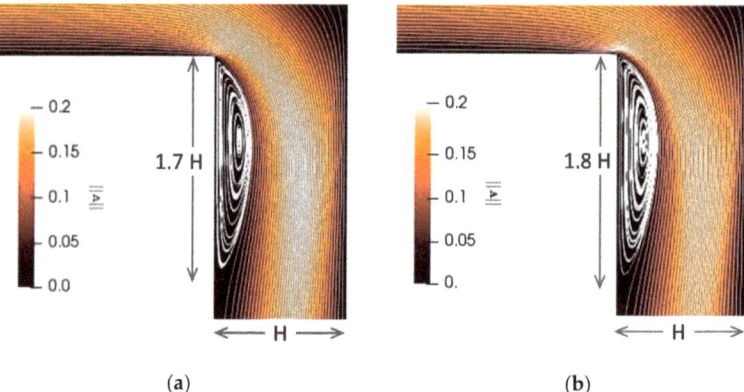

Figure 7. *Cont.*

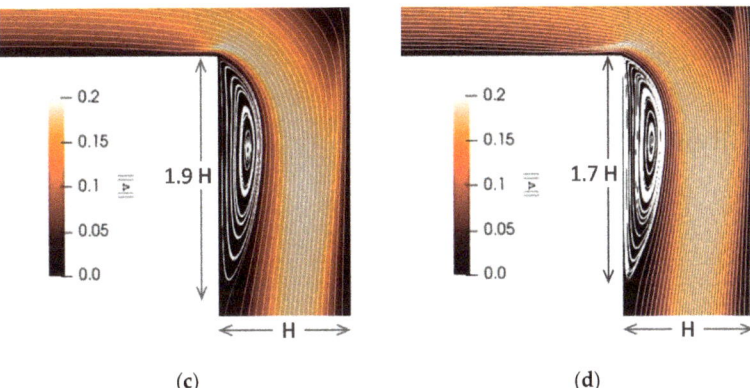

Figure 7. Downstream inertial vortices, Newtonian vs. polymer flow ($\beta = 0.1$.) for Re =100: (**a**) Newtonian flow; (**b**) polymer flow, (Wi = 0.5, El = 0.005); (**c**) polymer flow, (Wi = 1.0, El = 0.01); (**d**) polymer flow, (Wi = 1.5, El = 0.015).

The contour plots in Figure 8 show two inertial vortices both corresponding to a moderate Reynolds number of 50 also with $\beta = 0.1$. The first vortex, Figure 8a, has a Weissenberg number of 0.5, while the second, Figure 8b, corresponds to Wi = 1.0. Again, these inertial polymer vortices are comparable in size, in this case within 2%, indicating that the influence of the polymer in the size of the inertial vortex is also predicted to be moderate at a moderate Reynolds number.

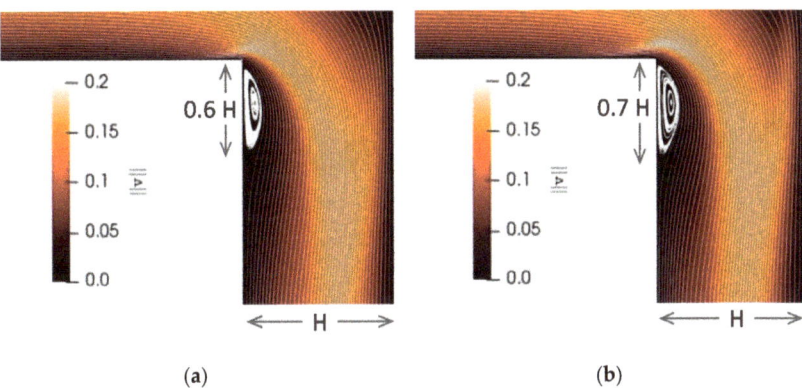

Figure 8. Downstream inertial vortices, Re = 50. Effect of Weissenberg Number (Polymer flow): (**a**) Wi = 0.5 and El = 0.01; (**b**) Wi = 1.0 and El = 0.02.

Figure 9 summarizes all inertial vortex length predictions for polymer flow at high Reynolds number. Both solutions for Newtonian flow, RheoTool and Fluent, are added for comparison. Predictions for Wi equal to 0.5 and 1.5 are similar to that of Newtonian Flow. At Wi = 1.5, polymer inertial vortices are predicted to be slightly smaller than the Newtonian counterpart as the Reynolds number is reduced.

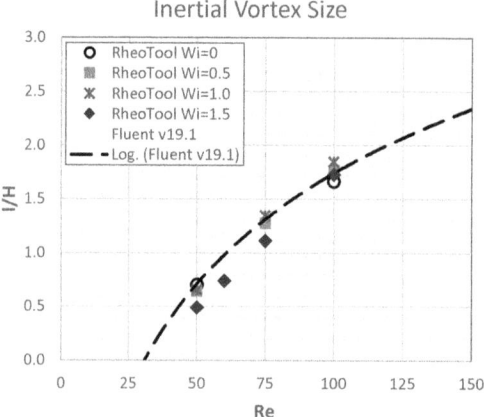

Figure 9. Non-dimensional inertial vortex length for polymer flow vs. Reynolds number.

From Figure 9 it could be inferred that adding polymers to the solution does not affect the size of the inertial vortices. Figure 10 presents the same data as Figure 9 but as a function of the elasticity number. At every Weissenberg number, the predictions clearly indicate that increasing elasticity leads to a rapid decrease in vortex size. This figure also highlights that it is the Reynolds number, indeed the inertial effects, that is predicted to drive the size of the inertial vortex, with the Weissenberg number playing a secondary role. This figure also seems to indicate that there is a maximum size for the inertial vortex at any given Reynolds number. This maximum vortex size is predicted for the cases where the inertial and elastic forces are balanced: Weissenberg equal to 1.0.

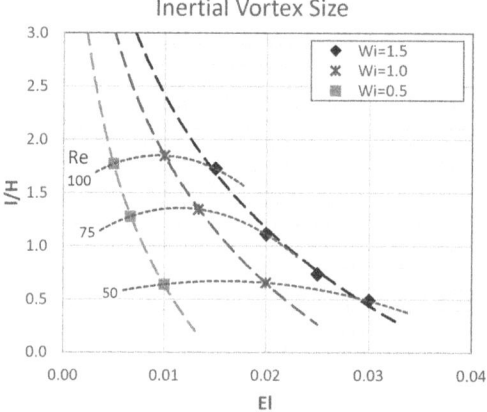

Figure 10. Non-dimensional inertial vortex length for polymer flow vs. elasticity.

3.2.2. Elastic Vortices

As the Reynolds number is reduced, the downstream or inertial vortices are predicted to disappear. However, as the Reynolds number is reduced further, vortices are predicted to form upstream of the corner bend. We refer to these vortices forming upstream of the bend corner as elastic vortices. At low Reynolds numbers, it is the Weissenberg number that is predicted to determine the size of the elastic vortex. To evaluate how changing the Weissenberg number affects vortices that form upstream of the bend corner, Figure 11 compares an elastic vortex for Wi = 1 and El = 1·10^6 to a vortex for Wi = 1.5 and El = 1.5·10^6. Both simulations correspond to a viscosity ratio of 0.1 and a Reynolds number of 1·10^{-6}. It is clear that the two vortices are predicted to be of different sizes, with the vortex for Wi = 1.5 being larger. Since the elasticity number and the Weissenberg number are linked, the vortices are predicted to be different in size, most likely due to the difference in Weissenberg numbers.

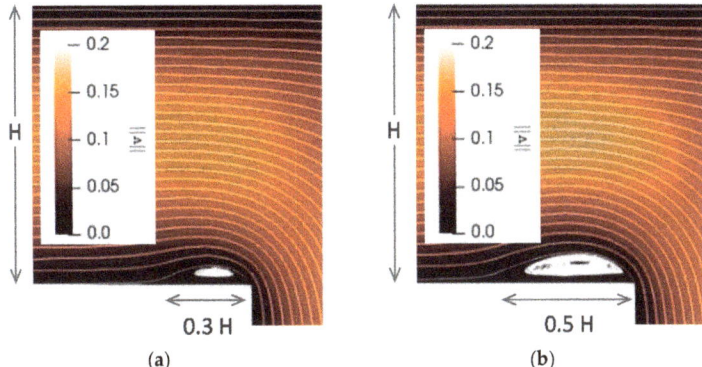

Figure 11. Upstream elastic vortices, effect of Weissenberg number (Polymer flow Re = 1×10^{-6}): (**a**) Wi = 1, El = 1×10^6; (**b**) Wi = 1.5, El = 1.5×10^6.

Lastly, we evaluate the effect of the viscosity ratio. As explained previously, this ratio can take any value between 0 and 1. Since a β value of 1 defines a purely Newtonian fluid, we expect that fluids with viscosity ratios less than 1 should exhibit polymer behavior similar to increasing the Weissenberg number in polymer flow. The elastic vortex should therefore be larger for smaller β values because of its reliance on elasticity to form.

The elastic vortices in Figure 12 show the effect of decreasing β; from β = 0.4 to β = 0.1. The second elastic vortex, lower viscosity ratio, is larger than expected, indeed, significantly larger, roughly three times. Reducing β is predicted to have a similar effect to that of increasing the Weissenberg number, increasing the size of the elastic vortex. However, the presence of elasticity is not sufficient to produce elastic vortices; it was found that elastic vortices are only predicted to form for viscosity ratios not exceeding a certain value. In other words, a dilute polymer solution, mostly solvent with a viscosity ratio near 1, is always predicted to have Newtonian-like flow behavior without a vortex forming upstream of the bend corner. Conversely, a concentrated polymer solution, mostly polymer with a viscosity ratio near 0, is predicted to have secondary flows, vortices upstream of the bend corner, for a large range of Reynolds numbers. These predictions are in line with the experiments of Gulati [5] who found that stable elastic vortices form within the flow of semidilute DNA solutions in a 90° micro bend channel above a certain threshold for the solution elasticity.

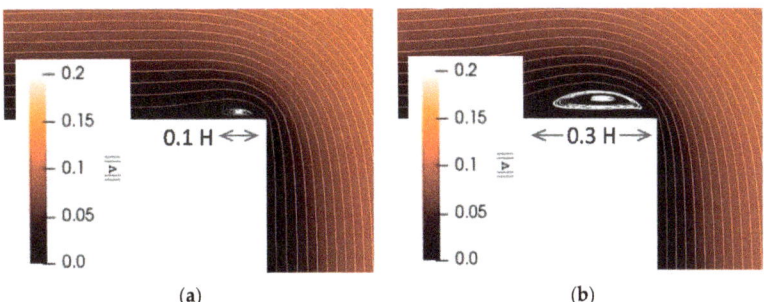

Figure 12. Upstream elastic vortices, effect of viscosity ratio (polymer flow) for Wi = 1 and El = $1 \cdot 10^6$: (**a**) elastic vortex for β = 0.4; (**b**) elastic vortex for β = 0.1.

Based on these findings we created a map to evaluate the effect of the three parameters, Wi, β, and El on the formation of elastic and inertial vortices. Figure 13 shows a phase diagram mapped for viscosity ratio versus elasticity number for Wi = 1.0. This value was chosen so the inverse of the elasticity number, El, is exactly the Reynolds number, Re. Three regions form in this phase diagram: the 'devoid region', no vortex forms; the 'inertia region', downstream or inertial vortices form; and the 'elastic region', upstream or elastic vortices form. As can be seen in the diagram, the boundary between the elastic region and the devoid region is a horizontal line at about β = 0.6. This is interesting because it suggests that there is a critical viscosity ratio needed for the elastic vortex to form. It might be possible that this boundary is not asymptotic, and that it is dependent on the Weissenberg and the Reynolds numbers. Take note of the logarithmic x-axis. The effect of the Weissenberg number on this predicted flow map will be the subject of a follow-up paper.

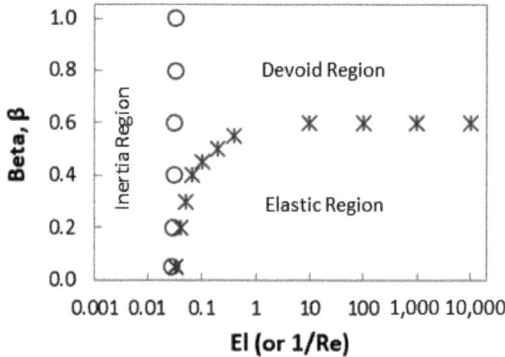

Figure 13. Phase space diagram, viscosity ratio versus elasticity number for Wi = 1.

The graph in Figure 14 is the same phase diagram as Figure 13, but the x-axis is now the Reynolds number (the inverse of the elasticity number for Wi = 1.0). In the figure, it is easier to see that elastic vortices only form at small Reynolds number values and viscosity ratios less than 0.6. As it has already been shown that viscosity ratio has a similar effect on vortex development to that of increasing the Weissenberg number, the question arises as to whether or not β will delay the onset of inertial vortices. Looking at the figure, the boundary of the inertial region is essentially a vertical line (Re = 29–37) indicating that β is predicted to have only a very subtle effect on the onset of inertial vortices. In other words, the Reynolds number is predicted to be the parameter with a primary role in determining the onset of inertial vortices.

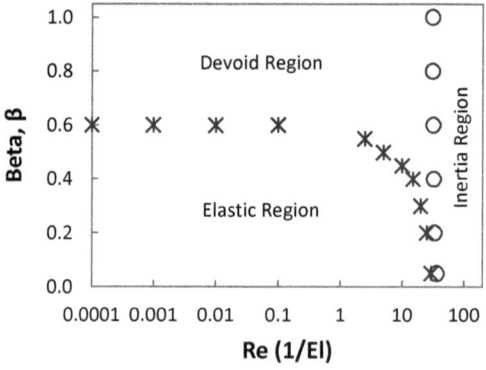

Figure 14. Phase space diagram, viscosity ratio versus Reynolds number for Wi = 1.

Lastly, Figure 15 plots the vortex length of elastic vortices versus the elasticity numbers at which they form for three values of Weissenberg number (Wi = 0.5, 1.0, 1.5) and a viscosity ratio $\beta = 0.1$. The figure shows that for elasticity numbers between 0.1 and 10 the size of the elastic vortex is predicted to depend on both the elasticity number, El, and the Weissenberg number, Wi. On the other hand, the figure also shows that for El = 10 and above, the elastic vortex size is predicted to be of the same size for a given Weissenberg number independently of the elasticity number. These predictions indicate that the Weissenberg number plays the primary role in determining the size of the elastic vortex that forms upstream of the bend corner for a fixed viscosity ratio. As discussed earlier, as the Weissenberg number increases so does the size of the elastic vortex. These predictions are also consistent with the experiments of Gulati [5] who found that the Weissenberg number is the parameter that determines the presence and size of the elastic vortices forming within the flow of semidilute DNA solutions in a 90° micro bend channel.

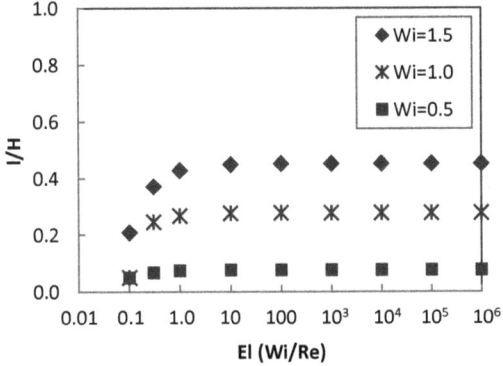

Figure 15. Non-dimensional elastic vortex length versus elasticity number for $\beta = 0.1$.

4. Discussion

It has been shown during this research paper that higher polymer concentrations and higher inertia are both predicted to lead to larger vortices in a sharp bend geometry. Higher polymer concentrations lead to elastic vortices located upstream of the bend. Conversely, higher Reynolds numbers lead to inertial vortices located downstream of the bend.

It was also found that after vortices are formed their size is predicted to be primarily determined by the properties of the polymer solution and the Reynolds number. In the case of elastic vortices, the vortex size is set by both the Weissenberg number and the viscosity ratio. In the case of inertial vortices, the vortex size is primarily set by the Reynolds number with the Weissenberg number playing a secondary role.

What is more interesting is the fact that predictions indicate vortices can be suppressed by adjusting the properties of the polymer solution, elasticity and viscosity ratio, for a given solution volumetric flow rate. The transition to inertia driven vortices is predicted to occur somewhat abruptly at elasticity numbers equivalent to the critical Reynolds number for Newtonian flow. Above this critical Reynolds number, inertia vortices are predicted to form downstream of the bend corner with polymer solution properties affecting the size only moderately. Below this critical Reynolds number, all polymer solution properties are predicted to affect not only the size but the formation of elastic vortices upstream of the bend corner. The size of the vortex is predicted to be determined by the Weissenberg number while the formation of the vortex itself is predicted to be determined by the viscosity ratio, i.e., the polymer concentration in the solution.

It should be noted that with our numerical approach we did not expect to predict or capture any secondary flow that might be induced or be present in the actual 3D geometry used as reference [5]. Similar works have found or predicted a minor effect of the aspect ratio of the channel in the size of the lip vortex [21] or the presence of secondary flows [28].

As the present work summarizes, our current primary interest was to evaluate the role that elasticity plays in the formation of lip vortices in the flow around a sharp corner. It is clear that modifying the underlying properties of the polymer solution predicts that the fluid dynamics can be significantly altered. Vortices, elastic and inertial, could both be suppressed or promoted by adjusting the polymer solution properties for a given volumetric flow situation.

5. Further Work

The phase diagrams presented in Figures 13 and 14 were created for a single Weissenberg number. The non-dimensional elastic vortex length behavior presented in Figure 15 was created for one single viscosity ratio. Additional diagrams for different values of both might confirm the universality of the predicted behavior.

In the present work, we used the FENE-P model to predict the behavior of the polymer solution. This model is computationally stiff. Extreme care has to be taken to find fully converged numerical solutions. Replicating the work with other polymer models would further verify the present findings.

In the long term, our goal is exploring the possibility of developing polymer and polymerlike solutions that respond to and change properties in a controlled manner based on self-induced flow instabilities. Ideally, developing an experimental setup where polymer flow could be tested to create a real phase diagram for the fluid dynamics would validate what the current work promises, controlling the formation of vortices based on the polymer solution's properties.

Author Contributions: B.W.: Software, Data Curation, Visualization. Writing—Initial Draft. J.L.: Software, Data Curation, Visualization. M.C.: Conceptualization, Methodology, Resources, Reviewing and Editing, Supervision. L.A.V.U.: Conceptualization, Validation, Formal Analysis, Investigation, Writing—Final Manuscript, Reviewing and Editing, Supervision, Project administration. All authors have read and agreed to the published version of the manuscript.

Funding: M.C. acknowledges the Donors of the American Chemical Society Petroleum Research Fund, for support of this research via Grant No. 56047-UNI9.

Institutional Review Board Statement: Not applicable.

Informed Consent Statement: Not applicable.

Conflicts of Interest: The authors declare no conflict of interest.

Appendix A

As discussed within the text, a 10:1 branches-to-channel height ratio for the geometry under analysis was found to be sufficient to predict fully developed flow before the bend corner. Figure A1 presents a comparison of the streamwise velocity component at two different locations. We are comparing profiles at a distance equivalent to 7H and 8H from the inlet.

Figure A1a presents the comparison for the same elastic vortex case discussed in the grid convergence analysis outlined in Figure 3; recall that an elastic vortex forms before the bend corner. The two profiles are nearly identical, indeed indistinguishable. The weighted average difference between the two profiles is 0.02%.

Figure A1b presents the comparison for the same inertial vortex case discussed in the grid convergence analysis section; recall that an inertial vortex forms after the bend corner. The two profiles are also nearly identical. The weighted average difference between the two profiles is 0.52%.

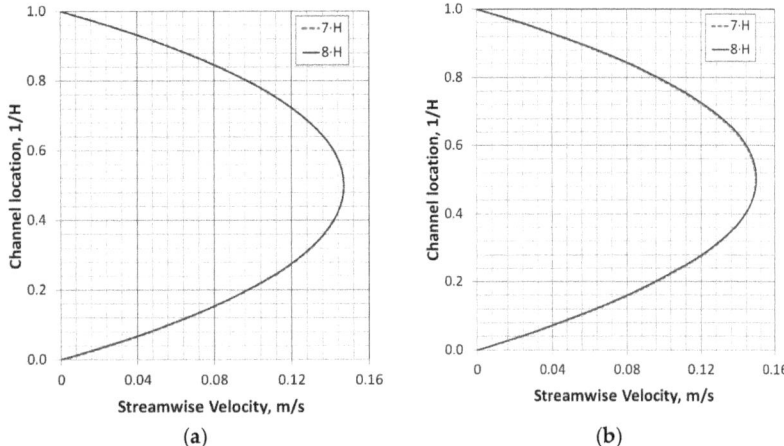

Figure A1. Streamwise velocity upstream of the corner bend: (**a**) elastic vortex for Wi = 1.0, El = 100, $\beta = 0.1$; (**b**) inertial vortex for Re = 100.

In both cases, it is clear the chosen length to height ratio is sufficient to obtain the fully developed flow condition before the bend corner. Hence, the inlet boundary condition should have no influence or effect on the size predictions of both elastic and inertial vortices.

As in Figure A1, Figure A2 presents a comparison of the streamwise velocity component at two different locations. We are now comparing profiles at a distance equivalent to 7H and 8H after the bend corner, or 3H and 2H far from the geometry outlet, respectively.

Figure A2a presents the comparison for the same elastic vortex case discussed above. The two profiles are nearly identical, again indistinguishable. The weighted average difference between the two profiles is 0.01%.

Figure A2b presents the comparison for the same inertial vortex case discussed above. The two profiles are again nearly identical. The weighted average difference between the two profiles is 0.65%.

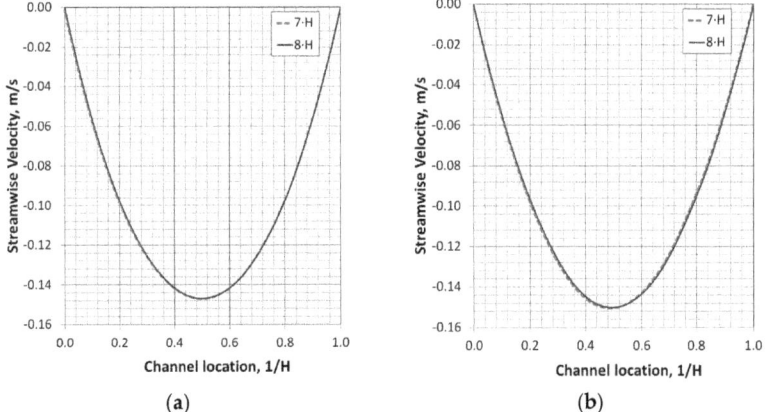

Figure A2. Streamwise velocity downstream of the corner bend: (**a**) elastic vortex for Wi = 1.0, El = 100, $\beta = 0.1$; (**b**) inertial vortex for Re = 100.

In both cases, it is clear the chosen length to height ratio is enough to predict that the flow will recover completely from the perturbation caused by the vortices before the geometry outlet, nearly reaching the fully developed flow condition. Hence, the outlet boundary condition should also have no influence or effect on the size predictions of either elastic or inertial vortices.

Appendix B

As discussed within the text, the size (length) of all predicted vortices was determined numerically by the change in direction of the tangential velocity near the wall. We indicated that doing this was equivalent to a change in sign of the shear stress or skin friction.

Figure A3 presents a comparison of the wall shear stress and the near wall streamwise (tangential) velocity. To allow a direct comparison, we normalized both stress and velocity by the maximum value along the respective walls.

Figure A3a presents the comparison for the same elastic vortex case discussed in Appendix A. The two profiles cross the zero line at nearly the same x-coordinate location. The length of the vortex determined by either method is the same with a 0.0% deviation.

Figure A3b presents the comparison for the same inertial vortex case discussed in Appendix A. The two profiles cross the zero line at almost the same y-coordinate location. The length of the vortex determined by either method is roughly the same, with a 0.5% deviation.

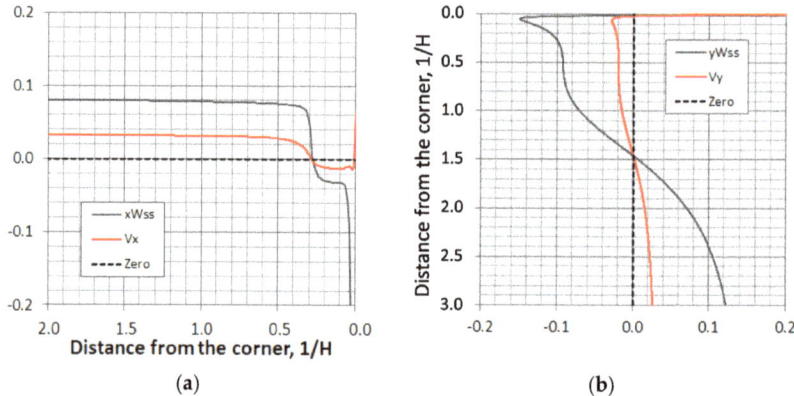

Figure A3. Normalized wall shear stress and streamwise velocity near the wall in the vortex region: (a) elastic vortex (xWss and Vx) for Wi = 1.0, El = 100, β = 0.1; (b) inertial vortex (yWss and Vy) for Re = 100.

In both cases, elastic and inertial vortices, it is clear that determining the vortex size by the change in direction of the tangential velocity near the wall is equivalent to a change in the sign of the wall shear stress. Any deviation between the methods that could be predicted is extremely small compared to the actual vortex sizes that were calculated.

References

1. Matsumoto, D.; Fukudome, K.; Wada, H. Two-dimensional fluid dynamics in a sharply bent channel: Laminar flow, separation bubble, and vortex dynamics. *Phys. Fluids* **2016**, *28*, 103602. [CrossRef]
2. Hayes, R.E.; Nandakumar, K.; Nasr-El-Din, H. Steady laminar flow in a 90 degree planar branch. *Comput. Fluids* **1989**, *17*, 537–553. [CrossRef]
3. Renardy, M. Current issues in non-Newtonian flows: A mathematical perspective. *J. Non-Newton. Fluid Mech.* **2000**, *90*, 243–259. [CrossRef]
4. Zhou, J.; Papautsky, I. Viscoelastic microfluidics: Progress and challenges. *Microsyst. Nanoeng.* **2020**, *6*, 113. [CrossRef]
5. Gulati, S.; Liepmann, D.; Muller, S. Elastic secondary flows of semidilute DNA solutions in abrupt 90° microbends. *Phys. Rev.* **2008**, *78*, 036314. [CrossRef] [PubMed]
6. Kim, J.; Hong, S.O.; Shimab, T.S.; Kim, J.M. Inertio-elastic flow instabilities in a 90° bent microchannel. *Soft Matter* **2017**, *13*, 5656. [CrossRef]
7. Hadri, F.; Guillou, S. Drag reduction by surfactant in closed turbulent flow. *Int. J. Eng. Sci. Technol.* **2010**, *2*, 6876–6879.

8. Munekata, M.; Matsuzaki, K.; Ohba, H. A Study on Viscoelastic Fluid Flow in a Square-Section 90-Degrees Bend. *J. Therm. Sci.* **2003**, *12*, 337–343. [CrossRef]
9. Keunings, R. On the Peterlin approximation for finitely extensible dumbbells. *J. Non-Newton. Fluid Mech.* **1997**, *68*, 85–100. [CrossRef]
10. Bird, R.B.; Curtiss, C.F.; Armstrong, R.C.; Hassager, O. Elastic Dumbbell Models. In *Dynamics of Polymeric Liquids*, 2nd ed.; John Wiley & Sons: New York, NY, USA, 1987; Volume 2.
11. Morrison, F.A. Newtonian Fluid Mechanics. In *Understanding Rheology*; Oxford University Press: New York, NY, USA, 2001.
12. Bird, R.B.; Armstrong, R.C.; Hassager, O. Differential Constitutive Equations. In *Dynamic of Polymeric Liquids*, 2nd ed.; John Wiley & Sons: New York, NY, USA, 1987; Volume 1.
13. RheoTOOL User Guide. Available online: https://devhub.io/repos/fppimenta-rheoTool (accessed on 20 May 2020).
14. Favero, J.L.; Secchi, A.R.; Cardozo, N.S.M.; Jasak, H. Viscoelastic flow analysis using the software OpenFOAM and differential constitutive equations. *J. Non-Newton. Fluid Mech.* **2010**, *165*, 1625–1636. [CrossRef]
15. Pimenta, F.; Alves, M.A. Stabilization of an open-source finite-volume solver for viscoelastic flows. *J. Non-Newton. Fluid Mech.* **2017**, *239*, 85–104. [CrossRef]
16. Fattala, R.; Kupferman, R. Constitutive laws for the matrix-logarithm of the conformation tensor. *J. Non-Newton. Fluid Mech.* **2004**, *123*, 281–285. [CrossRef]
17. Afonso, A.; Oliveira, P.J.; Pinho, F.T.; Alves, M.A. The log-conformation tensor approach in the finite-volume method framework. *J. Non-Newton. Fluid Mech.* **2009**, *157*, 55–65. [CrossRef]
18. Alves, M.A.; Oliveira, P.J.; Pinho, F.T. A convergent and universally bounded interpolation scheme for the treatment of advection. *Int. J. Num. Methods Fluids* **2003**, *41*, 47–75. [CrossRef]
19. Kalb, A.; Villasmil, U.; Larry, A.; Cromer, M. Role of chain scission in cross-slot flow of wormlike micellar solutions. *Phys. Rev. Fluids* **2017**, *2*, 071301. [CrossRef]
20. Kalb, A.; Villasmil-Urdaneta, L.A.; Cromer, M. Elastic instability and secondary flow in cross-slot flow of wormlike micellar solutions. *J. Non-Newton. Fluid Mech.* **2018**, *262*, 79–91. [CrossRef]
21. Hwang, M.Y.; Mohammadigoushki, H.; Muller, S.J. Flow of viscoelastic fluids around a sharp microfluidic bend: Role of wormlike micellar structure. *Phys. Rev. Fluids* **2017**, *2*, 043303. [CrossRef]
22. Xiong, R.; Chung, J.N. Effects of miter bend on pressure drop and flow structure in micro-fluidic channels. *Int. J. Heat Mass Transf.* **2008**, *51*, 2914–2924. [CrossRef]
23. Celik, I.B.; Ghia, U.; Roache, P.J.; Freitas, C.J.; Coleman, H.; Raad, P.E. Procedure for Estimation and Reporting of Uncertainty Due to Discretization in CFD Applications. *J. Fluids Eng.* **2008**, *130*, 078001.
24. Fernandes, C.; Araújo, M.; Ferrás, L.L.; Nóbrega, J.M. Improved Both Sides Diffusion (iBSD): A new and straightforward stabilization approach for viscoelastic fluid flows. *J. Non-Newton. Fluid Mech.* **2017**, *249*, 63–78. [CrossRef]
25. Ferrás, L.L.; Afonso, A.M.; Alves, M.A.; Nóbrega, J.M.; Carneiro, O.S.; Pinho, F.T. Slip flows of Newtonian and viscoelastic fluids in a 4:1 contraction. *J. Non-Newt. Fluid Mech.* **2014**, *214*, 28–37. [CrossRef]
26. Alves, M.A.; Oliveira, P.J.; Pinho, F.T. Benchmark solutions for the flow of Oldroyd-B and PTT fluids in planar contractions. *J. Non-Newton. Fluid Mech.* **2003**, *110*, 45–75. [CrossRef]
27. Matsui, T.; Hiramatsu, M.; Hanaki, M. Separation of Low Reynolds Number Flows Around a Corner. *Symp. Turbul. Liq.* **1975**, *3*, 283–288.
28. Poole, R.J.; Lindner, A.; Alves, M.A. Viscoelastic secondary flows in serpentine channels. *J. Non-Newton. Fluid Mech.* **2013**, *201*, 10–16. [CrossRef]

Article

Effects of Viscoelasticity on the Stress Evolution over the Lifetime of Filament-Wound Composite Flywheel Rotors for Energy Storage

Miles Skinner and Pierre Mertiny *

Department of Mechanical Engineering, University of Alberta, 9211-116 St., Edmonton, AB T6G 1H9, Canada; maskinne@ualberta.ca
* Correspondence: pmertiny@ualberta.ca

Abstract: High-velocity and long-lifetime operating conditions of modern high-speed energy storage flywheel rotors may create the necessary conditions for failure modes not included in current quasi-static failure analyses. In the present study, a computational algorithm based on an accepted analytical model was developed to investigate the viscoelastic behavior of carbon fiber reinforced polymer composite flywheel rotors with an aluminum hub assembled via a press-fit. The Tsai-Wu failure criterion was applied to assess failure. Two simulation cases were developed to explore the effects of viscoelasticity on composite flywheel rotors, i.e., a worst-case operating condition and a case akin to realistic flywheel operations. The simulations indicate that viscoelastic effects are likely to reduce peak stresses in the composite rim over time. However, viscoelasticity also affects stresses in the hub and the hub-rim interface in ways that may cause rotor failure. It was further found that charge-discharge cycles of the flywheel energy storage device may create significant fatigue loading conditions. It was therefore concluded that the design of composite flywheel rotors should include viscoelastic and fatigue analyses to ensure safe operation.

Keywords: viscoelasticity; polymer composite material; flywheel energy storage system; flywheel failure; Tsai-Wu criterion

Citation: Skinner, M.; Mertiny, P. Effects of Viscoelasticity on the Stress Evolution over the Lifetime of Filament-Wound Composite Flywheel Rotors for Energy Storage. *Appl. Sci.* **2021**, *11*, 9544. https://doi.org/10.3390/app11209544

Academic Editor: Luís L. Ferrás

Received: 21 September 2021
Accepted: 12 October 2021
Published: 14 October 2021

Publisher's Note: MDPI stays neutral with regard to jurisdictional claims in published maps and institutional affiliations.

Copyright: © 2021 by the authors. Licensee MDPI, Basel, Switzerland. This article is an open access article distributed under the terms and conditions of the Creative Commons Attribution (CC BY) license (https://creativecommons.org/licenses/by/4.0/).

1. Introduction

State-of-the-art high-speed flywheel energy storage systems (FESS) are recognized for several advantageous characteristics including a high charge and discharge rate, lifetimes ranging from 10 to 20 years and high specific energy up to 100 Wh·kg^{-1} [1]. Further, they are unaffected by depth of discharge or cycling effects common to electrochemical batteries and have a relatively high cycle efficiency—up to 95% depending on the electrical components [2]. While the high efficiency and long expected lifetime make FESS an attractive alternative over other short- and medium-term energy storage options, these same attributes pose design and operational challenges.

The majority of studies on fiber reinforced polymer (FRP) composite flywheel rotors have focused on instantaneous, or time-independent, behavior of composite rotors and hubs to optimize performance or minimize cost [3–5]. If rotor failure is considered, it is typically seen as a quasi-static process caused by excessive centrifugal loading exceeding material ultimate strengths [6]. While attempts have been made to predict rotor failure with progressive damage models [7], they largely neglect to incorporate viscoelasticity into the stress and failure analyses. It has been theorized that changes in the interfacial compressive forces could lead to rim separation or creep rupture [8], yet the number of studies on viscoelastic behavior in composite rotors supporting this notion are limited.

Some works presented solutions for the boundary-value problem presented by flywheel rotors constructed of viscoelastic materials and discussed creep effects [9,10]. Trufanov and Smetannikov [9] focused on flywheel rotors with an outer shell supporting an inner composite rim. Additionally, the rim is of non-uniform cross sections and features a variable winding angle, neither of which are commonly used in modern FESS [11,12]. Portnov [10] discussed a solution to the equilibrium equations to determine creep strain in rotating disks. Tzeng [13] expanded on previous works by simulating filament-wound composite flywheel rotors with uniform rotor cross section and discussing viscoelastic behavior at 10 years and infinite time (10^{10} h). Tzeng showed that viscoelastic stress relaxation of approximately 35% in the radial direction and a corresponding increase of approximately 9% in the circumferential direction can occur over the lifetime of the rotor. Emerson [14] conducted experimental investigations on flywheel rotors subjected to three temperatures and speed profiles over the course of 2 months using optoelectronic strain measurements. While rotor creep tests were inconclusive due to a mechanical failure, this work did not rule out any significant impact of creep on strains imposed for the press-fit assembly of the rotor. A similar conclusion was found elsewhere [15].

While there have been developments in the understanding of viscoelastic behavior of flywheel rotors, related insights do not necessarily translate well to typical use cases. For example, FESS in public transit [16] are installed with a vacuum enclosure that minimizes temperature fluctuations. Expected lifetimes are 10 to 20 years. In addition, load cycling occurs every few minutes, with viscoelastic effects effectively being negligible in between cycles. For other promising FESS applications, such as electric vehicle (EV) charging and renewable energy grid support, cycle times are likely much longer than for FESS in public transit, yet temperature conditions and timeframes would be similar.

The present study seeks to describe the viscoelastic behavior of composite flywheel rotors during their expected lifetimes using a computational algorithm to predict the stress evolution in the rotor. Additionally, the Tsai-Wu criterion is used to describe the total stress state, combining radial, circumferential, and axial stress to predict rotor failure. The simulated rotor material is a filament-wound carbon fiber reinforced polymer (CFRP) composite [17–19], similar to those typically used in flywheel rotor construction, making its application here appropriate. The rotor also includes an aluminum hub that facilitates the connection between the motor/generator unit and the bearing system. The effects of creep and viscoelastic stress relaxation on a flywheel rotor are examined with respect to two primary rotor failure modes: (i) separation between hub and rotor rims, and (ii) matrix cracking.

2. Composite Flywheel Rotor Modeling

2.1. Analytical Model Description

The analytical model has been discussed in several publications, therefore only a brief description will be provided here. While the present study focuses on the solution of a single-rim rotor, the analysis can be generalized to multi-rim rotors as described in [20], variable thickness rotors [21], and functionally graded materials [22]. The stress development in the thick composite rotor is assumed to be axisymmetric, meaning the resultant stresses and strains are independent of the circumferential coordinate. The material used for these rotors is a unidirectional filament-wound FRP composite where the winding angle is taken to be circumferential, i.e., 90°. Hence, the composite is assumed to be transversely isotropic. Additionally, it was assumed the aluminum hub and composite rim are permanently bonded, that is, the model is unable to simulate separation between hub and rim. However, the latter condition is indicated by interfacial radial stress being greater than or equal to zero. Due to axisymmetry, the rotor response must only satisfy the governing equation in the radial direction [23]. The stress equilibrium equation in cylindrical coordinates is given as [24]

$$\frac{\partial \sigma_r}{\partial r} + \frac{\sigma_r - \sigma_\theta}{r} + \rho r \omega^2 = 0, \tag{1}$$

where σ_r and σ_θ are the radial and circumferential hoop stresses, which are the only nontrivial terms in the stress matrix; ρ is the material density; and ω is the rotor angular velocity. The stress-strain relationship is defined as

$$\begin{Bmatrix} \sigma_\theta \\ \sigma_z \\ \sigma_r \\ \sigma_{\theta z} \end{Bmatrix} = \begin{bmatrix} Q_{11} & Q_{12} & Q_{13} & 0 \\ Q_{21} & Q_{22} & Q_{23} & 0 \\ Q_{31} & Q_{32} & Q_{33} & 0 \\ 0 & 0 & 0 & Q_{66} \end{bmatrix} \begin{Bmatrix} \varepsilon_\theta \\ \varepsilon_z \\ \varepsilon_r \\ \varepsilon_{\theta z} \end{Bmatrix}, \qquad (2)$$

where $[Q]$ is the stiffness matrix and $\{\varepsilon\}$ is the strain vector. Note that the z-coordinate is associated with the rotor axial direction. The $[Q]$ matrix is the inverse of the compliance matrix $[S]$ [25], such that

$$[Q] = [S]^{-1} = \begin{bmatrix} S_{11} & S_{12} & S_{13} & 0 \\ S_{21} & S_{22} & S_{23} & 0 \\ S_{31} & S_{32} & S_{33} & 0 \\ 0 & 0 & 0 & S_{66} \end{bmatrix}^{-1}. \qquad (3)$$

Considering Equation (2), the compliance matrix must define the behavior in the circumferential (parallel to fibers), radial (transverse to fibers), axial (transverse to fibers), and shear directions. Since the rotor material is assumed to be transversely isotropic with no applied shear forces, the symmetric matrix simplifies from 10 unique terms to seven. The strain in the circumferential and radial directions can be written as, respectively,

$$\varepsilon_\theta = \frac{u_r}{r} \text{ and } \varepsilon_r = \frac{\partial u_r}{\partial r}, \qquad (4)$$

where u_r is the displacement in the radial direction and r is an arbitrary location along the rotor radial direction. Invoking a plane strain assumption, strain in the axial and shear directions is defined correspondingly by Equation (5). The appropriateness of this assumption will be discussed later in this text.

$$\varepsilon_z = 0 \text{ and } \varepsilon_{\theta z} = 0. \qquad (5)$$

Combining Equations (1), (2), and (4) yields a second order inhomogeneous ordinary differential equation. Solving this equation gives the local displacement and local stress at an arbitrary radius defined as

$$u_r = -\rho \omega^2 \varphi_0 r^3 + C_1 \varphi_1 r^\kappa + C_2 \varphi_2 r^{-\kappa}, \qquad (6)$$

$$\sigma_r = -\rho \omega^2 \varphi_3 r^2 + C_1 r^{\kappa-1} + C_2 r^{-\kappa-1}. \qquad (7)$$

The C parameters are integration constants dependent on the boundary conditions and material properties. The κ and φ coefficients are intermediate terms dependent on the stiffness matrix, defined as follows:

$$\begin{gathered} \kappa = \sqrt{\frac{Q_{11}}{Q_{33}}}, \\ \varphi_0 = \frac{1}{(9-\kappa^2)Q_{33}}, \ \varphi_1 = \frac{1}{Q_{13}+\kappa Q_{33}}, \\ \varphi_2 = \frac{1}{Q_{13}-\kappa Q_{33}}, \ \varphi_3 = \frac{3Q_{33}+Q_{13}}{(9-\kappa^2)Q_{33}}. \end{gathered} \qquad (8)$$

Then, upon determining the integration constants, the radial displacement (Equation (6)) and radial stress (Equation (7)) can be found using Equation (8). Circumferential stress can be found by combining Equations (4), (6), and (8) in conjunction with the stress-strain relationship (Equation (2)).

Generalizing to a multi-rim flywheel rotor with an arbitrary number of rims, i.e., the rotor is constructed from N rims labeled j and can vary between $j = 1, 2, 3 \ldots N$, then the continuity condition at the interface states,

$$\sigma^j_{r,r_o} = \sigma^{j+1}_{r,r_i}; \quad u^j_{r,r_o} = u^{j+1}_{r,r_i}. \tag{9}$$

where σ^j_{r,r_o} is the radial stress at the outer radius, r_o, in the jth rim, and σ^{j+1}_{r,r_i} is the radial stress at the inner radius, r_i, of the next, $j + 1$, rim. The same notation is used to describe the radial displacements, u_r, at the interface.

2.2. Tsai-Wu Failure Criterion

The general Tsai-Wu failure criterion, described in [6,7,26,27], can be reduced to nine terms for a transversely isotropic material and considering the absence of shear stresses. This criterion finds a relationship, F, between the applied stress tensor and the material tensile strengths and predicts failure when $F \geq 1$. At failure, the stress tensor represents the maximum allowable stress and F equals unity. Hence, the Tsai-Wu failure criterion can be written as

$$F = F_{11}\left(\sigma_1^{all}\right)^2 + F_{22}\left(\sigma_2^{all}\right)^2 + F_{33}\left(\sigma_3^{all}\right)^2 + 2F_{12}\sigma_1^{all}\sigma_2^{all} + 2F_{13}\sigma_1^{all}\sigma_3^{all} + \\ 2F_{23}\sigma_2^{all}\sigma_3^{all} + F_1\sigma_1^{all} + F_2\sigma_2^{all} + F_3\sigma_3^{all} = 1, \tag{10}$$

where $\left(\sigma_i^{all}\right)$ is the allowable stress in the i = 1, 2, or 3 directions at an arbitrary point in the rotor. Note as applied herein, the 1 and 3 directions refer to the circumferential and radial stress, respectively, while the 2 direction refers to the axial direction. While the plane strain condition eliminates axial strain, it allows for axial stress; therefore, it is included in the failure criterion. Then,

$$F_{11} = \tfrac{1}{\sigma_{1t}\sigma_{1c}},\; F_{22} = F_{33} = \tfrac{1}{\sigma_{3t}\sigma_{3c}},\; F_1 = \tfrac{1}{\sigma_{1t}} - \tfrac{1}{\sigma_{1c}}, \\ F_2 = F_3 = \tfrac{1}{\sigma_{3t}} - \tfrac{1}{\sigma_{3c}},\; F_{12} = F_{13} = \tfrac{-1}{2\sqrt{\sigma_{1t}\sigma_{1c}\sigma_{3t}\sigma_{3c}}},\; F_{23} = F_{22} - \tfrac{1}{2\tau_{23}^2}, \tag{11}$$

where the subscripts t and c refer to the tensile and compressive ultimate strengths, respectively. The strength coefficients in the 2 and 3 directions are equal due to the transversely isotropic assumption, discussed further in Section 3.1.

It is common, and more valuable, to define the relationship between maximum allowable stress and the applied stress as the failure ratio (R) [7]. This relationship is found by combining the maximum allowable stress tensor, $\left(\sigma_i^{all}\right)$, with the applied stress tensor, $\left(\sigma_i^{app}\right)$, multiplied with R such that

$$\sigma_1^{all} = R\sigma_1^{app},\; \sigma_2^{all} = R\sigma_2^{app},\; \sigma_3^{all} = R\sigma_3^{app}. \tag{12}$$

Then, substituting Equation (12) into Equation (10) yields a quadratic equation, i.e.,

$$0 = \left[F_{11}\left(\sigma_1^{app}\right)^2 + F_{22}\left(\sigma_2^{app}\right)^2 + F_{33}\left(\sigma_3^{app}\right)^2 + 2F_{12}\sigma_1^{app}\sigma_2^{app} + 2F_{13}\sigma_1^{app}\sigma_3^{app} \right. \\ \left. + 2F_{23}\sigma_2^{app}\sigma_3^{app}\right]R^2 + \left[F_1\sigma_1^{app} + F_2\sigma_2^{app} + F_3\sigma_3^{app}\right]R - 1. \tag{13}$$

Solving this quadratic equation for R defines the failure ratio. When $\left(\sigma_i^{app}\right)$ equals $\left(\sigma_i^{all}\right)$, then R equals unity, indicating failure, whereas $R > 1$ indicates $\left(\sigma_i^{app}\right)$ is less than $\left(\sigma_i^{all}\right)$ and no failure is predicted. It is convenient to define a strength ratio (SR) to be $1/R$, as this is more intuitive conceptually and graphically [7]. Failure under this criterion is predicted when $SR \geq 1$.

2.3. Computational Methodology

The computational methodology has been discussed elsewhere [13,14], so only a brief description is provided here. The analytical model, described in Section 2.1, assumes constant loads, therefore the viscoelastic solution procedure requires approximating time-varying behavior through a number of discrete time and load steps. The response at each step is used to calculate stresses and SR for the flywheel throughout the simulation. First, the rotor dimensions, material properties, and simulation parameters—including time and velocity vectors of interest—are defined as inputs to the algorithm. Then, beginning at the first time and velocity of interest, the material stiffness matrix is calculated for each rim of the flywheel rotor (here, a single-rim rotor is considered). Next, the boundary conditions at each interface and at the inner and outer surface of the rotor are calculated. Using this information, the rotor response and SR are calculated for the given time and velocity. Finally, the algorithm iterates to the next time and velocity. This continues for all discrete times and velocities of interest, which yields the induced stresses and SR for all points in the flywheel rotor at all times and velocities of interest.

3. Modeling Parameters

The flywheel rotor simulated in this study is constructed from a single CFRP rim press-fitted to an aluminum hub. The hub and rim are simulated as cylinders with rectangular cross sections.

3.1. Materials

The aluminum and CFRP are both assumed to be uniform throughout the rotor and free of defects. Referring to [28], the aluminum exhibits negligible viscoelastic response at temperatures below 50 °C, therefore viscoelastic behavior in the aluminum is not considered. The material properties of the chosen 7075-T6 aluminum are found in [29].

The composite considered in this study is IM7 carbon fiber (Hexcel Corp., Stamford, CT, USA) with an 8552 epoxy resin system (Hexcel Corp., Stamford, CT, USA), as described by Tzeng et al. [30]. The filament winding process employed for fabricating CFRP flywheel rotors utilizes continuous unidirectional fiber reinforcement, which creates a transversely isotropic behavior [31].

In the CFRP rim, the fibers run circumferentially and display only subtle viscoelastic characteristics. The long-term behavior of a CFRP in the transverse direction is often described using a time-temperature superposition (TTSP) master curve. This curve is created by measuring short-term creep data at various elevated temperatures. Then, a shift factor is applied to the elevated temperature experimental data to shift them temporally, increasing the time axis while decreasing the temperature. Shifting all elevated temperature experimental data creates a smooth master curve representing the lifetime strain and compliance behavior, provided the applied stress from experimentation is known. Finally, curve fitting is performed on the master curve to generate empirical equations for the creep compliance. TTSP is applicable for modeling linear viscoelasticity, which is acceptable for this application as permanent damage, material aging, and other higher order effects are excluded from this simulation. The transverse compliance equations published by Tzeng et al. [30] are given in Table 1 and Equation (14), as are the aluminum properties, where variable t indicates time. The tensile strengths of the CFRP and yield strength of the aluminum necessary for the Tsai-Wu criteria are given in Table 2.

$$S_{11}^0 = 9.0 \times 10^{-12} \text{ Pa}^{-1}, \ S_{22}^0 = S_{33}^0 = 1.1 \times 10^{-10} \text{ Pa}^{-1}, \ S_{66}^0 = 2.0 \times 10^{-10} \text{ Pa}^{-1}. \quad (14)$$

Table 1. Material properties for aluminum 7075-T6 [29] and viscoelastic equations for CFRP [30] used in the present study.

Material	S_{11} [Pa^{-1}]	S_{22} [Pa^{-1}]	S_{33} [Pa^{-1}]	S_{66} [Pa^{-1}]	ν
Aluminum	1.39×10^{-11}	1.39×10^{-11}	1.39×10^{-11}	3.72×10^{-11}	0.33
CFRP	$S_{11}^0(t)^{0.01}$	$S_{22}^0(t)^{0.03}$	$S_{33}^0(t)^{0.03}$	$S_{66}^0(t)^{0.03}$	0.31

Table 2. Directionally dependent strengths of CFRP and yield strength of aluminum used to find SR from the Tsai-Wu failure criterion.

Material	σ_{1t} [MPa]	σ_{1c} [MPa]	σ_{3t} [MPa]	σ_{3c} [MPa]	τ [MPa]
CFRP	2720	1689	64.1	307	137
Aluminum	572	572	572	572	331

3.2. Flywheel Rotor Simulation Parameters

Two cases were considered to investigate the effects of viscoelastic behavior on the flywheel rotor. The first case simulates a worst-case scenario for creep and viscoelastic stress relaxation in the flywheel rotor. The second case more closely simulates a realistic scenario of an FESS experiencing daily charge/discharge cycles. The FESS capacity and flywheel rotor dimensions are identical between the two cases. Recent studies on appropriate sizing of FESS have identified various values ranging between 3 kWh and 20 kWh for residential applications, light rail transit, electric vehicle charging, and frequency regulation for microgrid applications [16,31,32]. For the present study, a capacity of 10 kWh was chosen as it is situated in the middle of the range for the applications mentioned above. Note that energy storage capacity scales linearly with rotor height (axial dimension), and scaling is not expected to affect creep behavior appreciably, so the chosen rotor configuration can easily be scaled up or down to adjust for a given application. This scaling could be done, as suggested in [33], by stacking individual composite disks on top of one another to form the rotor, in which case, the analysis for each individual disk is performed as described herein while capacity may be increased or decreased as needed. To illustrate the chosen capacity, a recent study [34] on residential photovoltaic (PV) potential in Lethbridge, Alberta, Canada, identified that the majority of residential homes had roof space for up to 10 kW of solar PV, meaning the FESS in this study could reasonably be expected to reach full capacity throughout the day even under less than ideal irradiation conditions, in order to provide power during high demand times such as in the evening. The simulated flywheel rotor dimensions and energy capacity used in this study are given in Figure 1 and Table 3. Note that changing power demand would necessarily require accelerating or decelerating the flywheel rotor, imposing shear stresses, which is not included in the current model, hence justifying the aforementioned biaxial stress condition.

Table 3. Flywheel rotor rim dimensions, press-fit interference, and energy capacity.

Parameter	Aluminum Hub	CFRP Rim	Complete Rotor
Inner radius	160 mm	200 mm	-
Outer radius	200 mm	330 mm	-
Press-fit interference	-	-	0.8 mm
Rotor height	-	-	430 mm
Energy capacity	-	-	11.19 kWh

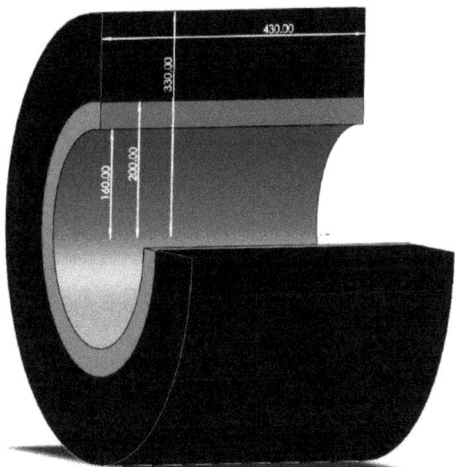

Figure 1. Simulated flywheel rotor showing dimensions of metallic hub and carbon fiber rim. The axis of rotation (AoR) is shown in blue.

Case 1: The worst-case scenario for viscoelastic effects is simulated by assuming the flywheel rotor to operate at its limit load at all times. The model was used to simulate 10 years of operation. Based on the Tsai-Wu failure criterion, a critical or limit velocity, ω_{limit}, was determined as 24,250 rpm. Note that failure is initially indicated at this velocity, for an SR lightly above unity (1.01); however, viscoelastic stress relaxation will improve the stress state, allowing for safe operation at this velocity over the long term.

Case 2: This scenario is intended to more realistically simulate the application of an FESS in solar PV electric grids or EV charging support. For these applications, the FESS is assumed to experience a single charge/discharge cycle every day as the system charges during off-peak hours and discharges during peak hours. Peak electricity demands are typically observed in the mornings and evenings [35], which are also when PV systems have low productivity; therefore, a household would rely on the FESS during these times to operate appliances or support the charging of a vehicle. (It should be mentioned here that the considered FESS capacity is not sufficient to fully charge typical EVs on the market. Rather, energy storage is seen as a means to support EV fast charging and associated peak loads.) Minimum demands are observed in the middle of the day when a PV system is most productive, thereby recharging the system. For this study, each day is divided into three 8-h phases and assigned a different average velocity for each period. These are referred to as the maximum phase, intermediate phase, and minimum phase. It is recognized that charging or discharging the FESS may occur over a period of hours; therefore, the intermediate phase represents the average velocity during the charge-discharge periods. Attempting to simulate a real-world scenario, the rotor is assumed to operate below ω_{limit}; therefore, the velocity during the maximum phase, ω_{Pmax}, is set at 0.9 ω_{limit}. For the minimum phase, the angular velocity, ω_{Pmin}, is chosen to be 0.25 ω_{limit}, as discussed in [36]. Finally, the intermediate phase angular velocity, ω_{Pint}, is halfway between ω_{Pmax} and ω_{Pmin}, i.e., 0.575 ω_{limit}. The rotor is simulated to rotate at each velocity, i.e., ω_{Pmax}, ω_{Pint} and ω_{Pmin}, for 8 h each day, for 365 days per year.

4. Results and Discussion

Filament-wound composite flywheel rotors may be subject to a variety of failure modes. Considering viscoelasticity and typical composite flywheel rotor construction, two failure modes are of primary concern. First, the rotor structural integrity is dependent on maintaining compressive loading at the interface between the hub and the composite rim, created by the press-fit during assembly. Therefore, in view of possible stress relaxation, a significant reduction or loss of this compressive loading may lead to rotor failure. Second, it has been shown that the prevalent polymer matrix materials for composite rotor fabrication, epoxy resins, experience creep embrittlement as they undergo viscoelastic stress relaxation, leading to an increased size and density of micro-cracks under subcritical loading conditions [37], i.e., applied loads which approach but do not exceed the matrix strength. Therefore, substantial viscoelastic stress increases the potential for micro-crack networks to substantially damage the matrix, which ultimately may lead to failure.

4.1. Algorithm Validation

The computational algorithm was validated by comparing simulation results with stress distributions for viscoelastic stress effects published by Tzeng et al. [13]. In their work, the rotor is constructed from two CFRP rims press-fitted together. The CFRP is an IM7/8552 transversely isotropic composite with no viscoelastic behavior in the fiber direction. Material properties are given in [13]. The simulation results are plotted alongside the published data in Figure 2. The close congruence that is observed between the published results and the current model provides validation that the present modeling approach is capable of accurately predicting the stress response in the flywheel rotor. Hence, model stress responses will herein be used in conjunction with the Tsai-Wu criterion to predict failure location and behavior.

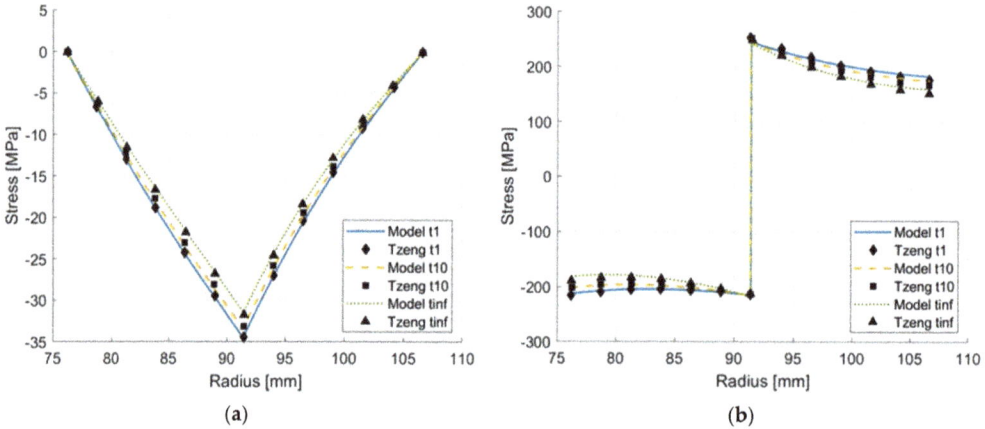

Figure 2. Radial (**a**) and circumferential (**b**) stress distribution comparison between the current model and Tzeng et al. [13] for a two-rim press-fit CFRP flywheel rotor.

Recalling the plane strain assumption made for the present analysis, modeling results validated the chosen approach, which simplified solving the radial inhomogeneous equilibrium equation. Contrasting present work with published literature, see e.g., [14,29,37–39], comparable results were achieved. It should be noted that some of these studies assume generalized plane strain. In addition, analyses that quantified axial stress [40] showed it to be an order of magnitude less than radial stress, and two orders less than the circumferential stress. Given the body of published works that impose and validate the plane strain assumption, and the comparatively small magnitude of axial stress, applying a plane strain assumption for the present analysis was seen as appropriate.

4.2. Viscoelastic Behavior

4.2.1. Case 1

Simulation results in terms of radial and circumferential stress are shown in Figure 3 for the flywheel rotor constructed from an aluminum hub with a thick CFRP rim. While the hub, located between $r = 160$ mm and $r = 200$ mm, was not inherently subject to viscoelastic behavior, its stress state changed as compressive loading from the composite rim decreased.

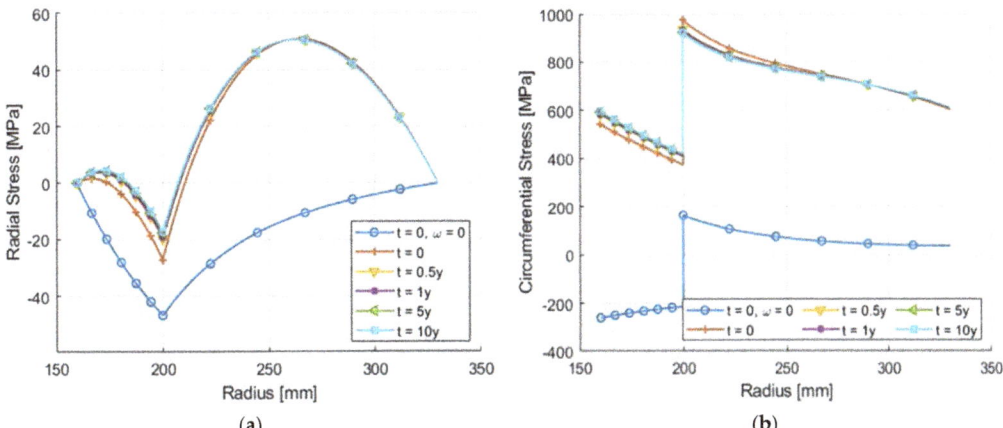

Figure 3. Radial (**a**) and circumferential (**b**) stress predictions for a flywheel rotor constructed of an aluminum hub and CFRP rim, after assembly (zero velocity), at startup, and at various times up to 10 years of continuous operation at the limit velocity of 24,250 rpm.

Considering the radial stress data depicted in Figure 3, the composite rim, in a pristine state post-manufacturing ($t = 0$ and $\omega = 0$), experiences high compressive loading, approximately −46.7 MPa, due to the press-fit assembly. After startup to ω_{limit}, the peak compressive load decreases to −27 MPa. This change is induced by the radial position of the hub and rim leading to differences in centrifugal loads, as well as differences in elastic modulus between the two materials. For comparison, the aluminum elastic modulus is 71 GPa while the CFRP longitudinal and transverse modulus is 111 GPa and 9.1 GPa, respectively. As a consequence, the outer rim deforms more than the aluminum hub, reducing the interfacial pressure. Note that this ability to compensate for differential deformation while maintaining rotor integrity is one advantage of a press-fit assembly.

It can further be seen in Figure 3 that circumferential stress in the aluminum hub increases over time. At the inner hub surface, circumferential stress increases from 542 MPa to 596 MPa after 10 years; an increase of 9.2%. (Note that even though this circumferential stress exceeds yield strength, the hub does not undergo failure because the stress coordinate for the given stress state still resides within the failure envelope, invoking, e.g., maximum distortion energy theory.) Additionally, the increased circumferential stress is coupled with a decrease in radial compressive stress, i.e., radial stress becomes less compressive in the aluminum hub. These changes in radial and circumferential stress are attributed to the increased compliance of the CFRP rim during this time period, allowing the hub to deform radially.

Regarding the composite rim, radial and circumferential peak stresses are predicted to decrease moderately between 1% and 5.5%, respectively, over the 10-year simulation period, which is to be expected based on previous research [14]. For greater clarity, peak stress values in the rotor over the simulated 10-year period were determined for (i) the interfacial pressure measured in the radial direction, (ii) the radial stress, and (iii) the circumferential stress. Corresponding values are given in Table 4. To illustrate their change over the simulated operation, they were normalized by their initial value at $t = 0$ and plotted in Figure 4. Within the first year of (continuous) operation, the rotor undergoes viscoelastic stress changes as the interfacial compressive stress decreases from −27 MPa to −19 MPa, or approximately 70.4% of the initial value. After 1 year, interfacial compressive stress decreases at a reduced rate, decreasing to 63.3% after 10 years. These results indicate that the composite material experiences rapid non-linear relaxation over the first year of operation, which to a large extent is ascribed to primary or transient creep (phase I). In the subsequent years of the rotor's service life, stress relaxation is significantly reduced, indicating the material has fully transitioned into secondary or steady-state creep (phase II). Conceivably, the initial rapid relaxation could be avoided by subjecting the composite rim to a suitable conditioning process prior to rotor assembly. In this case, only a relative minor decrease in interfacial pressure of about 7% would be expected.

Table 4. Rotor peak stress values for interfacial stress, radial peak stress, and circumferential peak stress at 0, 0.5, 1, 5, and 10 years of continuous flywheel operation at limit velocity.

Time [Year]	0	0.5	1	5	10
Interface pressure [MPa]	−26.98	−19.60	−19.01	−17.66	−17.08
Radial peak stress [MPa]	50.96	50.86	50.83	50.75	50.71
Circumferential peak stress [MPa]	975.51	936.54	933.20	925.40	922.01

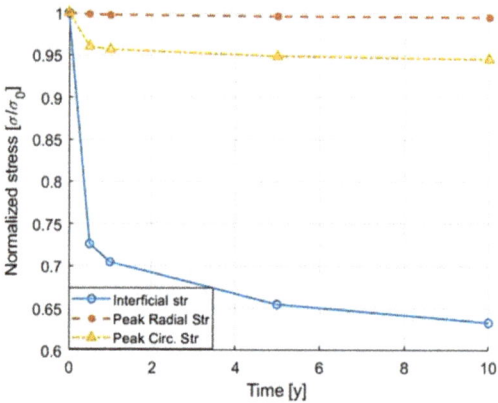

Figure 4. Normalized rotor peak stress for interfacial pressure, radial stress, and circumferential stress over time for continuous flywheel operation at limit velocity. Values are normalized by the corresponding stress at time $t = 0$.

Based on present findings, while viscoelastic stress relaxation leading to hub-rim separation is a conceivable scenario, it is controllable provided adequate interfacial pressure is achieved during assembly, or substantial initial creep effects can effectively be mitigated otherwise (e.g., by CFRP rim conditioning). It is interesting to note that for a reduced press-fit interference of 0.45 mm between the hub and rim (instead of 0.8 mm), creep effects are sufficient to cause zero interfacial pressure over the considered operating time, that is, separation between hub and rim would occur. Clearly, these results demonstrate that a viscoelastic analysis is warranted for the engineering design of FESS rotors.

The Tsai-Wu failure criteria were used to determine the SR data and predict the location and angular velocity associated with rotor failure. SR data facilitate predicting failure since these data are based on the combined stresses exceeding the strength threshold. In other words, the SR analysis provides an understanding of the total stress state of the rotor. SR data for the current rotor are provided in Figure 5.

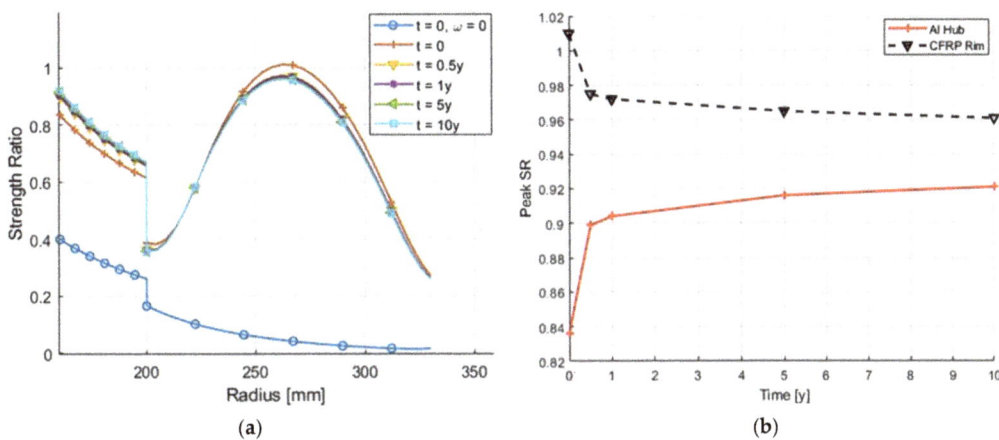

Figure 5. Strength ratio (*SR*) data for the flywheel rotor (**a**), and peak SR in the aluminum hub and CFRP rim at various times throughout the simulated lifetime (**b**).

The CFRP rim initially has the highest *SR* of 1.01 and is located at the midplane of the rim; however, viscoelastic stress relaxation improves the stress state of this rim, so *SR* decreases to 0.975 within 6 months, then continues to slowly decrease to about 0.96 after 10 years. The same cannot be said for the aluminum hub, which exhibits an evolution of stresses approaching failure. The *SR* for the hub, while initially substantially less than the composite rim, increases from 0.836 to 0.9 within 6 months, then continues to increase to 0.92 over the following 10 years. With an increase by 6.8%, changes in *SR* for the hub are rapid in the first year of operation, while the *SR* is predicted to increase by only another 2% over the next 9 years. Referring to Figure 3, this *SR* growth is caused by a rising circumferential stress in conjunction with a lessening of compression in the radial direction. This behavior can be understood recalling Equation (13), which is composed of radial, circumferential, axial, and coupled terms. The linear term for radial direction is $SR_{\text{radial}} = F_3 \sigma_3^{\text{app}} R$. Since the peak radial stress is compressive, a large negative value is introduced into Equation (13), thus reducing *SR* accordingly. As the CFRP rim undergoes viscoelastic deformation, the radial compressive load diminishes, so SR_{radial} diminishes as well, thus removing the negative term from the equation, causing the observed *SR* increase for the hub.

In summary, viscoelastic stress relaxation of the CFRP rim can improve its stress state to the detriment of the aluminum hub. It is reasonable to conclude that for a given rotor geometry, changing stresses may lead to damage of the aluminum hub and/or separation between hub and rim components. Viscoelastic effects should therefore be considered in flywheel rotor design. Nevertheless, based on the present observations, limited viscoelastic stress relaxation in the rotor may also be beneficial to the overall rotor performance. While substantial phase I creep may be a concern from a risk assessment point-of-view, mechanical conditioning and/or thermal aging could be a means to mitigate large initial viscoelastic effects after rotor assembly and operational startup while allowing for phase II creep to gradually evolve over the rotor's operational lifetime.

4.2.2. Case 2

As mentioned earlier, the load profile for case 2 is intended to more closely simulate the operation of an FESS in actual applications, such as for solar PV electric grids or EV charging. Graphs with the radial and circumferential stresses on day 1 of operation are given in Figure 6. These graphs serve as representative examples of the stress distribution for the simulated operation. While the magnitude of the radial and circumferential stresses was found to decrease in the rim and increase in the hub over time, the overall shape of the stress response at each velocity (minimum phase, ω_{Pmin}, intermediate phase, ω_{Pint}, maximum phase, ω_{Pmax}) was found to be similar for any day of the simulated operation, so only data for the first day are provided. Comparing case 1 (Figure 3) and case 2 (Figure 6), the stress responses at ω_{Pmin} and ω_{Pmax} closely resemble those from case 1 at $t = 0$, $\omega = 0$ and at $t = 0$, $\omega = \omega_{limit}$, respectively, as is expected given the similarity between angular velocities and the non-linear relationship between stress and velocity. Notably, for both cases, the circumferential stress in the hub is seen to change from initially having a positive slope ($\omega = 0$, $\omega = \omega_{Pmin}$) to having a negative slope at high-velocity operation (ω_{limit}, ω_{Pmax}). Between both extremes, stresses switch from tensile to compressive with the magnitude occurring at the hub inner surface. This loading scenario resembles fatigue loading with a negative stress ratio, positive mean stress, and a comparatively high stress range. The hub design should therefore include a fatigue analysis, especially for FESS that experiences high cycle rates, i.e., numerous cycles per day.

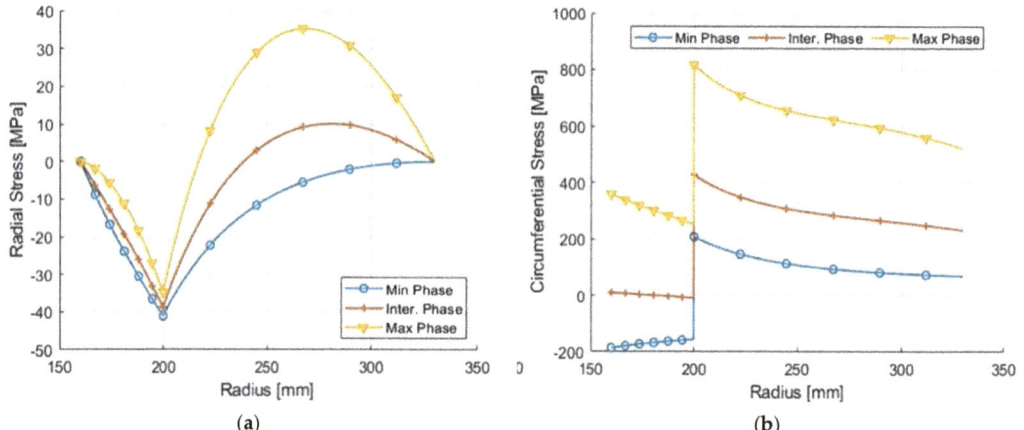

Figure 6. Radial (**a**) and circumferential (**b**) stress results at each velocity on day 1.

Figure 7 depicts SR data for all points along the rotor radial direction at each velocity for day 1 and day 365. Again, while magnitudes at each point are seen to vary for the different velocities, the overall shape of the SR curves at each velocity bear distinct similarities. Broadly, SR graphs exhibit similar trends as in case 1. After year 1, SR values at ω_{Pmax} increase in the hub but decrease in the composite rim. As one would expect, SR values in Figure 7 are lower compared to data in Figure 5, due to the overall lower stress levels and the reduced time that the rotor operates at high velocity.

During each phase, five key indicators are tracked throughout the simulation: (i) interface stress, (ii) peak radial tensile stress (i.e., neglecting compressive stresses, as these are found at the interface), (iii) peak circumferential stress in the CFRP rim, (iv) peak SR in the hub, and (v) peak SR in the CFRP rim. Values for each indicator recorded on day 1, 90, 180, 270, and 365 are given in Table 5. To facilitate comparisons with case 1, data from Table 5 were normalized using the day 1 value at each velocity and each location or component of interest, as depicted in Figure 8. Noting that since the peak radial tensile stress at ω_{Pmin} is negligible throughout the simulation (see Table 5), this dataset was omitted in Figure 8.

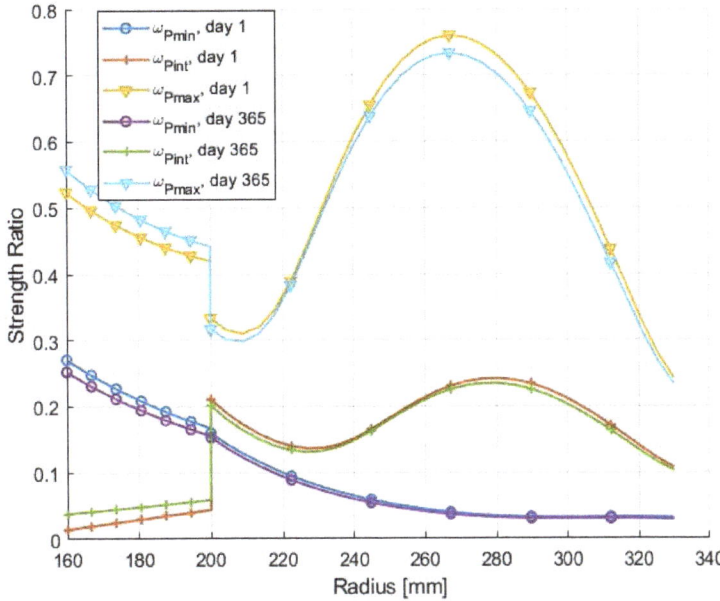

Figure 7. SR graphs for all points along the flywheel rotor radius at each velocity on day 1.

Table 5. Peak stress values at the hub-rim interface and radially and circumferentially in the CFRP rim, and peak SR values for the aluminum hub and the CFRP rim, at various times throughout the simulated one-year period.

	Phase	Day 1	Day 90	Day 180	Day 270	Day 365
Peak interface pressure [MPa]	ω_{Pmin}	−41.04	−39.26	−39.03	−38.90	−38.80
	ω_{Pint}	−38.76	−36.49	−36.17	−35.98	−35.84
	ω_{Pmax}	−34.78	−31.54	−31.05	−30.77	−30.55
Peak radial tensile stress in CFRP rim [MPa]	ω_{Pmin}	0.0	0.001	0.010	0.012	0.014
	ω_{Pint}	10.19	10.35	10.37	10.38	10.39
	ω_{Pmax}	35.33	35.26	35.24	35.23	35.22
Peak circumferential stress in CFRP rim [MPa]	ω_{Pmin}	207.9	207.3	207.2	207.1	207.1
	ω_{Pint}	426.8	420.2	419.3	418.7	418.3
	ω_{Pmax}	816.7	800.9	798.4	797.0	795.9
SR for aluminum hub [/]	ω_{Pmin}	0.271	0.256	0.255	0.253	0.253
	ω_{Pint}	0.044	0.056	0.057	0.058	0.059
	ω_{Pmax}	0.524	0.550	0.554	0.556	0.557
SR for CFRP rim [/]	ω_{Pmin}	0.160	0.155	0.154	0.153	0.153
	ω_{Pint}	0.242	0.236	0.235	0.235	0.234
	ω_{Pmax}	0.760	0.740	0.737	0.735	0.734

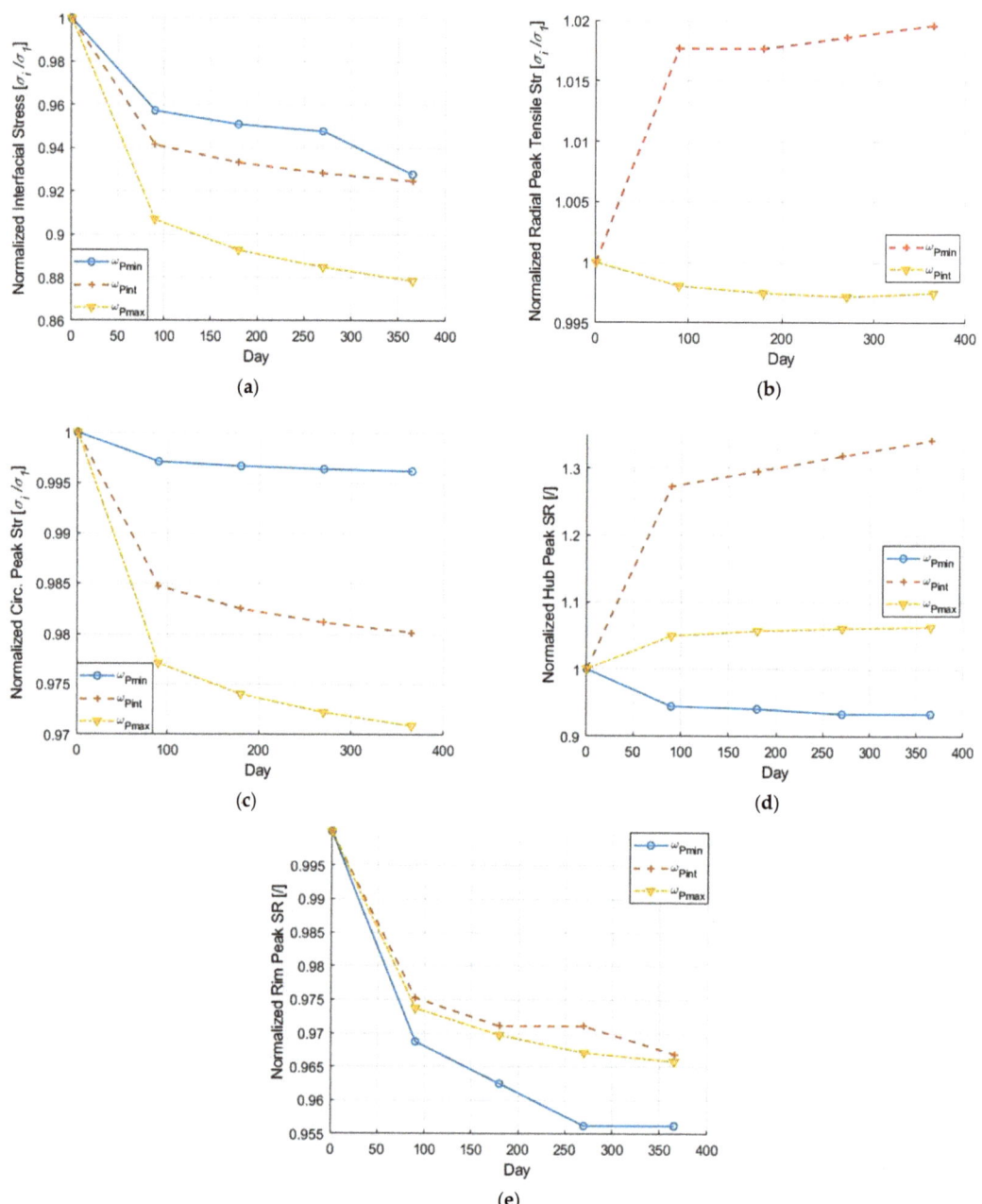

Figure 8. Stress results normalized by their initial value on day 1 for (**a**) interfacial pressure, (**b**) peak radial tensile stress, and (**c**) peak circumferential stress in the CFRP rim, (**d**) peak SR in the aluminum hub, and (**e**) peak SR in the CFRP rim.

Akin to case 1, values for the interfacial pressure exhibit the most significant change during the case 2 simulation, as shown in Figure 8a. This observation again substantiates conclusions in previous work [30] which suggested that stress relaxation at the interface could pose a risk to a flywheel rotor's structural integrity. After 1 year of operation, the interfacial pressure is predicted to decrease by approximately 7%, 7.5%, and 12% at the minimum, intermediate, and maximum velocities, respectively. However, these reductions are significantly less compared to case 1, for which a decrease of 29% is predicted after year 1. Again, lower predicted interface pressures in case 2 are due to the rotor being subjected to lower average angular velocities than in case 1, and the rotor spending less time subjected to high velocity loading conditions. Since for each charge-discharge cycle the rotor transitions from low to high velocity and vice versa, high cycle rates will typically subject the rotor to reduced viscoelastic effects, as it will spend less time at or near maximum velocities.

Considering Figure 8b,c, changes in peak stresses are rather benign. The peak radial tensile stress increases by a maximum of about 2%, which is comparable to the corresponding decrease seen in case 1. The peak circumferential stress shows a reduction of 0.5% at ω_{Pmin} and 3% at ω_{Pmax}. Reductions in both peak stress components are less than those for case 1 for the same reasons as discussed earlier.

In Figure 8d, peak SR values for the aluminum hub are indicated to decrease for ω_{Pmin} but to rise for the other two velocities. SR changes at ω_{Pmin} and ω_{Pint}, being seemingly high at the latter velocity, are largely irrelevant given the comparatively low absolute SR values for the hub at these operating conditions (see Figure 7). The rise in SR at ω_{Pmax} is considerable but is still confined to below 10% and remains uncritical. Referring to Figure 8e, the SR evolution for the CFRP rim is favorable, as observed for case 1, as values decrease over the considered operating period.

Considering relative SR changes between case 1 and case 2 at high velocity and at critical locations with respect to the rotor radial direction, i.e., the hub inner surface and the rim's cylindrical midsection, it is apparent that magnitudes in case 2 remain below those in case 1, which is to be expected given that the rotor is subjected to an overall reduced average velocity while also operating for less time under high velocity loading conditions. For example, in case 1, after the first year, the SR for the hub increases by 6.8%, while in case 2 (at ω_{Pmax}) over the same period, the increase is 3.3%.

5. Conclusions

The high-stress and long-lifetime operating conditions of modern composite flywheel rotors create the necessary conditions for viscoelastic failure modes not included in contemporary quasi-static failure analyses. In this study, a computational algorithm, based on an accepted analytical modeling approach, was developed to investigate the viscoelastic behavior of fiber reinforced polymer composite rotors during their lifetimes. Additionally, the Tsai-Wu failure criterion was used to compute strength ratios along the rotor radial direction. The values were used to assess the conditions for rotor failure. A composite flywheel rotor design was considered that meets capacity requirements to support an electrical vehicle charging system or solar PV residential electric grid. The rotor consists of a press-fit assembly of an aluminum hub with a carbon fiber polymer composite rim. The viscoelastic behavior of the flywheel rotor was studied for two cases: (i) a worst-case scenario of the rotor operating with an angular velocity at the failure threshold for a simulated lifetime of 10 years, and (ii) a charging/discharging cycle in which the rotor experiences a minimum, intermediate, and maximum velocity for 8 h each per day over a one-year period.

The case 1 simulation indicated that due to viscoelastic stress relaxation, the radial and circumferential stresses in the composite rotor reduce over time. After 10 years, peak stress in the radial and circumferential directions were found to decrease by approximately 1% and 5%, respectively. Given that rim stresses continually decrease over time, the risk of rim failure is diminishing during operation, provided no external factors, such as matrix cracking, affect the rotor's structural integrity. In contrast, circumferential stresses in the aluminum hub increase while radial stresses decrease. This behavior was attributed to an increasing compliance of the composite rim, allowing it to deform radially outward. Thus, radial compressive stresses in the rotor are reduced, and in turn, circumferential stresses in the hub are increased. The peak strength ratio for the composite rim decreases by approximately 4% compared to an 8% increase in the hub after 10 years. The latter is of concern, as the peak strength ratio in the aluminum hub converges toward unity, suggesting an increased risk of hub failure. The interfacial press-fit pressure is subject to the largest decrease, approximately 36%. Still, taken on its own, this behavior represents a low risk to the flywheel rotor's structural integrity as long as the rotor design prescribes a sufficient initial press-fit interference that upholds compressive interfacial pressure over the rotor's lifetime. Based on the observation in case 1, failure may occur in the rim at startup if the rotor reaches a critical velocity, but would decrease over time even if the critical velocity is maintained as viscoelastic stress relaxation improves the overall stress state in the composite rim. However, this behavior does not preclude possible failure due to other effects such as fatigue and matrix cracking, which warrants including such effects in flywheel rotor design and analysis.

The daily charge-discharge cycle considered in case 2 imposes cyclic loading conditions upon the rotor. In this scenario, strength ratios never reached unity, so failure is not predicted for any part of the rotor. Consistent with case 1, viscoelastic stress relaxation allows the radial and circumferential stress in the composite rim to decrease over time, creating a more favorable stress state regardless of angular velocity or time. However, the viscoelastic effects that improve the stress state for the rim are detrimental to the metallic hub. Moreover, charge-discharge cycles were found to impose fatigue loading with a negative stress ratio, positive mean stress, and a comparatively high stress range at the inner surface of the hub.

In summary, the present study conducted simulations on flywheel rotors of appropriate size and over appropriate time frames for applications such as in residential PV energy systems or EV charging stations. While previous studies began to explore this topic, the present study investigated the evolution of stresses in each principal direction between 6 months and 10 years of operation. Findings from these data are vital to consider when designing flywheel rotors for similar and other applications. The simulations conducted in the present study support the notion that viscoelastic effects reduce peak stresses in a composite rim over time. However, this study also showed that viscoelasticity may affect stresses in other parts of the rotor, i.e., the hub and the hub-rim interface, in ways that may lead to rotor failure. Moreover, it was noted that charge-discharge cycles of the flywheel energy storage device may create significant fatigue loading conditions. Therefore, it is concluded that flywheel rotor design should include viscoelastic and fatigue analyses to ensure safe operation, especially for devices experiencing high cycle rates and long-time operation near critical velocities.

Author Contributions: Conceptualization and methodology, M.S. and P.M.; validation, formal analysis, investigation, M.S.; resources, P.M.; data curation, M.S.; writing—original draft preparation, visualization, M.S.; writing—review and editing, M.S. and P.M.; supervision, project administration, funding acquisition, P.M. All authors have read and agreed to the published version of the manuscript.

Funding: This research and the APC was funded by the Canada First Research Excellence Fund with grant number Future Energy Systems T06-P03.

Institutional Review Board Statement: Not applicable.

Informed Consent Statement: Not applicable.

Data Availability Statement: Flywheel rotor simulation algorithm and data available from: mskin001/Flywheel_Rotor_Simulation at Cyclic_Simulation (github.com).

Conflicts of Interest: The authors declare no conflict of interest.

References

1. Hadjipaschalis, I.; Poullikkas, A.; Efthimiou, V. Overview of current and future energy storage technologies for electric power applications. *Renew. Sustain. Energy Rev.* **2009**, *13*, 1513–1522. [CrossRef]
2. Luo, X.; Wang, J.; Dooner, M.; Clarke, J. Overview of current development in electrical energy storage technologies and the application potential in power system operation. *Appl. Energy* **2015**, *137*, 511–536. [CrossRef]
3. Takkar, S.; Gupta, K.; Tiwari, V.; Singh, S.P. Dynamics of Rotating Composite Disc. *J. Vib. Eng. Technol.* **2019**, *7*, 629–637. [CrossRef]

4. Yang, L.; Crawford, C.; Ren, Z. A Fuzzy Satisfactory Optimization Method Based on Stress Analysis for a Hybrid Composite Flywheel. *IOP Conf. Ser. Mater. Sci. Eng.* **2018**, *398*, 012032. [CrossRef]
5. Mittelstedt, M.; Hansen, C.; Mertiny, P. Design and multi-objective optimization of fiber-reinforced polymer composite flywheel rotors. *Appl. Sci.* **2018**, *8*, 1256. [CrossRef]
6. Hartl, S.; Schulz, A.; Sima, H.; Koch, T.; Kaltenbacher, M. A Static Burst Test for Composite Flywheel Rotors. *Appl. Compos. Mater.* **2016**, *23*, 271–288. [CrossRef]
7. Corbin, C.K. *Burst Failure Prediction of Composite Flywheel Rotors: A Progressive Damage Approach via Stiffness Degredation*; Stanford University: Stanford, CA, USA, 2005.
8. Arnold, S.M.; Saleeb, A.F.; Al-Zoubi, N.R. Deformation and life analysis of composite flywheel disk systems. *Compos. Part B Eng.* **2002**, *33*, 433–459. [CrossRef]
9. Trufanov, N.A.; Smetannikov, O.Y. Creep of Composite Energy Accumulators. *Strength Mater.* **1991**, *23*, 671–675.
10. Portnov, G.G. Estimation of Limit Strains in Disk-Type Flywheels Made of Compliant Elastomeric Matrix Composite Undergoing Radial Creep. *Mech. Compos. Mater.* **2000**, *36*, 55–58. [CrossRef]
11. Levistor Boosting Forecourt Grid Power for the Next Generation of Fast Charging Electric Vehicles. Available online: https://levistor.com/#about (accessed on 14 August 2021).
12. Stornetic GmbH Powerful Storage System for Grid Services. Available online: https://stornetic.com/assets/downloads/stornetic_general_presentation.pdf (accessed on 14 August 2021).
13. Tzeng, J.T. Viscoelastic Analysis of Composite Cylinders Subjected to Rotation. *Trans. Ophthalmol. Soc.* **2001**, *101*, 200–202. [CrossRef]
14. Emerson, R.P. *Viscoelastic Flywheel Rotors: Modeling and Measurement*; Pennsylvania State University: State College, PA, USA, 2002.
15. Emerson, R.P.; Bakis, C.E. Relaxation of press-fit interference pressure in composite flywheel assemblies. In Proceedings of the 43rd International SAMPE Symposium and Exhibition, Anaheim, CA, USA, 31 May–4 June 1998; SAMPE: Anaheim, CA, USA, 1998; Volume 43, pp. 1904–1915.
16. Rupp, A.; Baier, H.; Mertiny, P.; Secanell, M. Analysis of a Flywheel Energy Storage System for Light Rail Transit. *Energy* **2016**, *107*, 625–638. [CrossRef]
17. Toray CARBON FIBER T700G. Available online: https://www.toraycma.com/wp-content/uploads/T700G-Technical-Data-Sheet-1.pdf (accessed on 14 August 2021).
18. Majda, P.; Skrodzewicz, J. A modified creep model of epoxy adhesive at ambient temperature. *Int. J. Adhes. Adhes.* **2009**, *29*, 396–404. [CrossRef]
19. Almeida, J.H.S.; Ornaghi, H.L.; Lorandi, N.P.; Bregolin, B.P.; Amico, S.C. Creep and interfacial behavior of carbon fiber reinforced epoxy filament wound laminates. *Polym. Compos.* **2018**, *39*, E2199–E2206. [CrossRef]
20. Thakur, P.; Sethi, M. Creep deformation and stress analysis in a transversely material disk subjected to rigid shaft. *Math. Mech. Solids* **2019**, *25*, 17–25. [CrossRef]
21. Debljinom, P.; Teorije, P.; Napona, P. Modelling of creep behaviour of a rotating disc in the presence of load and variable thickness by using SETH transition theory. *Struct. Integr. Life* **2018**, *18*, 153–160.
22. Zharfi, H. Creep relaxation in FGM rotating disc with nonlinear axisymmetric distribution of heterogeneity. *Theor. Appl. Mech. Lett.* **2019**, *9*, 382–390. [CrossRef]
23. Tzeng, J.T. *Viscoelastic Analysis of Composite Flywheel for Energy Storage*; Army Research Laboratory: Adelphi, MD, USA, 2001.
24. Lekhnitskiy, S.G. *Anisotropic Plates*; Air Force Systems Command: Moscow, Russia, 1957.
25. Ha, S.K.; Jeong, H.M.; Cho, Y.S. Optimum design of thick-walled composite rings for an energy storage system. *J. Compos. Mater.* **1998**, *32*, 851–873. [CrossRef]
26. Li, S.; Sitnikova, E.; Liang, Y.; Kaddour, A.S. The Tsai-Wu failure criterion rationalised in the context of UD composites. *Compos. Part A Appl. Sci. Manuf.* **2017**, *102*, 207–217. [CrossRef]
27. Tsai, S.W.; Wu, E.M. A General Theory of Strength for Anisotropic Materials. *J. Compos. Mater.* **1971**, *5*, 58–80. [CrossRef]
28. Rojas, J.I.; Nicolás, J.; Crespo, D. Study on mechanical relaxations of 7075 (Al-Zn-Mg) and 2024 (Al-Cu-Mg) alloys by application of the time-temperature superposition principle. *Adv. Mater. Sci. Eng.* **2017**, *2017*, 2602953. [CrossRef]
29. ASM International Handbook Committee. *Metals Handbook Properties and Selection: Nonferrous Alloys and Special-Purpose Materials*; ASM International: Geauga County, OH, USA, 1990; Volume 2, ISBN 978-0-87170-378-1.
30. Tzeng, J.T. Viscoelastic Modeling of Press-Fitted Composite Cylinders. *J. Compos. Tech. Res.* **2001**, *23*, 21–27.
31. Ding, H.; Chen, W.; Zhang, L. *Elasticity of Transversely Isotropic Materials*; Gladwell, G.M.L., Ed.; Springer: Dordrecht, The Netherlands, 2006; ISBN 9781119130536.
32. Sadananda, K.; Nani Babu, M.; Vasudevan, A.K. The unified approach to subcritical crack growth and fracture. *Eng. Fract. Mech.* **2019**, *212*, 238–257. [CrossRef]
33. Kheawcum, M.; Sangwongwanich, S. A Case Study on Flywheel Energy Storage System Application for Frequency Regulation of Islanded Amphoe Mueang Mae Hong Son Microgrid. In Proceedings of the 17th International Conference on Electrical Engineering/Electronics, Computer, Telecommunications and Information Technology, ECTI-CON 2020, Phuket, Thailand, 24–27 June 2020; pp. 421–426.
34. Amiryar, M.E.; Pullen, K.R. A review of flywheel energy storage system technologies and their applications. *Appl. Sci.* **2017**, *7*, 286. [CrossRef]

35. Mansouri Kouhestani, F.; Byrne, J.; Johnson, D.; Spencer, L.; Hazendonk, P.; Brown, B. Evaluating solar energy technical and economic potential on rooftops in an urban setting: The city of Lethbridge, Canada. *Int. J. Energy Environ. Eng.* **2019**, *10*, 13–32. [CrossRef]
36. Andersen, F.M.; Baldini, M.; Hansen, L.G.; Jensen, C.L. Households' hourly electricity consumption and peak demand in Denmark. *Appl. Energy* **2017**, *208*, 607–619. [CrossRef]
37. Peña-Alzola, R.; Sebastián, R.; Quesada, J.; Colmenar, A. Review of flywheel based energy storage systems. In Proceedings of the 2011 International Conference on Power Engineering, Energy and Electrical Drives, Malaga, Spain, 11–13 May 2011. [CrossRef]
38. Odegard, G.M.; Bandyopadhyay, A. Physical aging of epoxy polymers and their composites. *J. Polym. Sci. Part B Polym. Phys.* **2011**, *49*, 1695–1716. [CrossRef]
39. Ha, S.K.; Yang, H.-I.; Kim, D.-J. Optimum design of a hybrid composite flywheel with permanent magnet rotor. *J. Compos. Mater.* **1999**, *33*, 1544–1575. [CrossRef]
40. Pérez-Aparicio, J.L.; Ripoll, L. Exact, integrated and complete solutions for composite flywheels. *Compos. Struct.* **2011**, *93*, 1404–1415. [CrossRef]

Article

An Effect of MHD on Non-Newtonian Fluid Flow over a Porous Stretching/Shrinking Sheet with Heat Transfer

Angadi Basettappa Vishalakshi [1], Thippaiah Maranna [1], Ulavathi Shettar Mahabaleshwar [1] and David Laroze [2,*]

[1] Department of Mathematics, Shivagangotri, Davangere University, Davangere 577007, India; vishalavishu691@gmail.com (A.B.V.); marannat4@gmail.com (T.M.); u.s.m@davangereuniversity.ac.in (U.S.M.)
[2] Instituto de Alta Investigación, Universidad de Tarapacá, Casilla 7D, Arica 1000000, Chile
* Correspondence: dlarozen@uta.cl

Abstract: The current article explains the 3-D MHD fluid flow under the impact of a magnetic field with an inclined angle. The porous sheet is embedded in the flow of a fluid to yield the better results of the problem. The governing PDEs are mapped using various transformations to convert in the form of ODEs. The yielded ODEs momentum equation is examined analytically to derive the mass transpiration and then it is used in the energy equation and solved exactly by using various controlling parameters. In the case of multiple solutions, the closed-form exact solutions of highly non-linear differential equations of the flow are presented as viscoelastic fluid, which is classified as two classes, namely the second order liquid and Walters' liquid B fluid. The results can be obtained by using graphical arrangements. The current work is utilized in many real-life applications, such as automotive cooling systems, microelectronics, heat exchangers, and so on. At the end of the analysis, we concluded that velocity and mass transpiration was more for Chandrasekhar's number for both the stretching and shrinking case.

Keywords: Walters' liquid B; inclined MHD; similarity transformation; porous media; heat transfer; radiation

1. Introduction

The challenges on stretching sheets are helpful for engineering and industrial applications for manufacturing plastic, polymers, and more. In the present paper we are discussing the three-dimensional flow over a porous body on the non-Newtonian fluid in the presence of MHD and an inclined angle. Sakiadis [1] examined the behavior of the laminar and turbulent boundary layer flow of continuously moving solid surface and flat surface. This work is extended by Crane [2], considering fluid with a stretching sheet, after experiencing many challenges conducted on stretching sheet problems. Andersson [3,4] has examined the problem with viscous flow with uniform magnetic field; this work is properly valid for any Reynolds number. Wang [5], studied the stagnation point flow. Fang and Zhang [6] examined the heat transfer analysis on the basis of an analytical method. Miklavcic and Wang [7] discussed the asymmetric cases of two-dimensional flow in the presence of a suction parameter with multiple solutions. Turkyilmazoglu et al. [8,9] worked on Jeffrey fluid with a stagnation point. Mahabaleshwar et al. [10] examined the problems on a stretching surface by considering MHD Newtonian hybrid nanofluid flow due to superlinear stretching sheet. Very recently, Vishalakshi et al. [11] studied the stretching sheet problem by using Rivlin-Ericksen fluid by using mass transpiration and thermal communication. Mahabaleshwar et al. [12] investigated stretching sheet problems by considering different aspects of parameters, such as the Brinkman ratio, thermal radiation, porous medium parameter, and so on. Apart from these studies, some research was conducted on porous sheets while under the impact of magnetic parameter. Porous medium and magnetic parameters contributed a major role in the study of stretching sheet problems. There are many equations available to describe the porous medium. Many

investigations conducted on porous medium occurred under the impact of a magnetic field. Khan et al. [13] worked on the fluid flow with MHD, as well as the transfer of mass with a porous medium. Nadeem et al. [14] worked on the numerical results of MHD Casson nanofluid. Mahabaleshwar [15] conducted the work on magneto-convection electrically conducting micropolar liquids. Mahabaleshwar et al. [16–18] worked on fluid flow with heat transfer by considering different fluids using different parameters in the presence of porous medium. Mahabaleshwar et al. [19–21] reviewed the flow of Casson fluid, couple stress fluid, and nanofluid with heat transfer under the impact of MHD with various parameters. See some the recent investigations on MHD and porous medium in [22–27].

Inspired by the above literatures, this current work is the study of 3-D flow with transpiration and radiation. The novelty of the present work is to explain the three-dimensional flow of a fluid with heat transfer under the impact of magnetic field and in the presence of a porous medium. Resulting ODEs are obtained by changing PDEs by using suitable variables. Analytical results can be conducted by using different controlling parameters. Temperature equations can be examined analytically and exhibit in gamma functions. Results can be obtained with the help of different physical parameters. The results of skin friction and Nusselt number is also discussed. The present work contains many industrial applications as well as its argument with the work of Vishalakshi et al. [28].

2. Problem Statement and Solution

A 3-D fluid flow was named Walter's liquid B, due to a porous sheet with inclined angle, transpiration, and thermal radiation. Fluid flow moved towards the x-axis and y-axis and was placed normally to it. Let σ indicate electrical conductivity, assuming the flow of a fluid, along with strength, B_0. A porous medium was placed inside the flow of a fluid and schematically the present flow was indicated in Figure 1.

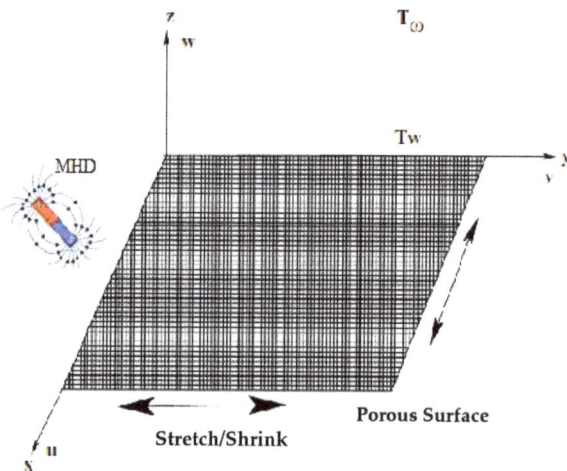

Figure 1. Schematic diagram of the three-dimensional flow.

Using these assumptions, the modelled governing equations are defined as follows [29–31]

$$u_x + v_y + w_z = 0, \tag{1}$$

$$uu_x + vu_y + wu_z = vu_{zz} - \left(\frac{v}{k_1} + \frac{\sigma B_0^2}{\rho}\sin^2(\tau)\right)u \\ -k\{uu_{xzz} + wu_{zzz} - (u_xu_{zz} + u_zw_{zz} + 2u_zu_{xz} + 2w_zu_{zz})\} \tag{2}$$

$$uv_x + vv_y + wv_z = vv_{zz} - \left(\frac{v}{k_1} + \frac{\sigma B_0^2}{\rho}\sin^2(\tau)\right)v \qquad (3)$$
$$-k\{vv_{xzz} + wv_{zzz} - (v_xv_{zz} + v_zw_{zz} + 2v_zv_{xz} + 2w_zv_{zz})\}$$

$$uT_x + vT_y + wT_z = \alpha T_{zz} - \frac{1}{\rho C_P}(q_r)_z, \qquad (4)$$

along with B. Cs (see [32])

$$\left.\begin{array}{l}u = ax + lu_z, \ v = by + lv_z, \ w = w_0, \ at \ z = 0 \\ u \to 0, \ u_z \to 0, \ v \to 0, \ as \ z \to \infty\end{array}\right\} \qquad (5)$$

where, u, v, and w indicate the velocities along the x, y, and z direction, respectively, and τ indicates the inclined angle; v is the kinematic viscosity, l indicates slip factor, ρ is the density, α is the thermal diffusivity, w_0 indicates wall transfer velocity, and k indicates permeability of the porous medium. Next we introduce the suitable variables as follows:

$$\eta = \sqrt{\frac{|a|}{v}}z, \ u = |a|xf_\eta(\eta), \ v = |a|yg_\eta(\eta), \ w = -\sqrt{|a|v}(f(\eta) + g(\eta)) \qquad (6)$$

by using the similarity transformation Equation (1) converted as follows:

$$f_{\eta\eta\eta} + (f+g)f_{\eta\eta} - f_\eta^2 - \left(Q\sin^2\tau + \frac{1}{Da}\right)f_\eta + \qquad (7)$$
$$K[(f+g)f_{\eta\eta\eta\eta} + (f_{\eta\eta} + g_{\eta\eta})f_{\eta\eta} - 2(f_\eta + g_\eta)f_{\eta\eta\eta}] = 0$$

$$g_{\eta\eta\eta} + (f+g)g_{\eta\eta} - g_\eta^2 - \left(Q\sin^2\tau + \frac{1}{Da}\right)g_\eta + \qquad (8)$$
$$K[(f+g)g_{\eta\eta\eta\eta} + (f_{\eta\eta} + g_{\eta\eta})g_{\eta\eta} - 2(f_\eta + g_\eta)g_{\eta\eta\eta}] = 0$$

Therefore, B. Cs defined in Equation (5) becomes:

$$f(0) = V_C, \ f_\eta(0) = d + \Gamma f_{\eta\eta}(0), \ g(0) = 0 \qquad (9)$$

$$f_\eta(\infty) \to 0, \ f_{\eta\eta}(\infty) \to 0, \ g_\eta(\infty) \to 0, \ g_{\eta\eta}(\infty) \to 0 \qquad (10)$$

where the $d = \frac{b}{|a|}$ indicates stretching/shrinking sheet parameter, mass flux velocity is given by $V_C = -\frac{w_0}{\sqrt{|a|v}}$, viscoelasticity is $K = \frac{|a|k}{v}$, Chandrasekhar's number is to be $Q = \frac{\sigma B_0^2}{|a|\rho}$, Darcy number is $Da^{-1} = \frac{v}{k_1|a|}$, and $\Gamma = l\sqrt{\frac{|a|}{v}}$ is the velocity slip parameter.

3. Exact Solutions of Momentum Equation

Let us consider the solution of Equations (7) and (8) are as follows:

$$f(\eta) = V_C + d\left(\frac{1 - \exp(-\lambda\eta)}{\lambda(1 + \Gamma\lambda)}\right), \ g(\eta) = d\left(\frac{1 - \exp(-\lambda\eta)}{\lambda(1 + \Gamma\lambda)}\right). \qquad (11)$$

where V_C indicates mass transpiration, if $V_C > 0$ indicates suction and $V_C < 0$ indicates injection.

By using the Equation (11) in Equations (7) and (8) to get the following resulting equations:

$$2K\lambda^2 - 1 = 0,$$
$$(1 + \Gamma\lambda)\left(\left(Q\sin^2\tau + \frac{1}{Da}\right) - \lambda(V_C - \lambda + KV_C\lambda^2)\right) - 2d(1 + K\lambda^2) = 0, \qquad (12)$$

After solving Equation (7) we get:

$$\lambda = \pm \frac{1}{\sqrt{2k_1}},$$

$$V_C = \frac{\left(Q\sin^2\tau + \frac{1}{Da}\right)(1+\Gamma\lambda) - 2d(1+K\lambda^2) + \lambda^2(1+\Gamma\lambda)}{\lambda(1+K\lambda^2)(1+\Gamma\lambda)}, \quad (13)$$

Skin friction co-officiants are also modified in the following form:

$$f_{\eta\eta}(0) = g_{\eta\eta}(0) = -\frac{d\lambda}{1+\Gamma\lambda}. \quad (14)$$

4. Exact Solutions of Energy Equation

This problem is essentially forced into a convection problem with the following boundary conditions:

$$T = T_w, \text{ at } z = 0$$
$$T \to T_\infty \text{ as } z \to \infty. \quad (15)$$

By using Rosseland's approximation, q_r is defined as follows (see Mahabaleshwar et al. [33–35]):

$$q_r = \frac{-4\sigma^*}{3k^*}\left(\frac{\partial T^4}{\partial z}\right). \quad (16)$$

where σ^* is the Stefan-Boltzmann constant, k^* is the coefficient of mean absorption, and T is the temperature of the fluid.

The term T^4 can be expanded as

$$T^4 = T_\infty^4 + 4T_\infty^3(T - T_\infty) + 6T_\infty^2(T - T_\infty)^2 + \ldots\ldots, \quad (17)$$

some higher order series ignore to get the result as:

$$T^4 = -3T_\infty^4 - 4T_\infty^3 T. \quad (18)$$

Using Equation (18) in Equation (16) to yield the result as:

$$\frac{\partial q_r}{\partial y} = -\frac{16\sigma^* T_\infty^3}{3k^*}\frac{\partial^2 T}{\partial y^2}. \quad (19)$$

By using the transformations defined in Equations (6) and (19) in Equation (4) to yield the following result:

$$\omega\theta_{\eta\eta}(\eta) + Pr(f(\eta) + g(\eta))\theta_\eta(\eta) = 0, \quad (20)$$

where $f(\eta)$ is given in Equation (11), we consider $\omega = \frac{3N+4}{3N}$, $N = \frac{-4\sigma^* T_\infty^3}{3k^* \kappa_f}$, and $Pr = \frac{\kappa_f}{\mu C_p}$.

Then the corresponding boundary conditions become:

$$\theta(0) = 1, \ \theta(\infty) \to 0\}, \quad (21)$$

To derive a homogeneous equation of Equation (19) by the use of power series method. The solution is $\theta(t) = \sum_{t=0}^{\infty} a_r t^{m+r}$, where a_r is the arbitrary constant and m is the constants to be determined.

Where:

$$t = \frac{2dk_1 Pr e^{-\lambda\eta}}{1+\Gamma\lambda} \quad (22)$$

On substituting t and also solving Equation (20) by using the B. Cs of Equation (21) to yield the following results:

$$\theta(\eta) = C_1 + C_2 \Gamma\left(\frac{2}{3\omega}\left(1 - 2K\left(Q\sin^2(\tau) + Da^{-1}\right)\right), \frac{4dKPre^{-\frac{\eta}{\sqrt{2}\sqrt{K}}}}{1 + \frac{\Gamma}{\sqrt{2}\sqrt{K}}}\right) \quad (23)$$

$$\theta(\eta) = \frac{\Gamma\left(\frac{2}{3\omega}(1 - 2K(Q\sin^2(\tau) + Da^{-1})), 0\right) - \Gamma\left(\frac{2}{3\omega}(1 - 2K(Q\sin^2(\tau) + Da^{-1})), \frac{4dKPre^{-\frac{\eta}{\sqrt{2}\sqrt{K}}}}{1 + \frac{\Gamma}{\sqrt{2}\sqrt{K}}}\right)}{\Gamma\left(\frac{2}{3\omega}(1 - 2K(Q\sin^2(\tau) + Da^{-1})), 0\right) - \Gamma\left(\frac{2}{3\omega}(1 - 2K(Q\sin^2(\tau) + Da^{-1})), \frac{4dKPr}{1 + \frac{\Gamma}{\sqrt{2}\sqrt{K}}}\right)} \quad (24)$$

5. **Results and Discussion**

In the current study, we emphasize the investigation on fluid flow with heat transfer under the impact of an inclined angle, Chandrasekhar's number transpiration, and radiation. The PDEs of the problem are mapped into ODEs using suitable transformations, then the resulting ODEs are solved analytically. Multiple solutions are used to analyse the present study. The analytical results of the momentum and energy equation is obtained at Equations (13) and (24), and the results of the momentum equation are obtained in terms of mass transpiration. The solution domain λ linked with another parameters through Equation (13). Analytical results of momentum and energy equation is, respectively, represented at Equations (13) and (24). By using graphical arrangements, the impact of different parameters can be performed.

Figure 2a,b exhibits the impact of $f(\eta)$ on η for various choices of Q for $d = 1$ and $d = -1$, respectively, and keeping other parameters as $\tau = 90°$, $k_1 = 1$, and $Da = 0.3$. Here, blue solid lines indicate the $\Gamma = 1$, and black dotted lines indicate the $\Gamma = 0$. From this graph, it is cleared that $f(\eta)$ is for values of Q for both $d = 1$ and $d = -1$. Figures 3 and 4 portray the effect of $f_\eta(\eta)$ on η for different choices of Γ and k_1, respectively. Figure 3a,b indicate the plots of $f_\eta(\eta)$ verses η for different choices of Γ for $d = 1$ and $d = -1$, respectively, in this $f_\eta(\eta)$ less for more values of Γ for $d = 1$. It is opposite if $d = -1$, i.e., $f_\eta(\eta)$ is for more values of Γ for $d = -1$. Figure 4a,b indicate the plots of $f_\eta(\eta)$ verses η for various values of k_1 for $d = 1$ and $d = -1$, respectively, in this t is observed that $f_\eta(\eta)$ is more for more choices of k_1 for $d = 1$. This impact is opposite if $d = -1$. i.e., $f_\eta(\eta)$ less for more values of k_1 for $d = -1$. In this problem we express the analytical method in terms of mass transpiration and the domain linked with other parameters through this equation.

Figure 5a,b portrays the plots of V_C verses k_1 for different choices of Q for $d = 1$ and $d = -1$, respectively, and keeps the other parameters as $\tau = 90°$, $Da = 0.3$. Here, blue solid lines indicate the $\Gamma = 2$ and black dotted lines indicate the $\Gamma = 0$. λ value connected with k_1 through Equation (13). In these graphs V_C is for values of Q for both $d = 1$ and $d = -1$.

Figure 6a,b demonstrated the impact of $\theta(\eta)$ on η for different values of Q for $d = 1$ and $d = -1$. In this $\theta(\eta)$ is for values of Q for both $d = 1$ and $d = -1$. Figure 7a,b demonstrated the impact of $\theta(\eta)$ on η for various choices of N for $d = 1$ and $d = -1$, in this it is observed that $\theta(\eta)$ is decreased for increasing the N for both $d = 1$ and $d = -1$. In these graphs it is observed that there is little difference between $d = 1$ and $d = -1$. In these figures, it is carefully observed that boundary value thickness is wider for the shrinking sheet case when compared to the stretching sheet case. Boundary value thickness is the velocity boundary layer; it is normally as the distance from the solid body.

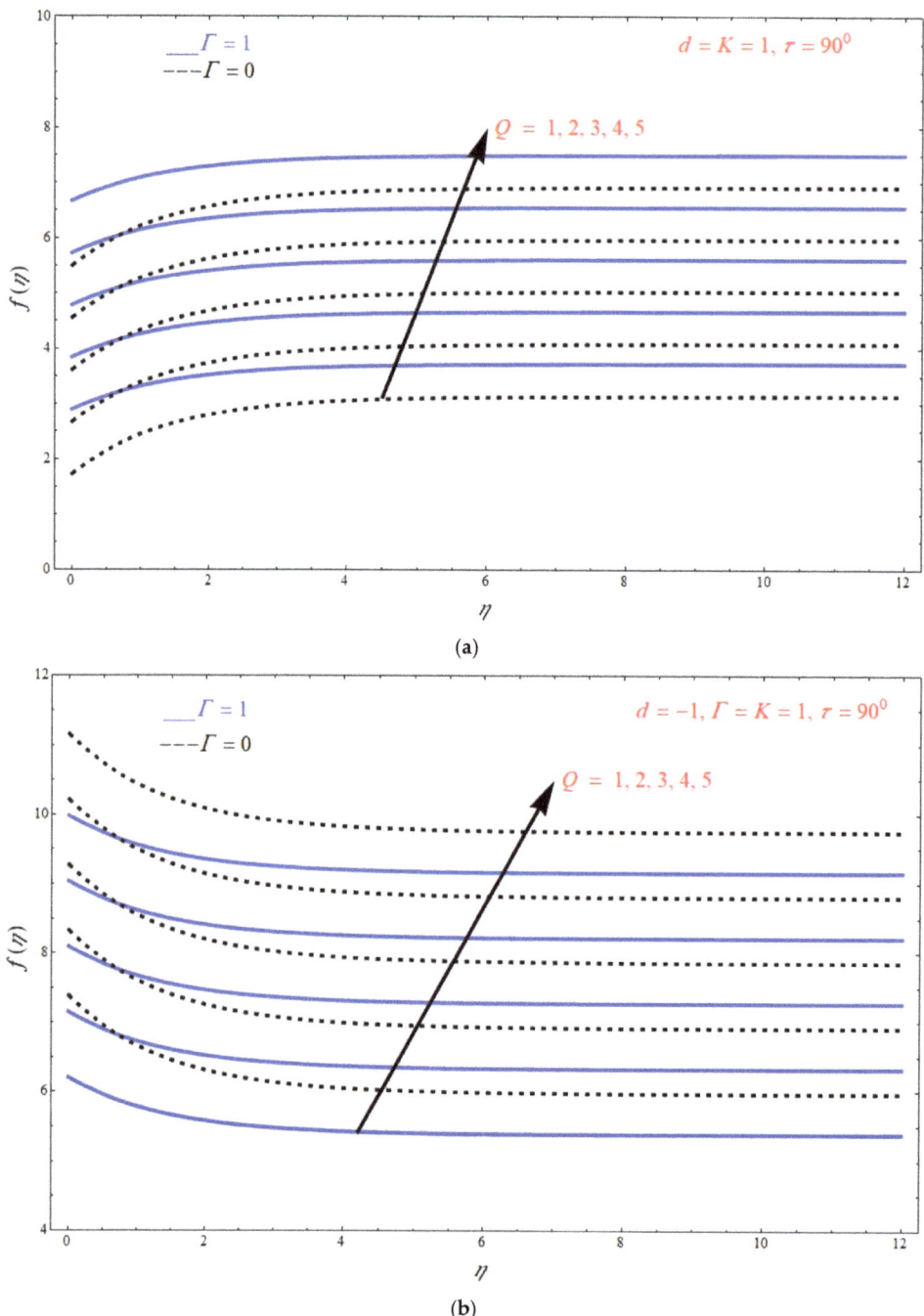

Figure 2. Impact of $f(\eta)$ on η for various choices of Q for (**a**) $d = 1$ and (**b**) $d = -1$.

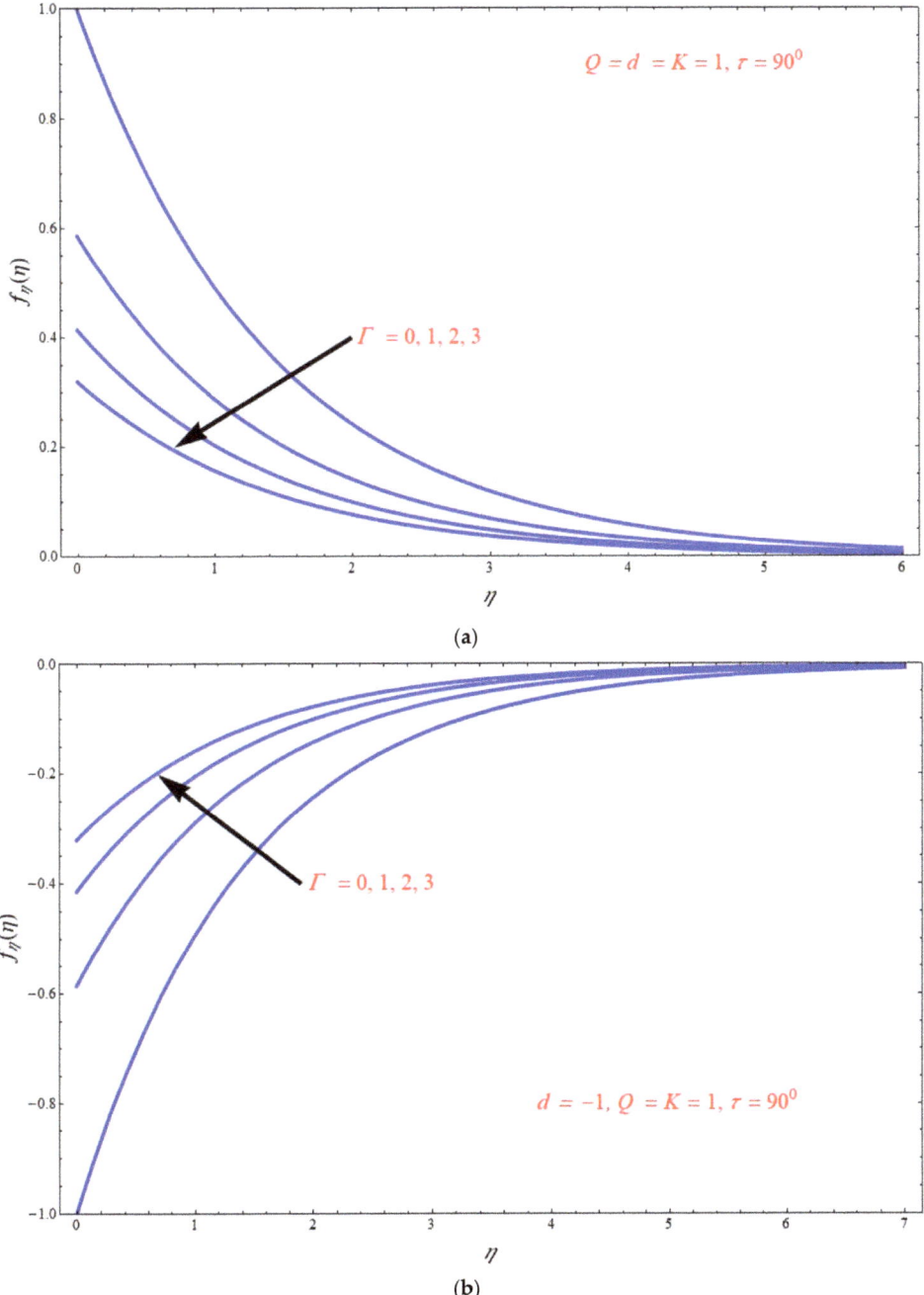

Figure 3. Plots of $f_\eta(\eta)$ verses η for different values of Γ for both (**a**) $d = 1$ and (**b**) $d = -1$.

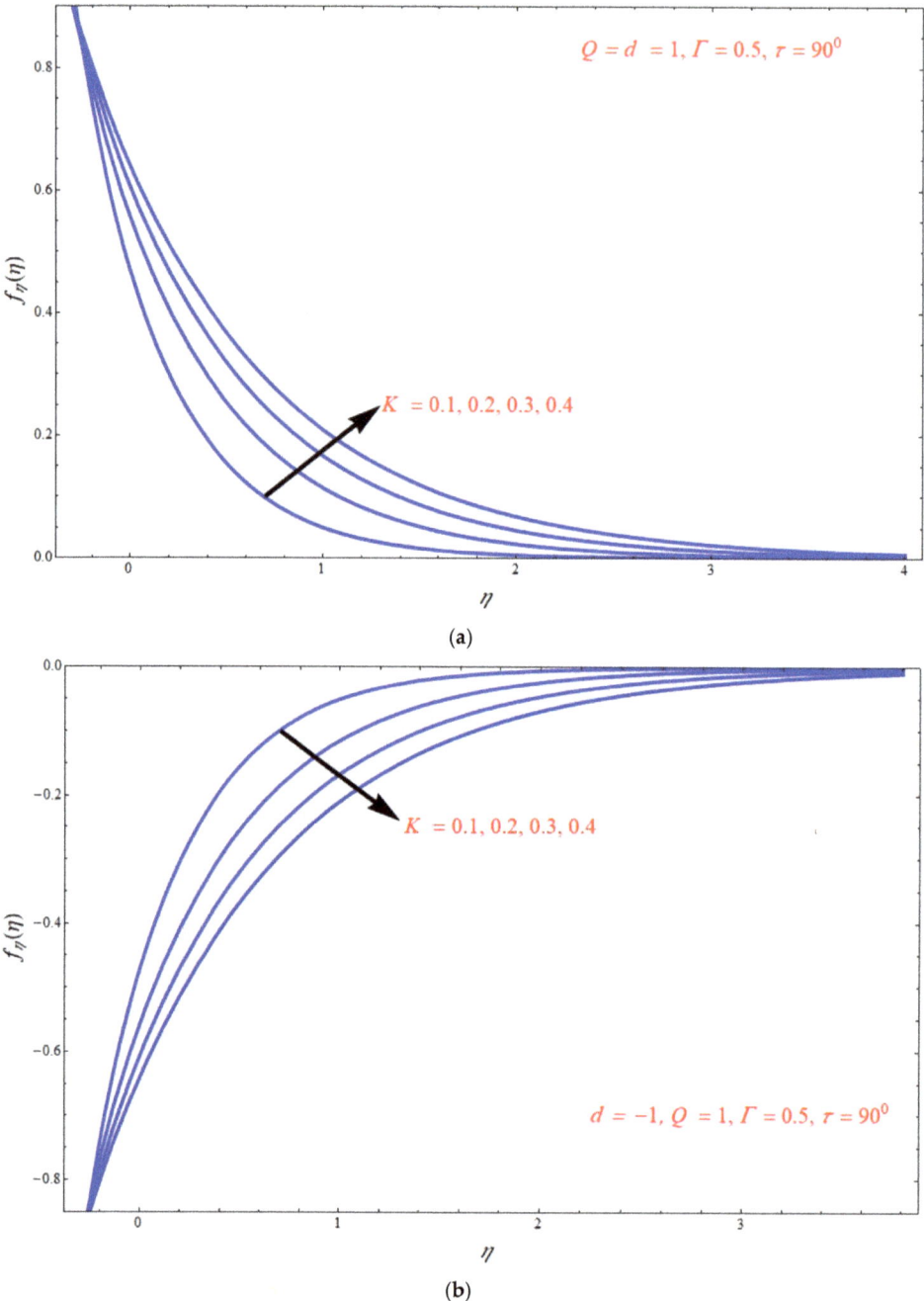

Figure 4. Plots of $f_\eta(\eta)$ verses η for different choices of k_1 for (**a**) $d = 1$ and (**b**) $d = -1$.

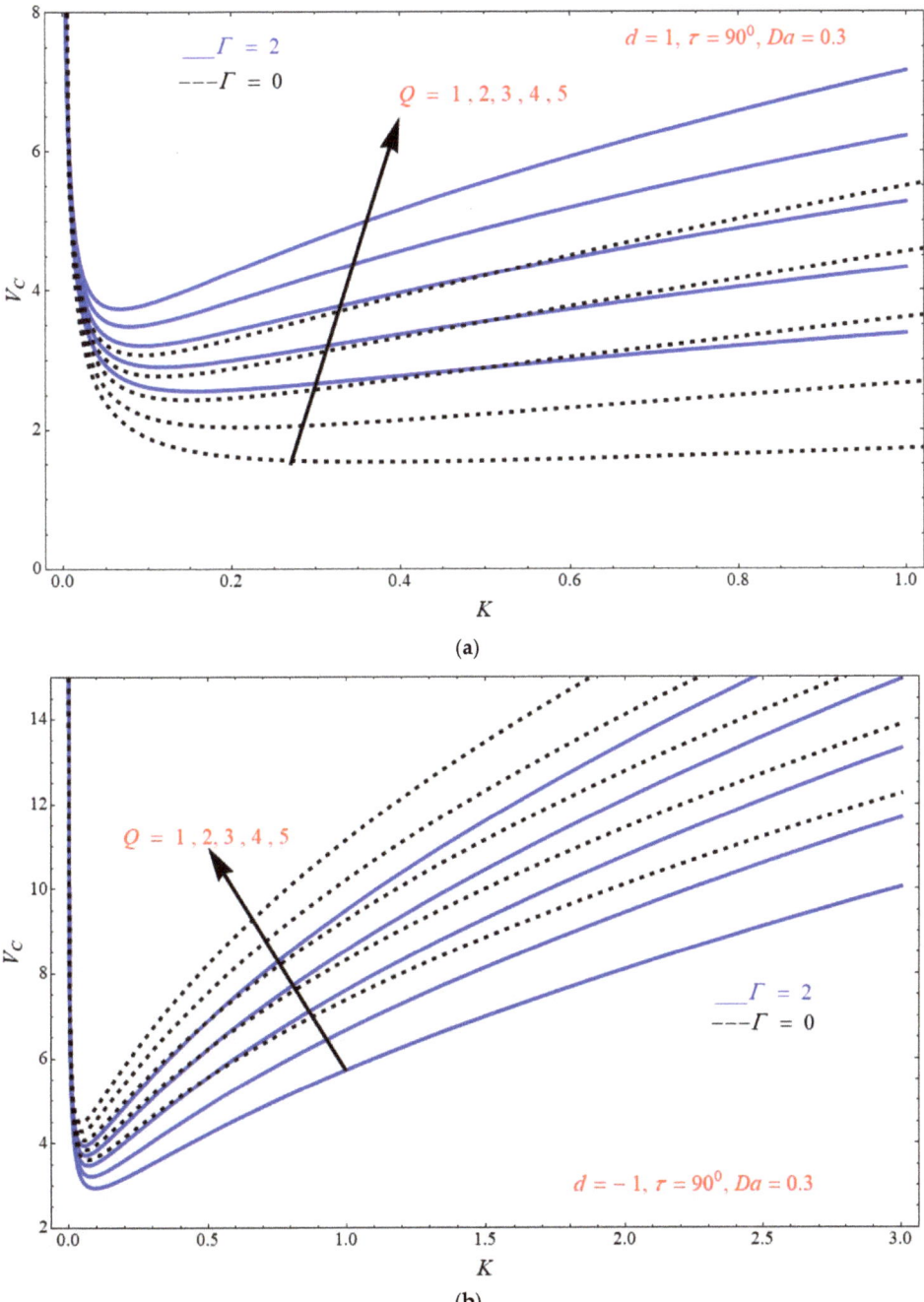

Figure 5. Impact of V_C on K for different values of Q for both (**a**) $d = 1$ and (**b**) $d = 1$.

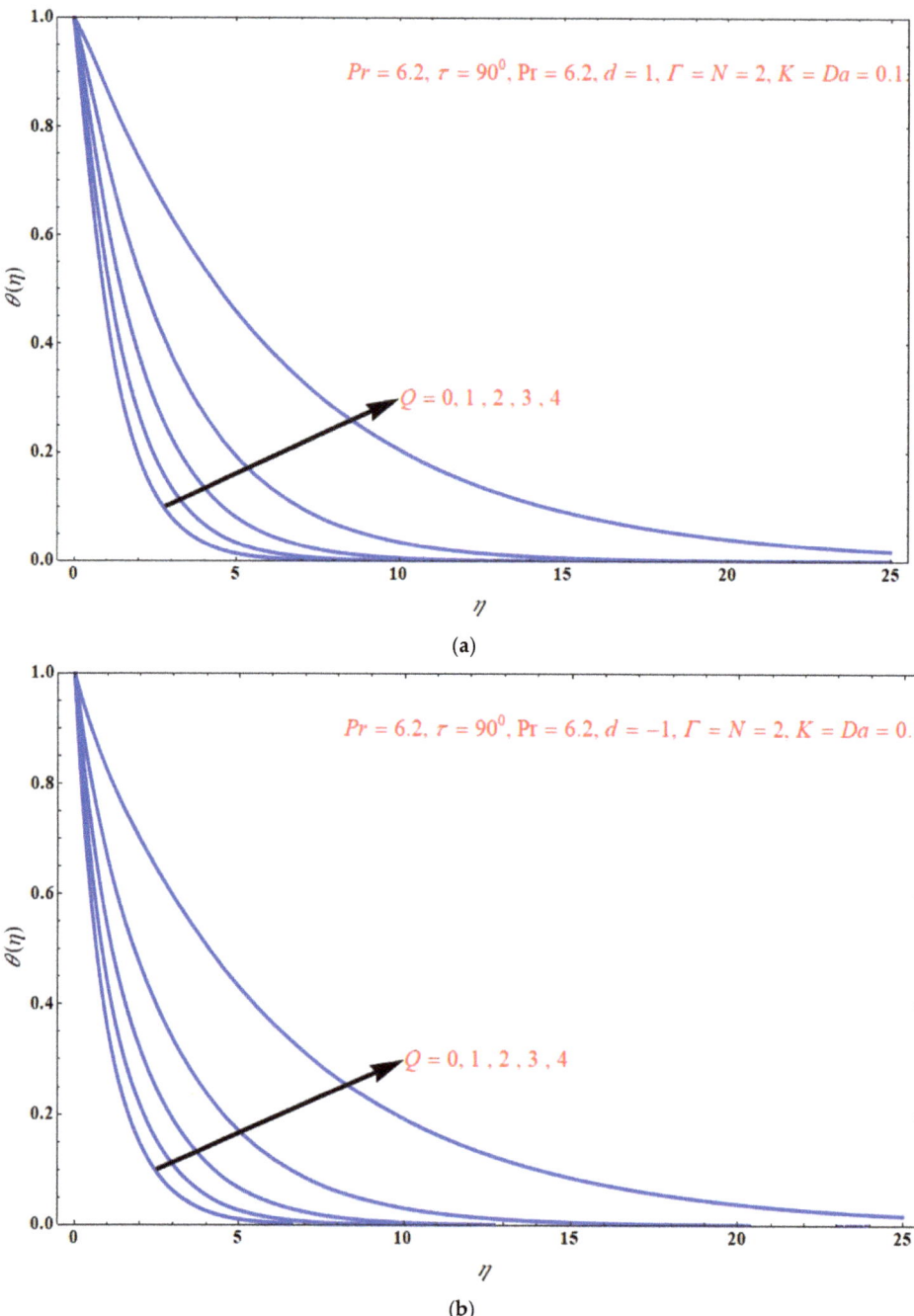

Figure 6. The plots of $\theta(\eta)$ verses η for different choices of Q for (**a**) $d = 1$ and (**b**) $d = -1$.

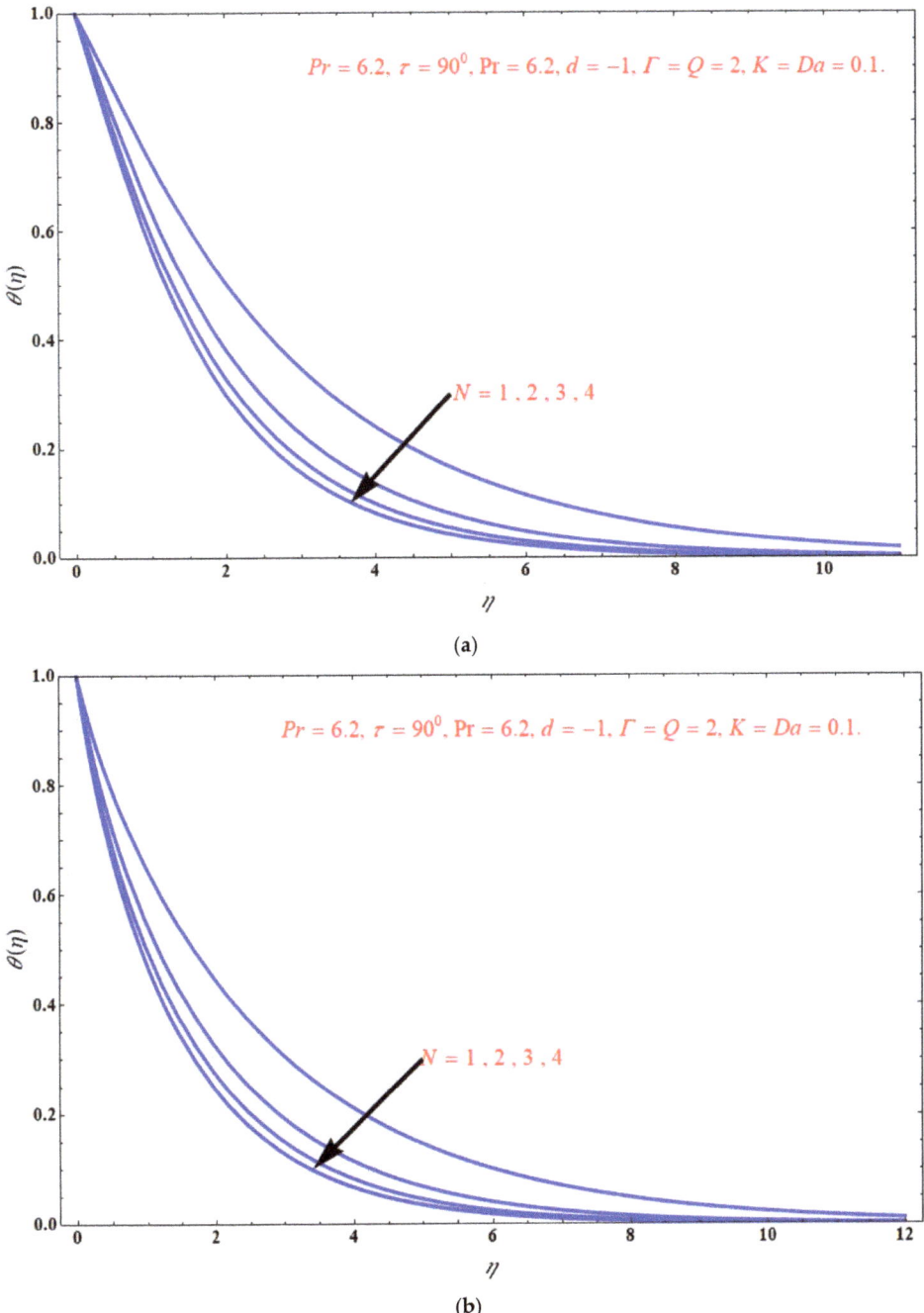

Figure 7. Impact of $\theta(\eta)$ on η for various choices of Q for both (**a**) $d = 1$ and (**b**) $d = -1$.

6. Concluding Remarks

A steady 3-D fluid flow over a porous sheet was taken to analyse the present study under the impact of inclined magnetic field. Multiple slips are considered in the current study to yield better results to the problem. The PDEs of the current problem were mapped into ODEs using suitable variables. Then, analytical solutions were obtained using various parameters. Graphical representations were achievable by using different parameters. With the graphical arrangements, the following results can be deduced:

$f(\eta)$ is for values of Q for both $d = 1$, and $d = -1$.

$f_\eta(\eta)$ less for values of Γ for $d = 1$. Also, it is for values of Γ for $d = -1$.

$f_\eta(\eta)$ increases with increased choices of k_1 for $d = 1$, but it decreases with increasing the values of k_1 for shrinking sheet condition.

V_C is for values of Q for both $d = 1$ and $d = -1$.

If $\tau = 0$, $\phi = 0$, $Bi \to \infty$ to get the results of Vishalakshi et al. [28].

If $Q = \beta = Da^{-1} = R = L = \tau = 0$. to get the results of classical Crane [2].

Author Contributions: Conceptualization: U.S.M.; methodology: U.S.M. and D.L.; software: A.B.V. and T.M.; formal analysis: A.B.V., T.M. and U.S.M.; investigation: A.B.V., T.M., U.S.M. and D.L.; writing—original draft preparation: U.S.M.; writing—review and editing: D.L. All authors have read and agreed to the published version of the manuscript.

Funding: D.L. acknowledges partial financial support from Centers of Excellence with BASAL/ANID financing, Grant Nos. AFB180001, CEDENNA.

Institutional Review Board Statement: Not available.

Informed Consent Statement: Not available.

Data Availability Statement: Data sharing is not applicable to this article.

Conflicts of Interest: The authors have no conflict to disclose.

Nomenclature

a and b	Stretching/shrinking sheet coefficient constant $[s^{-1}]$
B_0	Strength of the magnetic field $[wm^{-2}]$
C_P	Specific heat $[JKg^{-1}K^{-1}]$
d	Length scale $[-]$
Da	Darcy number $[-]$
Q	Chandrasekhar's number $[-]$
Pr	Prandtl number $[-]$
k_1	Permeability of porous medium m^2
k	Material constant of fluid $[-]$
K	Viscoelasticity $[-]$
l	Slip factor $[-]$
m	Constants to be determined $[-]$
N	Radiation parameter $[-]$.
q_r	Heat flux $[Wm^{-2}]$
T	Fluid temperature $[K]$
T_w	Wall temperature $[K]$
T_∞	For field temperature $[K]$
u v and w	Axial velocity towards x axis $[ms^{-1}]$
V_C	Mass transpiration $[-]$
w_0	Wall transfer velocity $[mg]$
x, y and z	Coordinates $[m]$

Greek symbols

α	Thermal diffusivity $[m^2 s^{-1}]$	
η	Similarity variable $[-]$	
Γ	Parameter of the analytical solution $[-]$	
λ	Constant domain $[-]$	
ν	Kinematic viscosity $[m^2 s^{-1}]$	
ρ	Density $[kgm^{-3}]$	
σ	Electrical conductivity $[S\ m^{-1}]$	
τ	Inclined angle [Rad]	
θ	Scaled fluid temperature [K]	
∞	Away from the sheet $[-]$	
γ_0	Porosity $[p \cdot u]$	

Abbreviations

BCs	Boundary conditions $[-]$
MHD	Magnetohydrodynamics
ODEs	Ordinary differential equations $[-]$
PDEs	Partial differential equations $[-]$

References

1. Sakiadis, B.C. Boundary layer behaviour on continuous solid surfaces: I boundary layer equations for two dimensional and axisymmetric flow. *AIChE J.* **1961**, *7*, 26–28. [CrossRef]
2. Crane, L.J. Flow past a stretching plate. *Z. Angew. Math. Phys.* **1970**, *21*, 645–647. [CrossRef]
3. Andersson, H.I. An exact solution of the Navier-Stokes equations for magnetohydrodynamics flow. *Acta Mech.* **1995**, *113*, 241–244. [CrossRef]
4. Andresson, H.I. Slip flow past a stretching surface. *Acta Mech.* **2002**, *158*, 121–125. [CrossRef]
5. Wang, C.Y. Stagnation flow towards a shrinking sheet. *Int. J. Non-Linear Mech.* **2008**, *43*, 377–382. [CrossRef]
6. Fang, T.G.; Zhang, J. Thermal boundary layers over a shrinking sheet: An analytical solution. *Acta Mech.* **2010**, *209*, 325–343. [CrossRef]
7. Miklavcic, M.; Wang, C.Y. Viscous flow due to a shrinking sheet. *Q. Appl. Math.* **2006**, *64*, 283–290. [CrossRef]
8. Turkyilmazoglu, M. Multiple solutions of heat and mass transfer of MHD slip flow for the viscoelastic fluid over a stretching sheet. *Int. J. Therm. Sci.* **2011**, *50*, 2264–2276. [CrossRef]
9. Turkyilmazoglu, M.; Pop, I. Exact analytical solutions for the flow and heat transfer near the stagnation point on a stretching/shrinking sheet in a Jeffrey fluid. *Int. J. Heat Mass Transfer.* **2013**, *57*, 82–88. [CrossRef]
10. Mahabaleshwar, U.S.; Anusha, T.; Hatami, M. The MHD Newtonian hybrid nanofluid flow and mass transfer analysis due to super-linear stretching sheet embedded in porous medium. *Sci. Rep.* **2021**, *11*, 22518. [CrossRef]
11. Vishalakshi, A.B.; Mahabaleshwar, U.S.; Sheikhnejad, Y. Impact of MHD and mass transpiration on Rivlin-Ericksen liquid flow over a stretching sheet in a porous media with thermal communication. *Transp. Porous Media* **2022**, *142*, 353–381. [CrossRef]
12. Mahabaleshwar, U.S.; Vishalakshi, A.B.; Azese, M.N. The role of Brinkmann ratio on non-Newtonian fluid flow due to a porous shrinking/stretching sheet with heat transfer. *Eur. J. Mech. B/Fluids* **2022**, *92*, 153–165. [CrossRef]
13. Nadeem, S.; Haq, R.U.I.; Akbar, N.S.; Khan, Z.H. MHD three dimensional Casson fluid flow past a porous linearly stretching sheet. *Alex. Eng. J.* **2013**, *52*, 577–582. [CrossRef]
14. Mahabaleshwar, U.S.; Pažanin, I.; Radulović, M.; Suarez-Grau, F.J. Effects of small boundary perturbation on the MHD duct flow. *Theor. Appl. Mech.* **2017**, *44*, 83–101.
15. Mahabaleshwar, U.S. Combined effect of temperature and gravity modulations on the onset of magneto-convection in weak electrically conducting micropolar liquids. *J. Eng. Sci.* **2007**, *45*, 525–540. [CrossRef]
16. Mahabaleshwar, U.S.; Sarris, I.E.; Lorenzini, G. Effect of radiation and Navier slip boundary of Walters' liquid B flow over stretching sheet in a porous media. *Int. J. Heat Mass Transf.* **2018**, *17*, 1327–1337. [CrossRef]
17. Mahabaleshwar, U.S.; Nagaraju, K.R.; Sheremet, M.A.; Baleanu, D.; Lorenzini, E. Mass transpiration on Newtonian flow over a porous stretching/shrinking sheet with slip. *Chin. J. Phys.* **2020**, *63*, 130–137. [CrossRef]
18. Mahabaleshwar, U.S.; Nagaraju, K.R.; Kumar, P.N.V.; Nadagouda, M.N.; Bennacer, R.; Sheremet, M.A. Effect of Dufour and Soret mechanisms on MHD mixed convective—Radiative non-Newtonian liquid flow and heat transfer over a porous sheet. *Therm. Sci. Eng. Prog.* **2020**, *16*, 100459. [CrossRef]
19. Mahabaleshwar, U.S.; Rekha, M.B.; Vinay Kumar, P.N.; Selimefendigil, F.; Sakanaka, P.H.; Lorenzini, G.; Ravichandra Nayakar, S.N. Mass transfer characteristics of MHD Casson fluid flow past stretching/shrinking sheet. *J. Eng. Thermophys.* **2020**, *29*, 285–302. [CrossRef]
20. Mahabaleshwar, U.S.; Sarris, I.E.; Hill, A.A.; Lorenzini, G.; Pop, I. An MHD couple stress fluid due to a perforated sheet undergoing linear stretching with heat transfer. *Int. J. Heat Mass Transf.* **2017**, *105*, 157–167. [CrossRef]

21. Mahabaleshwar, U.S.; Kumar, P.N.V.; Sheremet, M. *Magnetohydrodynamics Flow of a Nanofluid Driven by a Stretching/Shrinking Sheet with Suction*; Springer: Berlin/Heidelberg, Germany, 2016; p. 1901.
22. Sneha, K.N.; Mahabaleshwar, U.S.; Chan, A.; Hatami, M. Investigation of radiation and MHD on non-Newtonian fluid flow over a stretching/shrinking sheet with CNTs and mass transpiration. *Waves Random Complex Media* **2022**, 1–20. [CrossRef]
23. Anusha, T.; Mahabaleshwar, U.S.; Hatami, M. Navier slip effect on the thermal flow of Walter's liquid B flow due to porous stretching/shrinking with heat and mass transfer. *Case Stud. Therm. Eng.* **2021**, *28*, 101691. [CrossRef]
24. Anusha, T.; Mahabaleshwar, U.S.; Sheikhnejad, Y. An MHD of nanofluid flow over a porous stretching/shrinking plate with mass transpiration and Brinkman ratio. *Transp. Porous Media* **2021**, *142*, 333–352. [CrossRef]
25. Anusha, T.; Huang, H.; Mahabaleshwar, U.S. Two dimensional unsteady stagnation point flow of Casson hybrid nanofluid over a permeable flat surface and heat transfer analysis with radiation. *J. Taiwan Inst. Chem. Eng.* **2021**, *127*, 79–91. [CrossRef]
26. Kumar, P.N.V.; Mahabaleshwar, U.S.; Swaminathan, N.; Lorenzini, G. Effect of MHD and mass transpiration on a viscous liquid flow past porous stretching sheet with heat transfer. *J. Eng. Thermophys.* **2021**, *30*, 404–419. [CrossRef]
27. Li, Z.; Barnoon, P.; Davood, T.; Dehkordi, R.B.; Afrand, M. Mixed convection of non-Newtonian nanofluid in an H-Shaped cavity with cooler and heater cylinders filled by a porous material: Two phase approach. *Fac. Eng. Inf. Sci. Pap. Part B* **2019**, *30*, 2666–2685.
28. Vishalakshi, A.B.; Mahabaleshwar, U.S.; Sarris, I.E. An MHD fluid flow over a porous stretching/shrinking sheet with slips and mass transpiration. *Micromachines* **2022**, *13*, 116. [CrossRef]
29. Rostami, S.; Davood, T.; Shabani, B.; Sina, N.; Barnoon, P. Measurement of the thermal conductivity of MWCNCT-CuO/Water hybrid nanofluid using artificial neural networks (ANNs). *J. Therm. Anal. Calorim.* **2021**, *143*, 1097–1105. [CrossRef]
30. Makarim, D.A.; Suami, A.; Wijayanta, A.T.; Kobayashi, N.; Itaya, Y. Marangoni convection within thermosolutal and absorptive aqueous LiBr solution. *Int. J. Heat Mass Transf.* **2022**, *188*, 122621.
31. Wijayanta, A.T. Numerical solution strategy for natural convection problems in a triangular cavity using a direct meshless local petrov-Galerkin method combined with an implicit artificial-compressibility model. *Eng. Anal. Bound. Elem.* **2021**, *126*, 13–29.
32. Rahmati, A.R.; Akbari, O.A.; Marzban, A.; Davood, T.; Farzad, R.K.P. Simultaneous investigations effects of non-Newtonian nanofluid flow in different volume fractions of solid nanoparticles with slip and no-slip boundary conditions. *Therm. Sci. Eng. Prog.* **2018**, *5*, 263–277. [CrossRef]
33. Mahabaleshwar, U.S.; Sneha, K.N.; Huang, H.N. An effect of MHD and radiation on CNTS-water based nanofluid due to a stretching sheet in a Newtonian fluid. *Case Stud. Therm. Eng.* **2021**, *28*, 101462. [CrossRef]
34. Mahabaleshwar, U.S.; Vishalakshi, A.B.; Andersson, H.I. Hybrid nanofluid flow past a stretching/shrinking sheet with thermal radiation and mass transpiration. *Chin. J. Phys.* **2022**, *75*, 152–168. [CrossRef]
35. Mahabaleshwar, U.S.; Anusha, T.; Sakanaka, P.H.; Bhattacharyya, S. Impact of Lorentz force and Schmidt number on chemically reactive Newtonian fluid flow on a stretchable surface when Stefan blowing and thermal radiation and significant. *Arab. J. Sci. Eng.* **2021**, *46*, 12427–12443. [CrossRef]

Review

A Critical Review of Supersonic Flow Control for High-Speed Applications

Abdul Aabid [1,*], Sher Afghan Khan [2] and Muneer Baig [1]

[1] Department of Engineering Management, College of Engineering, Prince Sultan University, P.O. Box 66833, Riyadh 11586, Saudi Arabia; mbaig@psu.edu.sa
[2] Department of Mechanical Engineering, Faculty of Engineering, International Islamic University Malaysia, Kuala Lumpur 50728, Malaysia; sakhan06@gmail.com
* Correspondence: aaabid@psu.edu.sa or aabidhussain.ae@gmail.com

Citation: Aabid, A.; Khan, S.A.; Baig, M. A Critical Review of Supersonic Flow Control for High-Speed Applications. *Appl. Sci.* **2021**, *11*, 6899. https://doi.org/10.3390/app11156899

Academic Editors: Luís L. Ferrás and Alexandre M. Afonso

Received: 26 June 2021
Accepted: 21 July 2021
Published: 27 July 2021

Publisher's Note: MDPI stays neutral with regard to jurisdictional claims in published maps and institutional affiliations.

Copyright: © 2021 by the authors. Licensee MDPI, Basel, Switzerland. This article is an open access article distributed under the terms and conditions of the Creative Commons Attribution (CC BY) license (https://creativecommons.org/licenses/by/4.0/).

Abstract: In high-speed fluid dynamics, base pressure controls find many engineering applications, such as in the automobile and defense industries. Several studies have been reported on flow control with sudden expansion duct. Passive control was found to be more beneficial in the last four decades and is used in devices such as cavities, ribs, aerospikes, etc., but these need additional control mechanics and objects to control the flow. Therefore, in the last two decades, the active control method has been used via a microjet controller at the base region of the suddenly expanded duct of the convergent–divergent (CD) nozzle to control the flow, which was found to be a cost-efficient and energy-saving method. Hence, in this paper, a systemic literature review is conducted to investigate the research gap by reviewing the exhaustive work on the active control of high-speed aerodynamic flows from the nozzle as the major focus. Additionally, a basic idea about the nozzle and its configuration is discussed, and the passive control method for the control of flow, jet and noise are represented in order to investigate the existing contributions in supersonic speed applications. A critical review of the last two decades considering the challenges and limitations in this field is expressed. As a contribution, some major and minor gaps are introduced, and we plot the research trends in this field. As a result, this review can serve as guidance and an opportunity for scholars who want to use an active control approach via microjets for supersonic flow problems.

Keywords: flow control; de Laval nozzle; CD nozzle; microjet; supersonic flow; CFD; DOE

1. Introduction

In supersonic vehicles, the flow of exit from the rockets and missile engines has become a significant issue. It has been found that the loss of air inside the high-speed vehicle engines turns to increase the base drag. For example, a nozzle with sudden expansion ducts will form a recirculation zone, increasing base drag. When the base drag increases, the total amount of exit pressure will decrease, and this decrement will result in the loss of the forwarding force or thrust. Hence, many studies have reported controlling the high-speed flows as a passive and active control method. In a passive control method, the duct shape is modified with additional devices/shapes, such as ribs, cavities, cylinders, aerospikes, splitter plate, etc. In addition, researchers used different devices of flow formation, such as a nozzle as internal flow control and bluff body, non-circular cylinders, airfoil, and wings as external aerodynamics flow control. On the other hand, the active control of high-speed flow has been studied extensively over the last two decades. Researchers have used a high-speed nozzle with a sudden expansion duct and a microjet controller; a tiny hole in the base area is drilled to control the flow, which was found to be an excellent technique in a supersonic flow problem. Hence, this review is more focused on the active control approach, using a microjet in a CD nozzle.

The abrupt expansion of the external compressible flow over the back of the projectiles and its association to the base pressure has long been the focus of researchers' interest. The base drag, which accounts for a significant portion of the overall drag, is determined by the base pressure. Generally, the base pressure for a high-speed projectile is lower than the ambient pressure. The vast majority of the ballistics test data were supplied, leading us to presume that the base pressure ratio depends entirely on the flight Mach number. Compared to traditional ballistics testing processes, the experimental investigation of the internal flow apparatus provides several distinguishing benefits. A significant amount of air supply is lowered, which would generally be necessary for the wind tunnel test section to be large enough so that wall interference and other factors do not disrupt the model flow. Internal flows are free of stings and other support devices that are necessary for external flow investigations. The most significant benefit of the internal flow device is that static pressure and surface temperature measurements can be recorded as well as the entry to the expansion and the wake zone (Figure 1). These observations are crucial if the theoretical predictions are to be extensively investigated.

The sudden expansion problems in the subsonic and supersonic flow regimes are found in many applications. We discovered that previous researchers used a system to replicate high altitude conditions in jet and rocket engine test cells, jet discharge results in insufficient, sub-atmospheric discharge pressure. This was found by Khan et al. [1], who used microjets to control the sudden expansion flow (base pressure) from the CD nozzle as an active control method.

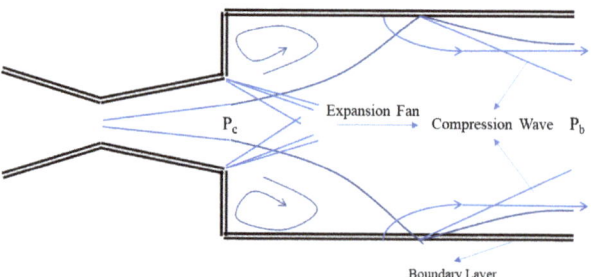

Figure 1. Sudden expansion flow field. Reprinted with permission from ref. [2]. Copyright 2021 Springer.

In this review, the following section explores the fundamental idea about flow development and its types, which gives a basic understanding of this work. Section three illustrates selected studies on the passive control method of high-speed flow, jet, and noise, which provides information about the passive devices and their benefits. Section four is the main objective of this work. Hence, for the detailed investigation, this is split into the different methodologies: experimental, computational fluid dynamics (CFD) and soft computing methods. We reviewed the most relevant papers related to the current work. Section five is the main contribution of this review; based on the existing data in this field, we explore and critically analyze the results. In addition, we propose some research gaps for future work and finally conclude, based on the current existing work.

2. High-Speed Flow Development in Nozzle

For supersonic flow development and investigations, a CD nozzle is utilized in most studies. A nozzle with no expanding part cannot produce supersonic air [3]; the flow is sonic at the throat; therefore, an asterisk denotes conditions at a sonic level. At the throat, the Mach number = 1, (V = a) and the throat area (a). Figure 2 illustrates a basic CD nozzle model [4] and the parameters with an asterisk are defined as critical values. If a high-pressure tank is connected to a pipe, then the velocity at the pipe exit changes depending on the backpressure. At any other region of the nozzle, the Mach number velocity and the local area can be derived by the continuity equation [5–7]. Even though it is possible to study one-dimensional flow behavior directly, it is only a particular case of two-dimensional flow. One of the known one-dimensional flow phenomena is the normal shock and formation of oblique shock waves. The occurrence of oblique shock waves in different flow fields occurs, such as flight at high Mach numbers, aircraft design, diffusers, and supersonic nozzles.

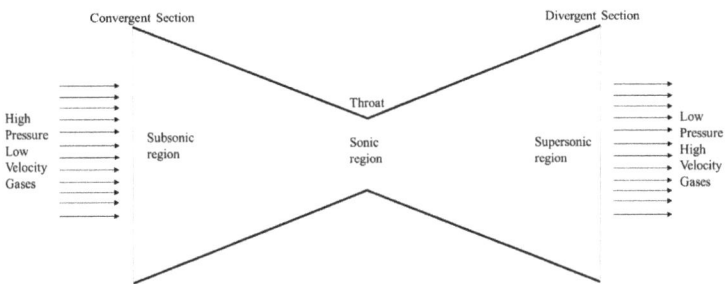

Figure 2. Convergent–divergent (CD) nozzle.

In previous studies, a particular flow rate was solved and analyzed mathematically to solve the energy equations, using a two-dimensional model of a nozzle. The measured iteration method improved the static tension, temperature, and velocity flows. The nozzle is produced and meshed with an automated technique and the sizing of the meshing processes and various values. The CD nozzle with sudden expansion was used to investigate the effect of expansion ratio for creeping expansion flows of fluid in the study of [8–11]. The CD nozzle was used to generate a supersonic flow and was simulated, using the ANSYS fluent software. The generic formula for the nozzle was determined manually, and the results were compared [3,12–14].

Nevertheless, the greatest difficulty in the CD nozzle outflow is in the subsonic and supersonic flow regimes. An abrupt expansion of the problem has various applications in the industry. In jet and rocket engine test cells, we observed that the system is used to simulate high-altitude conditions; the discharging of the jet results in inadequate discharge pressure, which is sub-atmospheric. Khan et al. [15] controlled the base pressure with the active control method, using microjets. In the active control method, the microjets were placed in the base region of the sudden expansion duct, and these microjets were directly connected to the settling chamber with the air directly passing to the duct. The inlet flow of the microjets reduced the recirculation zone, which resulted in decreasing the drag formation. A detailed study on these active flow control can be seen in Section 4.

Usually, the fluid flows are three-dimensional; the terms one, two, or three refer to the amount of coordinated space required to demonstrate the flow. The physical motion usually tends to be three-dimensional. At the same time, these are hard to evaluate and call for the most significant possible simplification. This is done by ignoring the flow variation in any direction, thus simplifying the problem. A three-dimensional problem can be reduced to a two-dimensional one and, subsequently, can be further simplified to one-dimensional one. The continuity equation, Bernoulli's equation, and momentum equation are used to study the one-dimensional flow case as methods of solution [5,16–19]. General classification and examples of flow fields can be seen in Refs. [20,21] which provided the information of 3D flows and sediment transport models for open channel flows [20]. Chong et al. [21] classified the 3D flow, considering the first-order coupled linear differential equation, using three matrix variations for both compressible and incompressible flows. The effect of nozzle geometry on the high-velocity oxygen fuel (HVOF) system was calculated [22]; the nozzle attached inside were adjusted, and they also proved that the cold spray technique influences the HVOF system and applies to the nozzle configuration.

3. Passive Control Methods

This section summarizes some of the passive control approaches of previous researchers, and it is split into three major concerns: flow, jet, and noise control, in order to extract the idea of these methods.

3.1. Flow Control

The examination of abrupt axi-symmetric expansion is a challenging subject that is becoming popular in a range of flow systems in the current scenario. The duct is employed with a smooth inner surface and a low base pressure in most cases. Sudden relief to the shear layer at the nozzle exit is due to the availability of a duct with a larger area. Due to a sudden increase in the duct area, the base pressure and flow-field may be articulated at the base area through the vortex dynamics produced, due to a sudden increase in the duct area. In contrast, the value of the base pressure is almost similar in the base area; however, the mean values are taken for the analysis. As a result of its wide application, abruptly increased flows have been thoroughly investigated.

The investigations were carried out in an abrupt expansion duct in several cases with passive devices such as a rib, cavity, step body, and boattail. Several studies were carried out experimentally to control the flow, and these results serve as the benchmark results. Viswanath [23] studied the impacts of riblets in pressure gradients and three-dimensionality on airfoils, wings, and wing–body combinations in different speed regimes to control the viscous drag in 3M riblets. Ishide and Itazawa [24] tweaked the leading edge flaps to enhance the delta wing design at a low Reynolds number of 1.9105 applied to a chord (geometric mean) of 286 mm as the typical length for micro and crewless air vehicles.

The flow past desired blunt-nosed body with and without spikes was tested in a primary 2D water channel, at a given Reynolds number [25], with spikes of various forms, such as conical, hemispherical, flat, and square nose. Khurana and Suzuki [26] used a forward-facing aerospike on the nose to examine the heat transfer and its control through the aerospikes for lifting the body configuration in a hypersonic flow. They used experimental techniques and a simplified model of the resulting basic symmetric-delta configuration. A similar study by the same authors [27] was carried out to assess the aerodynamic efficacy for aerospike on a hypothetical lifting-body in a hypersonic flow at various Mach numbers. In Ref. [28], the experimental research was carried out to determine the importance of the pressure hill height and the zone expansion impact for a flow-through in different forms of spikes with Reynolds numbers ranging from 2278 to 4405 to detect the vortex shedding process. In a prior work, the previous researchers [29] studied a theoretical estimate of the shock standoff distance in a supersonic range with the modified Newtonian impact theory. Based on this passive control strategy, the researchers [30] conducted the tests using a transonic wind tunnel of 38×30 cm in the Mach number ranges of 0.7 to 1.0 and controlled the entire afterbody drag of the outlet flow. In the absence of the jet flow at the base, the same author [31] examined the development of flow management with passive control devices for the axi-symmetric base and afterbody drag reduction in different speed regimes.

Experimentally, [32] studied the control of flow in a sonic jet in a circular convergent nozzle by locating two tabs at the nozzle exit. The effects of annular ribs as a passive control device on the base pressure control was examined, using an experimental technique in a subsonic and sonic nozzle with an abrupt expansion duct [33]. In addition, the researchers investigated the pressure fluctuations in a typical missile base area design at a freestream Mach number of 0.7, with and without a base cavity [34]. The study was discovered in resonant high-speed cavity flows controlled by high frequency pulsed supersonic microjets [35] in addition to CD nozzles. A compression corner calculation model was employed to perform extensive numerical investigations in the supersonic flow field with varied injection pressure ratios, actuation locations, and nozzle types [36]. Figure 3 illustrates the different devices that were used to control the high-speed supersonic flows, particularly in a CD nozzle.

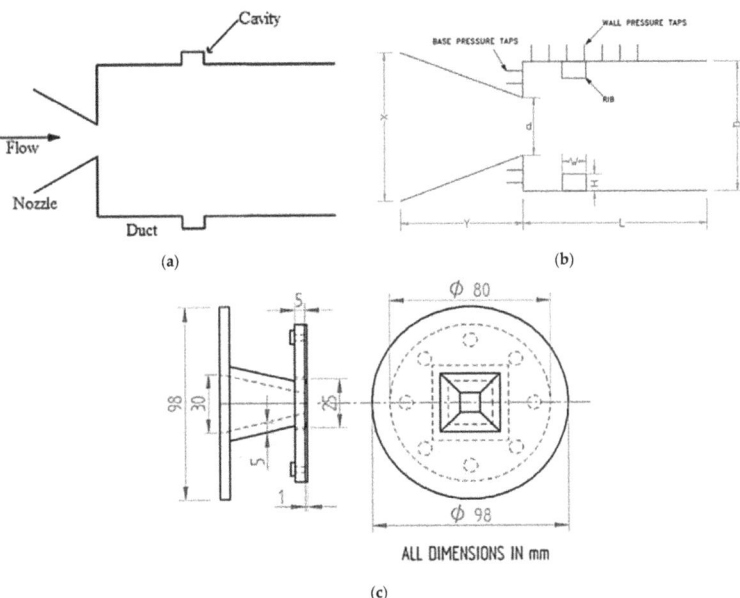

Figure 3. Passive control devices: (**a**) cavities [37], (**b**) ribs [38], and (**c**) cylinders [39,40]. Reprinted under the Creative Commons (CC) License (CC BY 3.0).

3.2. Jet Control

Mixing is required for efficient and effective jet performance in a variety of aerospace applications. Thrust vector control, missile engines, and aircraft propulsion systems are just a few of the technical applications based on jets. To manage the combustion chamber size and improve the vehicle's efficiency, effective mixing is necessary for air-breathing engines. Optimal mixing on a small and big scale is required for combustion cycles to work properly. Small-scale mixing is concerned with molecule mixing, whereas large-scale mixing is concerned with the dismantling of large-scale vortices. Acoustic radiations are caused by the existence of turbulent vortical structures and compressible waves (such as shock waves in supersonic flows), which frequently exceed acceptable levels in practical circumstances such as military and surveillance aircraft [41].

There are several studies that have reported on jet control, and we consider some cases in this review. It was observed that the high-speed flow controlled by the excitation of free shear-layer instabilities [42] uses localized arc filament plasma actuators in jets for different Mach numbers at a supersonic range. Focusing the aspect ratio of the nozzle and nozzle pressure ratio (NPR) with the Mach number variation, the elliptic jet control with limiting tab [43] was investigated and showed that each parameter is important to control the jet flows, similar to the overexpanded plug nozzle jet [44] controlled by the passive method. Additionally, some researchers used ventilated triangular tabs to control the jet [45], control the supersonic elliptic jet with ventilated tabs [46], and to measure the impact of the tab location relative to the nozzle exit on the shock structure of a supersonic jet [47].

Khan et al. [48] experimentally investigated the effect of the extended cowl on the flow field of planar plug nozzles for two different Mach number ranges (1.8 and 2.2) to observe the influence of the cowl length for the pressure distribution. Manigandan and Vijayaraja [49] experimentally investigated the flow-field and acoustic characteristics of the elliptical throat in the CD nozzle. According to the findings, switching from an elliptical to a circular throat alters the shock cell architecture, resulting in a substantial shift in the scream amplitude, owing to wave weakening. The jet controlled for mixing the flow was experimentally studied by Khan et al. [50] for the enhancement of the supersonic twin-jet mixing by vortex generator to observe the effects and the behavior of the daughter streams. Similarly, an impinging plug nozzle jet using a vortex generator [51] was studied experimentally. Figure 4 shows the nozzle and tab details, CAD drawing (Figure 4a), photographic view without tabs (Figure 4b), schematic sketch of nozzle exits with triangular tabs along the major and minor axes (Figure 4c), and triangular tab dimensions and the photograph of the nozzle with triangular tabs along the major axis (Figure 4d).

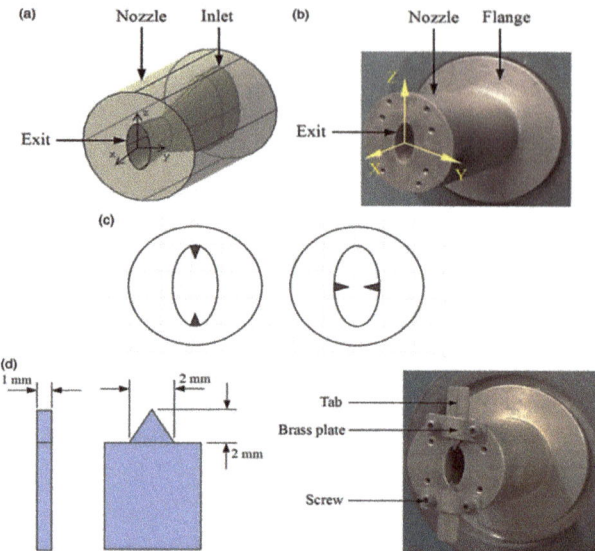

Figure 4. Jet control device, (**a**) Nozzle drawing, (**b**) Nozzle image without tabs, (**c**) Both minor and major tabs on the nozzle drawing (**d**) Triangular tabs dimension and nozzle image with major tabs [52].

3.3. Noise Control

In the past, the decrease in shock-related noise from aviation engines operating at supercritical nozzle pressure ratio received attention. The fact that shock-related noise occurs on many engines, including the turbofan engines utilized in today's commercial aircrafts, emphasizes its relevance. It has long been known and shown in model scale tests that employing a proper CD nozzle instead of a convergent nozzle, as used in most supersonic aircraft engines, may minimize/remove the shock noise component [53]. However, due to many unsolved issues concerning the noise characteristics of CD nozzles operating under non-design circumstances, the actual application of this idea has been avoided [54]. Although an appropriate CD nozzle may be used to produce a shock-free jet flow at a certain design pressure ratio, the same nozzle can be used for off-design pressure ratios during takeoff and landing operations, resulting in shock noise [55].

The bulk of these noise reduction systems are referred to be passive since they cannot be switched off or changed while in flight and might result in performance losses. Penn State [56] is developing a fluid insert technique for supersonic jet noise reduction. The fluid insert method aims to reduce noise in low bypass ratio turbofans while having minimal impact on engine performance. The fluid inserts blast air into the diverging portion of the nozzle on demand, which may be turned off or adjusted depending on the flight regime. Although significant research has been conducted in the form of noise measurements and Reynolds-averaged Navier–Stokes (RANS) calculations to enhance the fluid insert technology [57,58], the reason why these inserts work is still not understood completely. The correlation of changes in the flow field with corresponding changes in the noise is inadequate, using only existing RANS data [59,60]. It was suggested that unstable scale-resolving simulations be used to obtain more insight into the flow field and to better understand noise reduction techniques [56]. Additionally, when properly structurally supported, acoustic reflectors of an adequate scale are a suitable noise reduction solution for the high-pressure venting typical of blowdown operations [61]. Due to disadvantages, such as delayed convergence and the complexity of the phase shift mechanism, a unique technique was used that does not use secondary path modeling [62] and the sensitivity of noise to system uncertainties [63]. Fluid inserts reduce the convection speed of wave packets in the jet shear layer, resulting in a reduced Mach wave radiation angle [64].

This interaction is discussed here with a focus on noise creation and reduction when a jet is parallel to or impinging on a solid wall. Various researchers have presented computational methods [65,66] with large-eddy simulations [67–71], high fidelity simulations [72] and 3D simulation [73]. The noise reduction of supersonic jets by nozzle trailing-edge changes was investigated experimentally [74], and hemispherical noise reduction reflectors on transonic jet flows [75] were conducted. Different injection sites, angles, and circumstances were also investigated, resulting in distinct acoustic behavior and flow-field changes [76]. Using steady fluidic injection, researchers conducted an empirical scaling analysis of supersonic jet control [77]. Aft and lateral wall inclinations for a cavity [78] supersonic cavity flow utilizing high-speed upstream injection [79] and cavity dynamics to the introduction of various storage configurations installed at different positions inside the cavity [80] were all numerically modeled for noise reduction. The employment of a single injector as a fluid insert helps break up the large-scale structures of the flow, according to direct cross-correlations of near-field data with far-field microphone signals [81]. Pipe-jet noise is reduced via geometric changes in the form of trailing edge castellations. The interaction between the streamwise vortices is determined by the number of castellations, which changes the sound generated [82]. Figure 5 shows the noise reduction configurations, which are adopted in the nozzle exit region, and represents the baseline nozzle (Figure 5a), nozzle (Figure 5b), schematic of a single fluid inserts in the nozzle (Figure 5c), and designation of different azimuthal planes for the 3FC-2FI nozzle (Figure 5d).

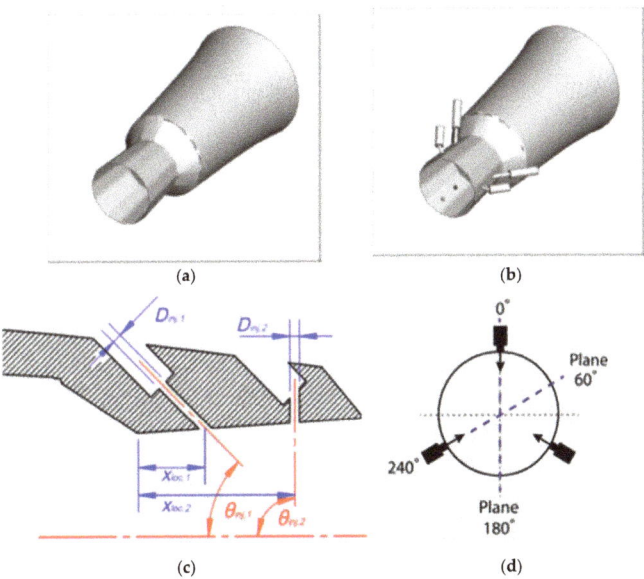

Figure 5. Details of nozzle configurations. Reprinted with permission from ref. [64]. Copyright 2020 Springer.

4. Active Control Methods

The major consideration of this review work is to explore the active control methods of supersonic flows. Therefore, this section is split into the methodologies employed by the researchers: experimental, CFD, and soft computing approaches.

4.1. Experimental Investigation

The wall and base pressure in the suddenly extended duct were measured by changing the geometrical and flow parameters of the CD nozzle by Khan et al. [15]. The Mach number can vary in any supersonic range, as seen in Figure 6 (CD nozzle with a larger duct). To build a CD nozzle, the exit diameter of the nozzle may be kept constant, as the base pressure results for nozzles with an exit diameter are accessible in the literature [15]. To calculate the throat diameter, the isentropic relations were utilized [5]. The nozzles are calibrated after manufacture in order to determine the exact Mach number at the nozzle exit.

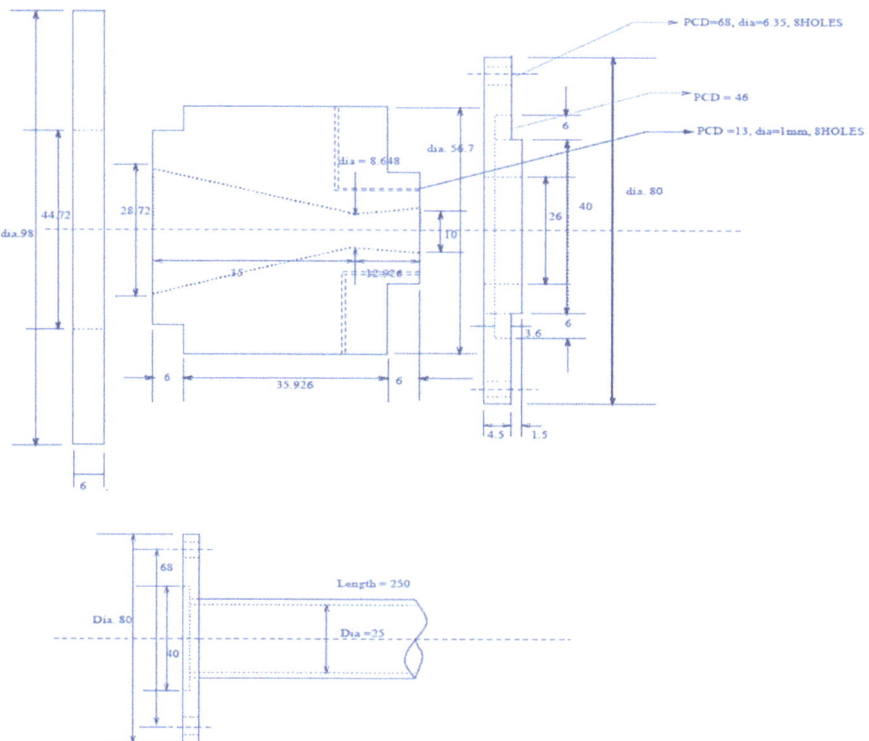

Figure 6. Flow expansion, nozzle, and enlarged duct design for Mach number 1.87.

The experimental test model was an axisymmetric CD nozzle that was attached to a concentric axisymmetric tube with an area ratio, and it could be adjusted. Figure 7 depicts a perspective of the experimental setup that was used for experimentations. The researchers used a pitch circle diameter of 13 mm; there were eight micro-holes, four of which (designated as c) were used for blowing, and the other four (labeled as m) were used to record the base pressure. Controlling the base pressure was accomplished by blowing through control holes (c) with air from the primary settling chamber [15]. The pressure taps are shown in Figure 7 together with the rapid expansion duct. A perspective of the experimental apparatus is shown in Figure 8. Brass was utilized for the investigations' dramatically enlarged ductwork. Lower L/Ds were then achieved by machining the duct after measurements for a certain length of pipe were completed. Pressure taps made of stainless steel were used to record the static wall pressure. Because the flow field soon after departing the nozzle is so important, the first nine holes were spaced four millimeters apart, and the remaining holes were spaced eight millimeters apart. The moisture content of the compressed air is absorbed in a dual-tower semi-automatic silica gel dryer. Later the dried air is heated and used to remove the moisture from the air altogether. A regulatory pressure valve operated by a pressure relief pilot permits the dryer to operate; three tanks were used to store the dry air with a total capacity [1].

In previous experimental methods, the NPR was 3 to 11 at various expansion levels and showed that the L/D could be defined for a given Mach number and NPR, resulting in a cumulative increase/decrease in the base pressure [1,83–88]. The experiments were also conducted by [89] for area ratio 2.56 and area ratio 4.84 [90]. The statistical approach aims to verify the experimental data, obtained at the VTI Žarkovo Institute during supersonic wind tunnel tests: free escape and flow with one chosen type of obstacle [91]. The tests were carried out at design NPR with sudden expansion for the duct length L = 10D to 1D. The considered cross-sectional areas of the tube were 2.56, 3.24, 4.84, and 6.25 [92]. Figure 7 shows the experimental setup for the base and wall pressures.

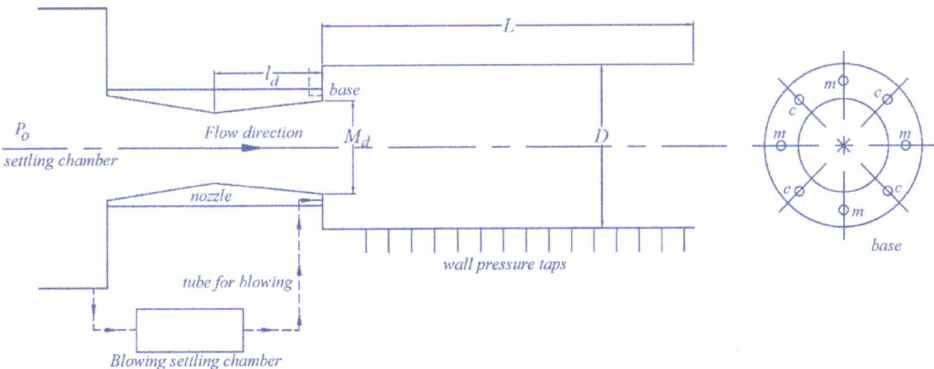

Figure 7. Experimental setup [93]. Reprinted under the Creative Commons (CC) License (CC BY 4.0).

In the presence and absence of a base cavity, the experiments were conducted to analyze the pressure variations in the base region of a conventional missile configuration at a freestream Mach number of 0.7. The authors attempted to identify the variations in pressure and illustrate the influence base cavities have on their behavior. Experiments consisted of unstable calculations of pressure at six azimuthal locations. The disparity in the pressure fluctuations in the azimuthal direction was observed due to a standard axisymmetric, which is significant [35]. For the Mach numbers ranging from 1 to 3, the effect of the microjet control in the CD nozzle with a sudden expansion duct was experimentally tested at supersonic Mach numbers. The research varied nozzle parameters, such as an area ratio of 2.56; NPR ranged from 3 to 11; and L/D ranged from 10 to 1 [94]. They extended their work for Mach 2.5 and 3.0, and cross-sectional areas of 2.56, 3.24, 4.84, and 6.25; the L/D ratio of the duct was 10 to 1, and the NPR used were from 3 and 11 [95].

To control the drag, the researchers [96] worked with base drag and experimentally studied the aerospikes behind the base of bluff bodies. A 1 mm thick plate with two spikes at 11.5 mm was mounted between the nozzle as a passive device. The efficacy of the continuous grooved cavity was tested experimentally to control the base flows with specific parameters [97]; in the transonic system, the aerospikes effectively control the base pressure [98] without disrupting the main flow, respectively.

Figure 8. A view of the extended duct with pressure tapings and setup site [97]. Reprinted under the Creative Commons (CC) License (CC BY 3.0).

In summary, Khan et al., investigated the active control of the base pressure [1,15,82–85,99] in which they considered the differed Mach number at various supersonic flow ranges, such as Mach numbers 1.25, 1.30, 1.48, 1.6, 1.8 and 2.0 in [84,85]; 1.87, 2.2 and 2.58 in [15]; and 2.0, 2.5 and 3.0 in [82], respectively. The test area ratio was from 2.56, 3.24, 4.84, and 6.25; the NPR was from 3 to 9; and the duct length for L = 10D until 1D. More attention was given in the recent study to wall pressure distribution for different inertia levels, relief to the flow, NPR, and the L/D ratio [93,100–107]. After monitoring the flow from the active microjet control, it is also necessary to know whether the flow changes in the duct or not. Hence, recent findings have also shown that several studies are conducted with equivalent or variable Mach number and area ratios with the same NPR and L/D ratios. The majority of wall pressure flows were considered with varying the area ratio, such as 2.56 [93,100], 3.24 [101,102,108,109], 4.84 [103–105,110], and varying the Mach number for the same area ratio 2.56 with Mach 2 and 3 [106], 1.3. 1.9, and 2.4 [111] and for the area ratio 6.25 with Mach 1.1 and 1.5 [112] and 2.1 and 2.8 [113].

4.2. Computational Fluid Dynamics Approach

As the second method of this study on high-speed flow control, a CFD approach was chosen, and a relative work is overviewed to examine the objective and outcomes of the previous investigation.

Apart from experimental studies, a numerical method was employed by various researchers; such studies can be seen in this section. A fluid-structure analysis was found for the delta wing (cropped) [114] based on an aeroelastic solver in the time domain. Two fluidic thrust modulation methods were employed—shock thrust and throat shifting thrust modulation [115]—for the investigation. Initially, the CFD approach examines the supersonic flow through a de Laval nozzle and obtains complete isolation of the thrust flow due to shock waves. The flow was studied for friction, and the temperature of gases at the exit of the combustion chamber [116]. The finite volume model was developed for the estimated two-dimensional and three-dimensional flow formations, using turbulence model efficacies via ANSYS simulation [117]; it was used to investigate the extensive flow field within the supersonic ejector and improved the ejectors mixing chamber wall structure to attain an optimum entrainment ratio to obtain the highest possible capability the ejector can achieve. The ANSYS fluent and ICEM meshing tool was used to conduct the simulation and analyze the ejector performance: k-epsilon realizable and k-ω SST [118] and k-ω [119]. To find the micro-Laval nozzle performance, it was primarily investigated by its machined surface topology, and a circular cross-section micro-Laval nozzle was modeled [120].

Modeling of the nozzle geometry and generating mesh was carried out using the GAMBIT 2.4 program and validated with the findings of the experiment taken from the literature, which are well known [121]. Patel et al. [122] derived the principles of the de Laval nozzle, using the nozzle operating theory, and examined the experimental approach of evaluating flow for various nozzles; the variance of flow parameters, such as pressure, temperature, velocity, and density, was visualized [123] to investigate the effects of geometry and flow parameters on the thrust force created by the flow from the CD nozzles to a circular duct with a larger cross-section. The study was carried out for all Mach numbers using various area ratios (2–12) and NPR (3 to 11) [124–127]. The researchers optimized the rocket air ejector configuration in AutoCAD, then analyzed it using Ansys CFX using the numerical approach with 3D models [128]. The primary purpose was to optimize the CD nozzle location for each operating circumstance and the influence of reflected shock waves and boundary layers on the ejector performance in the mixing chamber constant zone [129]. To compute the mass flow rate and multiphase sound velocity for a CD nozzle, the equilibrium and homogeneous model that gives rise to no-slip in temperature and velocity between the particle and gas phases was used [130]. Additionally, the mixing noise and shock-related wideband noise [131] in a nozzle was discussed. The CFD methodology was also employed in this definition for varied area ratios, the Mach number, NPR, and L/D, and analyzed the flow region surrounding the base [132].

Meanwhile, it was found that ANSYS fluent was the most appropriate tool to solve this problem. The CFD approach was a dominant concept in ANSYS software and a foundational analysis to settle the problem. It is essential to know the governing equations and the turbulent modeling of the current flow. In this respect, the studies examined and illustrated the testing approach and identified an issue for situations considering the various nozzle parameters and the flow range. In the analysis, the L/D of 10 was used for the conduit. An area ratio between the exit area of the CD nozzle and the duct area was also found and observed in the flow formations (Figure 9). The idea is used by the CFD method for various area ratios, the inertia level, NPR, and the duct length [14,132–137] for the compressible flow.

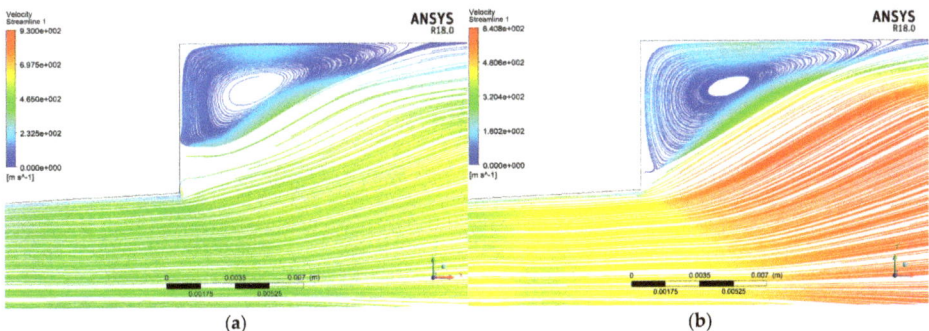

Figure 9. Velocity streamline for area ratio 6.25 (**a**) with control (**b**) without control [132]. Reprinted under the Creative Commons (CC) License (CC BY 4.0).

The CFD procedure also used to explore the flow through the wedge [138,139], non-circular cylinder [137,140], splitter plate [141–143], airfoil [144,145] and powered submarine [146] for an incompressible flow range. From the fluent, more studies were found in recent years in which some studies have been utilized with the pressure-based solver, and K-epsilon turbulent modeling was used [134,136]. The fluent results were continued with a change in parameters to investigate the effect of active microjet controller [2,147]. Figure 10 shows the perfect 2D model with the contour received as a result.

4.3. Soft Computing Methods

Soft computing is an emerging technology for discovering these kinds of problems, which parallel the remarkable capacity of the human imagination to target and study in an atmosphere of ambiguity and imprecision. It contains many computer models, including experimental system architecture, fuzzy set theory, neural networks, and approximate reasoning, and it requires intensive computing for learning and reworking.

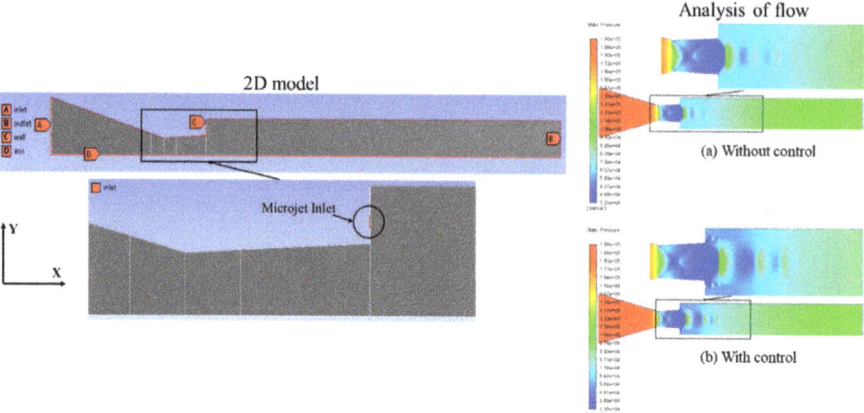

Figure 10. D nozzle with sudden expansion duct and their results. Reprinted with permission from ref. [2]. Copyright 2021 Springer.

4.3.1. Design of Experiments

If one needs to verify scientific predictions properly, these measurements are valuable. In several engineering and medicine implementations, the design of experiments (DOE) method was used to define the variables that are important in the production and optimization phase to accomplish a helpful target. This approach was then used to find the most prevalent strategy for the solution of the current object. The DOE plays a vital role in identifying relevant variables in the development process for industrial uses, such as planning and experimentation [148–150]. These kinds of elements were used successfully with the influence of designer control and changes over two or three stages. The experiments were carried out based on the orthogonal array (OA) to obtain each possible response factor affecting the input variables.

Researchers used the DOE principle based on the current problem and found that this could be an efficient way to achieve the base pressure control with an appropriate parameter. Therefore, to refine the base pressure control, the DOE technique was used. The impact of microjets on control was achieved to obtain base pressure differences of various parameters. The general DOE approach method can be seen in Figure 11.

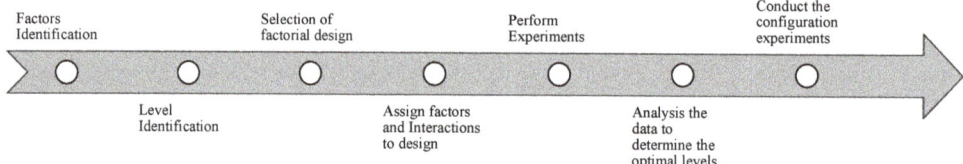

Figure 11. Steps in implementation of DOE approach. Reprinted with permission from ref. [2]. Copyright 2021 Springer.

The experimental study was used for various parametric combinations using microjets to control the base pressure, using a CD nozzle. The data were used to refine the optimal mix of parameters employed to provide precise control of base pressure for improved performance with DOE [151]. A Taguchi design L_9-OA and variance analysis (ANOVA) was used to analyze the influence of nozzle parameters affecting the base pressure. Multiple linear regression models, confirmation checks, and linear regression equations were performed for accuracy in an optimization. The ANOVA method was also used to obtain the individual parameters statistical significance on the total base pressure variability [152]. The observation on the control becomes effective for lower area ratio, compared to the higher area ratio with the aid of 15 arbitrary test cases; two linear regression model presentations were tested for their estimated accuracy [153]. To optimize the response surface methodology (RSM) of experimental data, non-linear regression models based on the central composite design (CCD) and box-Behnken design (BBD) were developed to simplify the input–output relationships [154]. The DOE with L_{27}-OA and ANOVA was used to determine the feedback (in percentage terms) of various process parameters and their correlations with and without control on the base pressure [155,156]. The optimum nozzle parameters were targeted, such as convergent angle, divergent angle, and the throat radius of the nozzle; the best values were assessed based on the flow parameters [157]. Jaimon et al. [158] used the DOE method to predict the suddenly expanded flow with and without microjets as an active control. To develop the linear model, they used a complete factorial design of the L_{16} orthogonal array (OA). Using Taguchi's L_{27} orthogonal array, a regression analysis was made [159], and optimized results investigated the suitable parameters for base pressure control.

4.3.2. Fuzzy Logic

Jagannath et al. [160] discovered a fuzzy logic methodology for investigating pressure loss in a sudden expansion duct. The authors aimed to notice minor pressure loss when the L/D was 1, as evidenced by the fuzzy logic formulation. According to the authors, this can be a qualitative examination of internal fluid flow through a nozzle with a sudden expansion duct using the fuzzy logic methodology [161]. Because of all other characteristics, such as wall static pressure and loss pressure, including the base pressure, they discovered the best value of L/D. When the Mach number is changed to 1.58 or 2.06, all other parameters remain the same as in the sudden expansion duct with cavities [162]. They found that the fuzzy logic technique L/D of 1 is sufficient for smooth flow growth based on de Laval nozzle and wall static pressure changes in both circumstances. Quadros et al. [163] described the critical aspects of the fuzzy logic technique in turbulent supersonic flow simulations as a cost-effective methodology. The Mamdani-based fuzzy logic methodology was utilized to build connections between input and output in the CFD findings. Triangular, generalized bell shape and Gaussian membership functions were employed in this technique.

4.3.3. Machine Learning

A neural network was employed to examine the predicted floor pressure in a convergent nozzle with shallow cavity internal flows at various subsonic Mach numbers [164]. The authors used feedforward neural networks (NNs) to highlight the modeling problem. A mixed density network was trained using an artificial neural network (ANN) methodology using the updated database of the future flow profile. The ANN [165] was used to learn and train flow characteristics over a transport aircraft configuration to estimate the aerodynamic coefficients using different network sizes. The K-means technique was used to cluster this massive amount of data. According to the RSM, microjets are efficient when a favorable pressure gradient influences the flow. The effect of a lower area ratio and a longer duct length on the base pressure is minimal. Furthermore, the random forest strategy was applied, which belongs to this group and is employed in Bootstrap aggregation employment regression and classification research [166]. Heat maps are used to visualize the massive amount of experimental data generated. Six back-propagation neural network models (BPMs) based on input and output possibilities are built to forecast pressure in high-speed flows for the first time. The six BPMs with two hidden layers containing four neurons were determined to be the best suited for regression analysis. The very non-linear values of the base and wall pressure are correctly predicted by BPM 5 and BPM 6 [167]. Figure 12 depicts a broad schematic depiction of the back-propagation model (BPM).

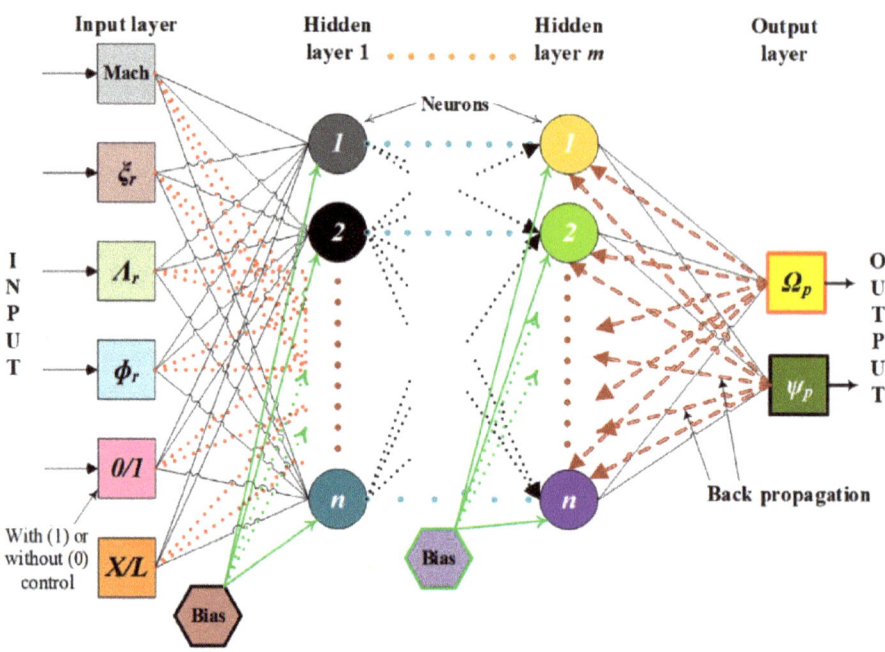

Figure 12. Back-propagation for base pressure modeling. Reprinted with permission from ref. [167]. Copyright 2020 AIP.

5. Critical Analysis of Literature

After conducting an exhaustive review on active flow management by blowing compressed air at the duct base, it was observed that the researchers utilized experimental, CFD and soft computing methods for determining the flow formation inside the nozzle (Figure 13). The experimental method was used in maximum cases for which it was challenging to obtain high-speed flow formations inside the duct, with shadowgraph for clear demonstration. Hence, the researchers utilized the CFD methods; it is easier to obtain the contours for the different variations of the flow parameters. Multiple high-speed supersonic flows were studied well by the researchers. In the CFD, different turbulence models were used to investigate the effectiveness of the aerodynamic flow variations; the limitation found was that there is a lack of three-dimensional modeling and analysis, which can be explored well in future investigation.

Next, a soft computing approach was recently found to optimize the parameters on pressure flows; several studies have been reported via DOE. However, in few cases, machine learning and fuzzy logic methods stated that the optimum results can predict from the standard statical methods. A valuable combination of parameters in the development and optimization of the flow process examines the necessary factors to achieve. Therefore, soft computing minimizes the number of experimental and theoretical workflows for a given situation. That can be more on future work for the researchers to measure the microjet effect on the nozzle pressure control and to find the optimum parameters for controlling the base pressure. However, there are no studies have been reported in the literature on either theoretical or analytical approaches with fundamental governing equations.

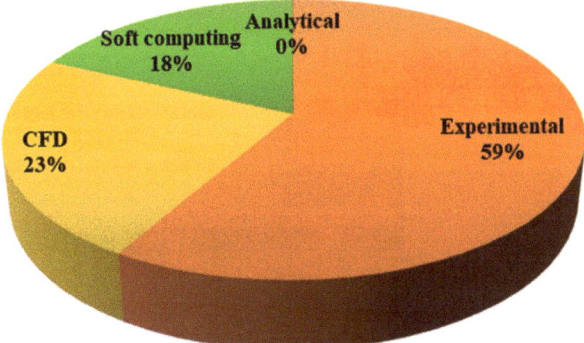

Figure 13. Methodologies used for active flow control.

In contrast, some of the passive control devices are shown in this review to know another type of control method of high-speed flow, jet, and noise. One of the critical issues in passive control is the need for additional energy sources to activate the control mechanism. Indeed, the present literature focuses more on active control and their scenarios. Some of the significant gaps/observations between the methodology related to the present work are shown in Table 1.

Table 1. Major observation for active flow control.

S. No.	Control	Experimental	CFD	Optimization	Remark
1	Active	✓	✓	✓	Very recently, CFD and optimization studies were found, and still a lot of scope in optimization
2	Passive	✓	✓	✓	Several studies have been reported in all approaches
3	Without Control	✓	✓	✓	Several studies have been reported in all approaches

For in-depth analysis of methodologies adopted for the active control method shown by the trends plot, we consider the past two decades. From the graphical view, it has clearly shown that many works have been done with the experimental method in the early years as discussed previously. However, due to recent advanced technologies, such as simulation and data optimization tools, experimental work was reduced in recent years and the soft computing approaches have increased, which have found advantages in research work. Indeed, until now, no mathematical model was made to predict the exact pressure values before and after the microjet controller. The researchers utilized different techniques to predict the results in a flow object. This gives an idea of how the methodologies are increasing in the current scenarios as a scientific approach to solving the respective problem, such as flow control. Figure 14 illustrates the last 20 years of work related to the current study in percentages, considering different techniques used to evaluate the pressure values.

In another consideration, trends in the control of flow are considered a flow model, such as nozzle/jets, bluff body, and airfoil (Figure 15). Nozzle/jets are a type of high-speed configuration that was found in several studies from the past two decades for aerospace applications. Next, the bluff body is also considered to know flow formation, and it is controlled with passive devices, such as splitter plates, for the application of both automotive and aerospace industries. Lastly, airfoil is found in many studies but in most of the cases, this has been only developed to investigate the aerodynamic forces in some studies where the outer flow control was found.

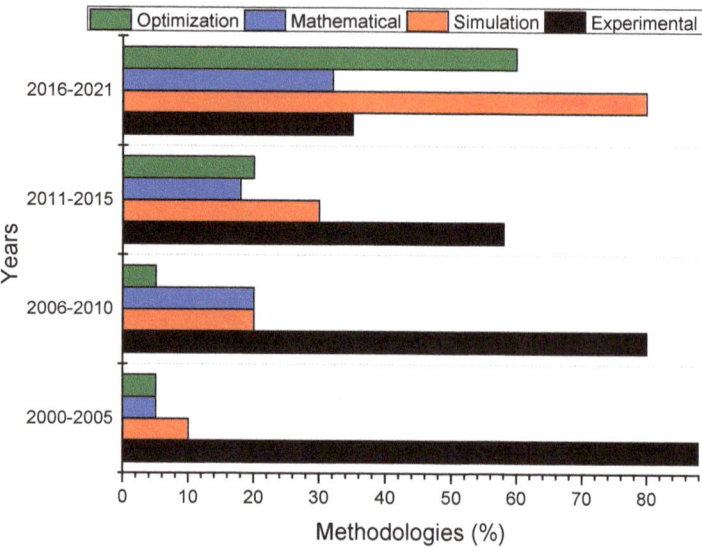

Figure 14. Trends in research methodologies in active control of flow.

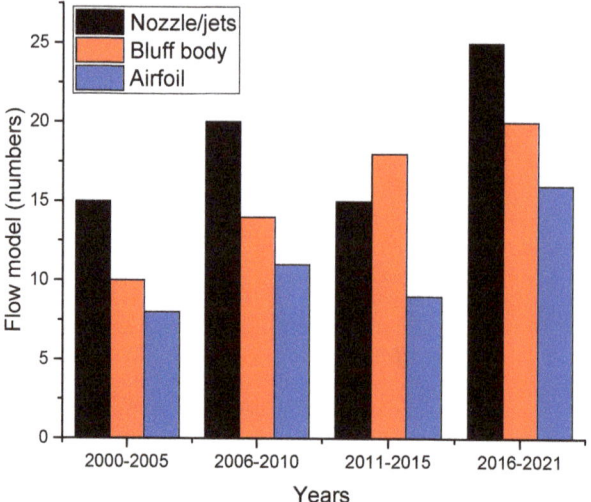

Figure 15. Trend's inflow model and its control.

Apart from the microjet controller for the control of high-speed flow, the researcher utilized different objects, such as smart material (actuator) to control the high-speed flow when focused on the depth of the active control approach. In addition, the devices for the flow control were found in the form of the cavity, corner model, diffuser, bluff body, cylinders, airfoil, wedge, etc. In all studies, some outer flow controlled, and some inner flow controlled was investigated. Table 2 illustrate the summary of the previous work with the limitations in their studies.

Table 2. Summary and limitations of the flow control method.

Type of Object	Technique Adopted	Focused Parameter	Limitations	Reference
Compression corner calculation model	ANSYS Fluent with k–ε turbulence model	Numerical investigations in the supersonic flow field with different injection pressure ratios, actuation positions, and nozzle types	The only simulation was used for investigation	[36]
CD nozzle with sudden expansion duct	Experimental, CFD and DOE	Effect of microjet control in a sudden expansion duct with the parametric investigation	Numerical investigation done by 2D model and microjet location is fixed to a single point	[2]
CD nozzle with sudden expansion duct	RSM, clustering, and random forest regression	Pressure in suddenly expanded high-speed aerodynamic flow	Predicting the pressure values using the optimization method with and without microjet controller	[166]
Resonant high-speed cavity	Wind tunnel test with a blowdown-type facility	High-frequency pulsed supersonic microjets to control resonant high-speed cavity flows	The fact that the REM/ SmartREM actuator performance can be enhanced	[35]
Cylindrical cavity structure takes	A computational method based on high-order numerical techniques	Flow physics of a pulsed microjet actuator for high-speed flow control	Efficient and geometrically complicated pulsed actuators were developed for various high-speed flow and noise control	[168]
Ultra-compact serpentine inlet	Experimental work with high-pressure air supply and CFD	Microjet flow control in an ultra-compact serpentine inlet	Mach number of inlet throat between 0.2 to 0.5, and it is varied to a higher number	[169]
Crewless Arial Vehicle	Experimental wind tunnel test	Microjet-based active flow control on a fixed-wing UAV	Experimental results possible to simulate with the CFD model	[170]
Impinging Jet	Experimental work with supersonic jet facility	Control of impinging supersonic jet flows using microjets	A fluent model will be helpful in such an example	[171]
Supersonic Crossflow model	Experimental work in a supersonic wind tunnel	Properties of resonance-enhanced microjets in supersonic crossflow	Limited to experimental work and also the Mach number can be varied	[172]
Ahmed body and NACA 0015 airfoil	An experimental study was conducted in a subsonic wind tunnel	Active flow control by micro-blowing and effects on aerodynamic performances	Experimental work only	[173]
Flap's pressure-side	Experimental approach	Microjet configuration sensitivities for active flow control on multi-element high-lift systems	CFD and data optimization are helpful techniques to predict the multi-element high-lift system	[174]
Flap's pressure-side	Experimental approach	Surface-normal active flow control for lift enhancement and separation mitigation for high-lift standard research model	Other aerodynamic forces	[175]
Spaced jet configurations	Experimental investigation	Control of compression-ramp-induced interaction with steady microjets.	It is advantageous to utilize a control design (MJ6) that functions well	[176]
Double-offset diffuser	Experimental technique	Flow dynamics affected by active flow control in an offset diffuser	Parametric study to predict more accurate results from the experimental data	[177]
Series of cavity flow	Experimental technique	The effects of high-frequency, supersonic microjet injection on a high-speed cavity flow	Specific Mach 1.5	[178]
cylindrical cavity	Simulation and Experimental investigation	Simulations of pulsed actuators for high-speed flow control	The parametric investigation will be effective in changing flow control results	[179]

Table 2. *Cont.*

Type of Object	Technique Adopted	Focused Parameter	Limitations	Reference
CD Nozzle with sudden expansion duct	CFD approach with 2D model	Base pressure control using microjets	Limited to two-dimensional model	[133,137,147,180]
Wedge	CFD approach with 2D model	Aerodynamics flow control	Limited to two-dimensional model	[137,138]
Bluff body	CFD approach with 2D model	Splitter plate to control flow and non-circular cylinder	Limited to a two-dimensional model	[137,140–142]
Airfoil	CFD approach with 2D model	Analysis of flows and prediction of CH10 airfoil for unmanned aerial vehicle wing design	Only simulation	[150]

Challenges in the Flow Control Method

Some major issues can be found when the active flow control method is used:

- **Experimental test:** One of the major issues during the test is how to reduce the flow disturbances in the connecting pipe from the main settling chamber.
- **Instrumentation:** With its sensing hole facing the flow, the pitot probe mounted on a rigid 3D traverse with a resolution of 0.1 mm was used for pressure measurement. Deficient Reynolds numbers based on the probe diameter significantly influenced the pressures measured by pitot probes. However, this effect is seldom a problem in supersonic streams because a probe of reasonable size will usually have a Reynolds number above 500, which is above the range of troublesome Reynolds numbers.
- **Data accuracy:** Pressure distribution along the jet centerline with a normal to the tab is difficult to measure and needs a proper location and observation.
- **The nozzle calibration:** The measured pitot pressure can determine the Mach number by treating the flow through the nozzle to be isentropic, thus leading to the total pressure at the nozzle exit being the same as the settling chamber pressure.
- **Simulation:** The 2D model was developed in several studies but this can only be suitable when it is uncontrolled or passive controlled. Indeed, a microjet controller needs a 3D model for more accurate results and it is difficult to design, also requiring a supercomputer to simulate/analyze.
- **Base pressure:** Based on the previous result, the base pressure either increases or decreases when it is controlled and also it varies by varying the nozzle parameter. However, it is critical to increasing the supersonic or hypersonic ranges due to high-speed formation in the setup that has a chance to incur breaks/damages; hence, it needs a very high configured/quality setup for testing.
- **Microjet location:** It was found that the microjets were located in the PCD of 13 mm but there is no other location that is reported in the literature; this can be explored well in future studies.
- **Soft computing:** As compared to DOE, only a few studies are reported with machine learning algorithms and the Fuzzy logic approach. These methods can be explored well in future investigations.

6. Conclusions

Throughout this review, the previous work was presented and discussed. The concept of CD nozzle and flow-field was addressed with a fundamental governing equation. The study shows the critical point of view of how researchers utilized the fundamental concepts in the problem solving of high-speed flows, considering active and passive methods. For the passive method, flow, jet, and noise control studies were considered and critically reviewed. Most of the studies in this literature were presented for the active control method; this was the major work of this review. Active control investigation was divided into three powerful techniques: experimental, numerical, and soft computing methods. It was found over the last two decades, the study on active control was done widely with an experimental approach. In contrast, the CFD study was conducted very recently but needs improvement on its performance with the three-dimensional approach and the proper finite volume method. Moreover, soft computing is well utilized by the design of experiments methods as compared to machine learning and fuzzy logic. Indeed, no analytical investigation has been made on such a problem. Furthermore, a critical analysis and research gaps in this field were discussed.

In summary, guidelines for scientists seeking to control flow with microjets in any high-speed flow development object were introduced. Descriptions, findings, and analyses of the critical literature on supersonic vehicle applications are included in these guidelines. The categorization can provide a quick overview of the microjet controller study topics. Furthermore, researchers may provide comprehensive viewpoints and benchmarks for particular study topics via the difficulties and possibilities. In brief, these recommendations can assist researchers in developing new ideas, especially in the early phases of this field of study.

Author Contributions: Conceptualization, methodology, investigation, data curation, writing—original draft preparation, A.A.; resources, writing—review and editing, supervision, S.A.K. and M.B.; project administration, A.A, S.A.K. and M.B.; funding acquisition, M.B. All authors have read and agreed to the published version of the manuscript.

Funding: This research received no external funding.

Acknowledgments: This research is supported by the Structures and Materials (S&M) Research Lab of Prince Sultan University. Furthermore, the authors acknowledge the support of Prince Sultan University for paying the article processing charges (APC) of this publication.

Conflicts of Interest: The authors declare no conflict of interest.

References

1. Khan, S.A.; Rathakrishnan, E. Control of Suddenly Expanded Flows with Micro-Jets. *Int. J. Turbo Jet Engines* **2003**, *20*, 63–82. [CrossRef]
2. AAabid, A.; Khan, S.A. Investigation of High-Speed Flow Control from CD Nozzle Using Design of Experiments and CFD Methods. *Arab. J. Sci. Eng.* **2021**, *46*, 2201–2230. [CrossRef]
3. Ramanjaneyulu, S. Design and flow analysis of Convergent Divergent nozzle using CFD. *Int. J. Res. Appl. Sci. Eng. Technol.* **2019**, *7*, 4020–4029. [CrossRef]
4. Singh, J.; Zerpa, L.E.; Partington, B.; Gamboa, J. Effect of nozzle geometry on critical-subcritical flow transitions. *Heliyon* **2019**, *5*, 19. [CrossRef]
5. Rathakrishnan, E. *Applied Gas Dynamics*; Wiley: Hoboken, NJ, USA, 2019.
6. Greyvenstein, G.P. An implicit method for the analysis of transient flows in pipe networks. *Int. J. Numer. Methods Eng.* **2001**, *53*, 1127–1143. [CrossRef]
7. Keir, A.S.; Ives, R.; Hamad, F. CFD analysis of C-D nozzle compared with theoretical & experimental data. *INCAS Bull.* **2018**, *10*, 53–64. [CrossRef]
8. Ferrás, L.L.; Afonso, A.M.; Alves, M.A.; Nóbrega, J.M.; Pinho, F.T. Newtonian and viscoelastic fluid flows through an abrupt 1:4 expansion with slip boundary conditions. *Phys. Fluids* **2020**, *32*, 043103. [CrossRef]
9. Poole, R.; Pinho, F.; Alves, M.; Oliveira, P. The effect of expansion ratio for creeping expansion flows of UCM fluids. *J. Non Newton. Fluid Mech.* **2009**, *163*, 35–44. [CrossRef]
10. Poole, R.; Alves, M.; Oliveira, P.J.; Pinho, F. Plane sudden expansion flows of viscoelastic liquids. *J. Non Newton. Fluid Mech.* **2007**, *146*, 79–91. [CrossRef]
11. Dhinakaran, S.; Oliveira, M.; Pinho, F.; Alves, M. Steady flow of power-law fluids in a 1:3 planar sudden expansion. *J. Non Newton. Fluid Mech.* **2013**, *198*, 48–58. [CrossRef]
12. Francisco, A.R.L. Nozzles. *J. Chem. Inf. Model.* **2013**, *53*, 1689–1699. [CrossRef]
13. Kumar, R.R.; Devarajan, Y. CFD simulation analysis of two-dimensional convergent-divergent nozzle. *Int. J. Ambient. Energy* **2018**, *41*, 1505–1515. [CrossRef]
14. Khan, A.; Aabid, A.; Khan, S.A. CFD Analysis of Convergent-Divergent Nozzle Flow and Base Pressure Control Using Micro-JETS. 2018. Available online: www.sciencepubco.com/index.php/IJET (accessed on 26 July 2021).
15. Khan, S.A.; Rathakrishnan, E. Active Control of Suddenly Expanded Flows from Overexpanded Nozzles. *Int. J. Turbo Jet Engines* **2002**, *19*, 119–126. [CrossRef]
16. Raman, R.K.; Dewang, Y.; Raghuwanshi, J. A review on applications of computational fluid dynamics. *Int. J. LNCT* **2018**, *6*, 8.

17. Xu, G.; Luxbacher, K.D.; Ragab, S.; Xu, J.; Ding, X. Computational fluid dynamics applied to mining engineering: A review. *Int. J. Min. Reclam. Environ.* **2016**, *31*, 1–25. [CrossRef]
18. Alobaid, F. Computational Fluid Dynamics. *Mater. Intern. Struct.* **2018**, *57*, 87–204. [CrossRef]
19. Tapasvi, V.; Gupta, M.S.; Kumaraswamy, T. Designing and Simulating Compressible Flow in a Nozzle. *Int. J. Eng. Adv. Technol.* **2015**, *6*, 46–54.
20. Lai, Y.G.; Wu, K. A three-dimensional flow and sediment transport model for free-surface open channel flows on unstructured flexible meshes. *Fluids* **2019**, *4*, 18. [CrossRef]
21. Chong, M.S.; Perry, A.E.; Cantwell, B.J. A general classification of three-dimensional flow fields. *Phys. Fluids A Fluid Dyn.* **1990**, *2*, 765–777. [CrossRef]
22. Sakaki, K.; Shimizu, Y. Effect of the Increase in the Entrance Convergent Section Length of the Gun Nozzle on the High-Velocity Oxygen Fuel and Cold Spray Process. *J. Therm. Spray Technol.* **2001**, *10*, 487–496. [CrossRef]
23. Viswanath, P.R. Aircraft viscous drag reduction using riblets. *Prog. Aerosp. Sci.* **2002**, *38*, 571–600. [CrossRef]
24. Ishide, T.; Itazawa, M. Aerodynamic improvement of a delta wing in combination with leading edge flaps. *Theor. Appl. Mech. Lett.* **2017**, *7*, 357–361. [CrossRef]
25. Khurana, S.; Suzuki, K.; Rathakrishnan, E. Flow Field around a Blunt-nosed Body with Spike. *Int. J. Turbo Jet Engines* **2012**, *29*, 217–221. [CrossRef]
26. Khurana, S.; Suzuki, K. Towards Heat Transfer Control by Aerospikes for Lifting-Body Configuration in Hypersonic Flow at Mach 7. In Proceedings of the 44th AIAA Thermophysics Conference, San Diego, CA, USA, 24–27 June 2013.
27. Khurana, S.; Suzuki, K. Assessment of Aerodynamic Effectiveness for Aerospike Application on Hypothesized Lifting-Body in Hypersonic Flow. In Proceedings of the Fluid Dynamics and Co-Located Conferences, San Diego, CA, USA, 24–27 June 2013; pp. 24–27.
28. Khurana, S.; Suzuki, K.; Rathakrishnan, E. Flow field behavior with Reynolds number variance around a spiked body. *Mod. Phys. Lett. B* **2016**, *30*, 1650362. [CrossRef]
29. Sinclair, J.; Cui, X. A theoretical approximation of the shock standoff distance for supersonic flows around a circular cylinder. *Phys. Fluids* **2017**, *29*, 026102. [CrossRef]
30. Viswanath, P.R. Passive devices for axisymmetric base drag reduction at transonic speeds. *J. Aircr.* **1988**, *25*, 258–262. [CrossRef]
31. Viswanath, P. Flow management techniques for base and afterbody drag reduction. *Prog. Aerosp. Sci.* **1996**, *32*, 79–129. [CrossRef]
32. Singh, N.K.; Rathakrishnan, E. Sonic Jet Control with Tabs. *Int. J. Turbo Jet Engines* **2002**, *19*, 107–118. [CrossRef]
33. Vijayaraja, K.; Senthilkumar, C.; Elangovan, S.; Rathakrishnan, E. Base Pressure Control with Annular Ribs. *Int. J. Turbo Jet Engines* **2014**, *31*, 111–118. [CrossRef]
34. Vikramaditya, N.S.; Viji, M.; Verma, S.B.; Ali, N.; Thakur, D.N. Base Pressure Fluctuations on Typical Missile Configuration in Presence of Base Cavity. *J. Spacecr. Rocket.* **2018**, *55*, 335–345. [CrossRef]
35. Kreth, P.A.; Alvi, F.S. Using High-Frequency Pulsed Supersonic Microjets to Control Resonant High-Speed Cavity Flows. *AIAA J.* **2020**, *58*, 3378–3392. [CrossRef]
36. Liu, Y.; Zhang, H.; Liu, P. Flow control in supersonic flow field based on micro jets. *Adv. Mech. Eng.* **2019**, *11*. [CrossRef]
37. Rathakrishnan, E.; Ramanaraju, O.; Padmanaban, K. Influence of cavities on suddenly expanded flow field. *Mech. Res. Commun.* **1989**, *16*, 139–146. [CrossRef]
38. Sethuraman, V.; Khan, S.A. Effect of sudden expansion for varied area ratios at subsonic and sonic flow regimes. *Int. J. Energy Environ. Econ.* **2016**, *24*, 99–112.
39. Asadullah, M.; Khan, S.A.; Asrar, W.; Sulaeman, E. Active control of base pressure with counter clockwise rotating cylinder at Mach 2. In Proceedings of the 2017 IEEE 4th International Conference on Engineering Technologies and Applied Sciences, Salmabad, Bahrain, 29 November–1 December 2017.
40. Khan, S.A.; Asadullah, M.; Sadhiq, J. Passive Control of Base Drag Employing Dimple in Subsonic Suddenly Expanded Flow. *Int. J. Mech. Mechatron. Eng.* **2018**, *8*, 69–74.
41. Gutmark, E.J.; Grinstein, F.F. Flow control with noncircular jets. *Annu. Rev. Fluid Mech.* **1999**, *31*, 239–272. [CrossRef]
42. Samimy, M.; Webb, N.; Crawley, M. Excitation of Free Shear-Layer Instabilities for High-Speed Flow Control. *AIAA J.* **2018**, *56*, 1770–1791. [CrossRef]
43. Rathakrishnan, E.; Ethirajan, R. AR 4 elliptic jet control with limiting tab. *Fluid Dyn. Res.* **2017**, *50*, 025505. [CrossRef]
44. Khan, A.; Kumar, R. Experimental Study and Passive Control of Overexpanded Plug Nozzle Jet. *J. Spacecr. Rocket.* **2018**, *55*, 778–782. [CrossRef]
45. Jacksi, K.; Ibrahim, F.; Ali, S. Scholars Journal of Engineering and Technology (SJET). *Sch. J. Eng. Technol.* **2018**, *9523*, 49–53. [CrossRef]
46. Akram, S.; Rathakrishnan, E. Control of Supersonic Elliptic Jet with Ventilated Tabs. *Int. J. Turbo Jet Engines* **2017**, *37*, 267–283. [CrossRef]
47. Kumar, P.A.; Aileni, M.; Rathakrishnan, E. Impact of tab location relative to the nozzle exit on the shock structure of a supersonic jet. *Phys. Fluids* **2019**, *31*, 076104. [CrossRef]
48. Khan, A.; Panthi, R.; Kumar, R.; Ibrahim, S.M. Experimental investigation of the effect of extended cowl on the flow field of planar plug nozzles. *Aerosp. Sci. Technol.* **2019**, *88*, 208–221. [CrossRef]

49. Manigandan, S.; Vijayaraja, K. Flow field and acoustic characteristics of elliptical throat CD nozzle. *Int. J. Ambient. Energy* **2017**, *40*, 57–62. [CrossRef]
50. Khan, A.; Akram, S.; Kumar, R. Experimental study on enhancement of supersonic twin-jet mixing by vortex generators. *Aerosp. Sci. Technol.* **2020**, *96*, 105521. [CrossRef]
51. Panthi, R.; Krishna, T.V.; Nanda, S.R.; Khan, A.; Kumar, R.; Sugarno, M.I. Experimental Study of Impinging Plug Nozzle Jet Using a Vortex Generator. *J. Spacecr. Rocket.* **2020**, *57*, 1414–1418. [CrossRef]
52. Kumar, S.A.; Rathakrishnan, E. Elliptic jet control with triangular tab. *Proc. Inst. Mech. Eng. Part G J. Aerosp. Eng.* **2017**, *231*, 1460–1477. [CrossRef]
53. Tam, C.; Tanna, H. Shock associated noise of supersonic jets from convergent-divergent nozzles. *J. Sound Vib.* **1982**, *81*, 337–358. [CrossRef]
54. Liu, J.; Ramamurti, R. Numerical Study of Supersonic Jet Noise Emanating from an F404 Nozzle at Model Scale. *AIAA Scitech 2019 Forum* **2019**, 1–28. [CrossRef]
55. Akatsuka, J.; Hromisin, S.; Falcone, J.; McLaughlin, D.K.; Morris, P.J. Mean Flow Measurements in Supersonic Jets with Noise Reduction Devices. *AIAA Scitech 2019 Forum* **2019**, 1–16. [CrossRef]
56. Prasad, C.; Morris, P.J. Unsteady Simulations of Fluid Inserts for Supersonic Jet Noise Reduction. *AIAA Scitech 2019 Forum* **2019**, 1–22. [CrossRef]
57. Gao, J.; Xu, X.; Li, X. Numerical simulation of supersonic twin-jet noise with high-order finite difference scheme. *AIAA J.* **2018**, *56*, 290–300. [CrossRef]
58. Chen, B.; Wang, Y. Active aerodynamic noise control research for supersonic aircraft cavity by nonlinear numerical simulation. *Int. J. Electr. Eng. Educ.* **2021**. [CrossRef]
59. Zhu, W.; Xiao, Z.; Fu, S. Numerical modeling screen for flow and noise control around tandem cylinders. *AIAA J.* **2020**, *58*, 2504–2516. [CrossRef]
60. Patel, T.K. Analysis of Supersonic Jet Noise in the Sideline and Upstream Directions Using the Navier-Stokes Equations. Ph.D. Thesis, University of Florida, Gainesville, FL, USA, 2020.
61. Coombs, J.; Schembri, T.; Zander, A. The effect of hemispherical surface on noise suppression of a supersonic jet. In Proceedings of the Acoustics 2019—Sound Decisions: Moving forward with Acoustics, Mornington Peninsula, Australia, 13–14 May 2019; pp. 1–10.
62. Mondal, K.; Das, S.; Hamada, N.; Abu, A.; Das, S.; Faris, W.; Thiam, H.; Toh, T.; Ahmas, A. An improved narrowband active noise control system without secondary path modelling based on the time domain. *Int. J. Veh. Noise Vib.* **2019**, *15*, 110–132. [CrossRef]
63. Berton, J.J.; Huff, D.L.; Geiselhart, K.; Seidel, J. Supersonic Technology Concept Aeroplanes for Environmental Studies. *AIAA Scitech 2020 Forum* **2020**. [CrossRef]
64. Prasad, C.; Morris, P.J. A study of noise reduction mechanisms of jets with fluid inserts. *J. Sound Vib.* **2020**, *476*, 115331. [CrossRef]
65. Salehian, S.; Mankbadi, R.R. A review of aeroacoustics of supersonic jets interacting with solid surfaces. *AIAA Scitech 2020 Forum* **2020**, 1–31. [CrossRef]
66. Rahmani, S.K.; Alhawwary, M.A.; Wang, Z.J.; Phommachanh, J.; Hill, C.; Hartwell, B.; Collicott, B.; Swim, J.; Farokhi, S.; Taghavi, R.; et al. Noise Mitigation of a Supersonic Jet Using Shear Layer Swirl. *AIAA Scitech 2020 Forum* **2020**, 1–30. [CrossRef]
67. Horner, C.; Sescu, A.; Afsar, M.; Collins, E.; Azarpeyvand, M. Passive Noise Control Strategies for Jets Exhausting over Flat Surfaces: An LES Study. *AIAA Aviation 2020 Forum* **2020**, 1–17. [CrossRef]
68. Prasad, C.; Morris, P.J. Steady active control of noise radiation from highly heated supersonic jets. *J. Acoust. Soc. Am.* **2021**, *149*, 1306–1317. [CrossRef] [PubMed]
69. Pourhashem, H.; Kumar, S.; Kalkhoran, I.M. Flow field characteristics of a supersonic jet influenced by downstream microjet fluidic injection. *Aerosp. Sci. Technol.* **2019**, *93*, 105281. [CrossRef]
70. Li, B.; Ye, C.-C.; Wan, Z.-H.; Liu, N.-S.; Sun, D.-J.; Lu, X.-Y. Noise control of subsonic flow past open cavities based on porous floors. *Phys. Fluids* **2020**, *32*, 125101. [CrossRef]
71. Martin, R.; Soria, M.; Rodriguez, I.; Lehmkuhl, O. On the Flow and Passive Noise Control of an Open Cavity at Re = 5000. *Flow Turbul. Combust.* **2021**, 1–26. [CrossRef]
72. Ye, C.C.; Zhang, P.J.Y.; Wan, Z.H.; Sun, D.J.; Lu, X.Y. Numerical investigation of the bevelled effects on shock structure and screech noise in planar supersonic jets. *Phys. Fluids* **2020**, *32*. [CrossRef]
73. Su, Z.; Liu, E.; Xu, Y.; Xie, P.; Shang, C.; Zhu, Q. Flow field and noise characteristics of manifold in natural gas transportation station. *Oil Gas Sci. Technol. Rev. l'IFP* **2019**, *74*, 70. [CrossRef]
74. Wei, X.; Mariani, R.; Chua, L.; Lim, H.D.; Lu, Z.; Cui, Y.; New, T.H. Mitigation of under-expanded supersonic jet noise through stepped nozzles. *J. Sound Vib.* **2019**, *459*, 1–18. [CrossRef]
75. Coombs, J.; Zander, A.; Schembri, T. Influence of a Hemispherical Noise Reduction Reflector on Transonic Jet Flows. In Proceedings of the 22nd Australasian Fluid Mechanics Conference AFMC2020, Brisbane, Australia, 7–10 December 2020.
76. Cuppoletti, D.; Gutmark, E.J.; Hafsteinsson, H.E.; Eriksson, L.-E. Elimination of Shock Associated Noise in Supersonic Jets by Destructive Wave Interference. *AIAA J.* **2018**, *57*, 720–734. [CrossRef]
77. Kumar, P.A.; Kumar, S.M.A.; Mitra, A.S.; Rathakrishnan, E. Empirical scaling analysis of supersonic jet control using steady fluidic injection. *Phys. Fluids* **2019**, *31*, 056107. [CrossRef]

78. Mancini, S.; Kolb, A.; Gonzalez-Martino, I.; Casalino, D. Effects of wall modifications on pressure oscillations in high-subsonic and supersonic flows over rectangular cavities. In Proceedings of the 25th AIAA/CEAS Aeroacoustics Conference, Delft, The Netherlands, 20–23 May 2019; pp. 1–20. [CrossRef]
79. Xiansheng, W.; Dangguo, Y.; Jun, L.; Fangqi, Z. Control of Pressure Oscillations Induced by Supersonic Cavity Flow. *AIAA J.* **2020**, *58*, 2070–2077. [CrossRef]
80. Robertson, G.; Kumar, R. Effects of a Generic Store on Cavity Resonance at Supersonic Speeds. *AIAA J.* **2020**, *58*, 4426–4437. [CrossRef]
81. Prasad, C.; Morris, P. Effect of fluid injection on turbulence and noise reduction of a supersonic jet. *Philos. Trans. R. Soc. A Math. Phys. Eng. Sci.* **2019**, *377*, 20190082. [CrossRef] [PubMed]
82. Anureka, R.; Srinivasan, K. Passive control of pipe-jet noise using trailing-edge castellations. *Appl. Acoust.* **2020**, *170*, 107516. [CrossRef]
83. Khan, S.A.; Rathakrishnan, E. Active Control of Suddenly Expanded Flows from Underexpanded Nozzles. *Int. J. Turbo Jet Engines* **2004**, *21*, 233–254. [CrossRef]
84. Khan, S.A.; Rathakrishnan, E. Control of Suddenly Expanded Flows from Correctly Expanded Nozzles. *Int. J. Turbo Jet Engines* **2004**, *21*, 255–278. [CrossRef]
85. Khan, S.A.; Rathakrishnan, E. Active Control of Suddenly Expanded Flows from Underexpanded Nozzles—Part II. *Int. J. Turbo Jet Engines* **2005**, *22*, 163–183. [CrossRef]
86. Khan, S.; Rathakrishnan, E. Control of suddenly expanded flow. *Aircr. Eng. Aerosp. Technol.* **2006**, *78*, 293–309. [CrossRef]
87. Khan, S.A.; Rathakrishnan, E. Nozzle Expansion Level Effect on Suddenly Expanded Flow. *Int. J. Turbo Jet Engines* **2006**, *23*, 233–258. [CrossRef]
88. Rehman, S.; Khan, S.A. Control of base pressure with micro-jets: Part I. *Aircr. Eng. Aerosp. Technol.* **2008**, *80*, 158–164. [CrossRef]
89. Baig, M.A.A.; Al-Mufadi, F.; Khan, S.A.; Rathakrishnan, E. Control of Base Flows with Micro Jets Control of Base Flows with Micro Jets. *Int. J. Turbo Jet Engines* **2011**, *28*, 59–69. [CrossRef]
90. Biag, M.A.A.; Khan, S.A.; Rathakrishnan, E. Active control of base pressure in suddenly expanded flow for area ratio 4.84. *Int. J. Eng. Sci. Technol.* **2012**, *4*, 1892–1902.
91. Kostic, O.; Stefanović, Z.; Kostič, I.; Olivera, K.; Zoran, S.; Ivan, K. CFD modeling of supersonic airflow generated by 2D nozzle with and without an obstacle at the exit section. *FME Trans.* **2015**, *43*, 107–113. [CrossRef]
92. Fharukh, M.; Asadullah, M.; Khan, S.A. Experimental Study of Suddenly Expanded Flow from Correctly Expanded Nozzles. *ARPN J. Eng. Appl. Sci.* **2016**, *11*, 10041–10047.
93. Khan, S.A.; Aabid, A.; Chaudhary, Z.I. Influence of control mechanism on the flow field of duct at mach 1.2 for area ratio 2.56. *Int. J. Innov. Technol. Explor. Eng.* **2019**, *8*, 1135–1138. [CrossRef]
94. Chaudhary, Z.I.; Shinde, V.B.; Khan, S.A. Investigation of base flow for an axisymmetric suddenly expanded nozzle with micro JET. *Int. J. Eng. Technol.* **2018**, *7*, 236–242. [CrossRef]
95. Ahmed, F.; Khan, S.A. Investigation of efficacy of low length-to-diameter ratio and nozzle pressure ratio on base pressure in an abruptly expanded flow. In Proceedings of the MATEC Web of Conferences, Chennai, India, 31 July 2018; Volume 01004, pp. 1–6. [CrossRef]
96. Khan, S.A.; Asadullah, M.; Jalaluddeen, A.; Baig, M.A. Flow Control with Aerospike behind Bluff Body. *Int. J. Mech. Prod. Eng. Res. Dev.* **2018**, *8*, 1001–1008.
97. Khan, S.A.; Al Robaian, A.A.; Asadullah, M.; Khan, A.M. Grooved Cavity as a Passive Controller behind Backward Facing Step. *J. Adv. Res. Fluid Mech. Therm. Sci.* **2019**, *53*, 185–193.
98. Khan, S.A.; Alrobaian, A.A.; Asadullah, M. Threaded Spikes for Bluff Body Base Flow Control. *J. Adv. Res. Fluid Mech. Therm. Sci.* **2019**, *2*, 194–203.
99. Khan, S.A.; Rathakrishnan, E. Active Control of Base Pressure in Supersonic Regime. *J. Inst. Eng. Aerosp. Eng. J.* **2006**, *87*, 1–10.
100. AAabid Fharukh, A.; Sher Afghan, K. Experimental Investigation of Wall Pressure Distribution in a Suddenly Expanded Duct from a Convergent-Divergent Nozzle. In Proceedings of the 2019 6th IEEE International Conference on Engineering Technologies and Applied Sciences (ICETAS), Kuala Lumpur, Malaysia, 20–21 December 2019; p. 6. [CrossRef]
101. Khan, S.A.; Aabid, A.; Mokashi, I.; Ahmed, Z. Effect of Micro Jet Control on the Flow Filed of the Duct at Mach 1.5. *Int. J. Recent Technol. Eng.* **2019**, *8*, 1758–1762. [CrossRef]
102. Khan, S.A.; Mokashi, I.; Aabid, A.; Faheem, M. Experimental Research on Wall Pressure Distribution in C-D Nozzle at Mach number 1.1 for Area Ratio 3.24. *Int. J. Recent Technol. Eng.* **2019**, *8*, 971–975. [CrossRef]
103. Faheem, M.; Kareemullah, M.; Aabid, A.; Mokashi, I.; Khan, S.A. Experiment on of nozzle flow with sudden expansion at mach 1.1. *Int. J. Recent Technol. Eng.* **2019**, *8*, 1769–1775. [CrossRef]
104. Akhtar, M.N.; Bakar, E.A.; Aabid, A.; Khan, S.A. Effects of micro jets on the flow field of the duct with sudden expansion. *Int. J. Innov. Technol. Explor. Eng.* **2019**, *8*, 636–640. [CrossRef]
105. Azami, M.H.; Faheem, M.; Aabid, A.; Mokashi, I.; Khan, S.A. Experimental Research of Wall Pressure Distribution and Effect of Micro Jet at Mach 1.5. *Int. J. Recent Technol. Eng.* **2019**, *8*, 1000–1003.
106. Aabid, A.; Khan, S.A. Determination of wall pressure flows at supersonic Mach numbers. *Mater. Today Proc.* **2021**, *38*, 2347–2352. [CrossRef]

107. Saleel, A.; Baig, M.A.A.; Khan, S.A. Experimental Investigation of the Base Flow and Base Pressure of Sudden Expansion Nozzle. *IOP Conf. Ser. Mater. Sci. Eng.* **2018**, *370*, 012052. [CrossRef]
108. Khan, S.A.; Ahmed, Z.; Aabid, A.; Mokashi, I. Experimental research on flow development and control effectiveness in the duct at high speed. *Int. J. Recent Technol. Eng.* **2019**, *8*, 1763–1768. [CrossRef]
109. Azami, M.H.; Faheem, M.; Aabid, A.; Mokashi, I.; Khan, S.A. Inspection of Supersonic Flows in a CD Nozzle using Experimental Method. *Int. J. Recent Technol. Eng.* **2019**, *8*, 996–999.
110. Akhtar, M.N.; Bakar, E.A.; Aabid, A.; Khan, S.A. Control of CD nozzle flow using microjets at mach 2.1. *Int. J. Innov. Technol. Explor. Eng.* **2019**, *8*, 631–635. [CrossRef]
111. Aabid, A.; Khan, S.A. Studies on Flows Development in a Suddenly Expanded Circular Duct at Supersonic Mach Numbers. *Int. J. Heat Technol.* **2021**, *38*, 185–194. [CrossRef]
112. Aabid, A.; Khan, S.A.; Baig, M.A.A.; Reddy, A.R. Investigation of Flow Growth in aDuct Flowsfor Higher Area Ratio. *IOP Conf. Ser. Mater. Sci. Eng.* **2021**, *1057*, 10. [CrossRef]
113. Aabid, A.; Khan, S.A.; Baig, M.A.A.; Rao, K.S. Effect of Control on the Duct Flow at High Mach Numbers. *IOP Conf. Ser. Mater. Sci. Eng.* **2021**, *1057*, 9. [CrossRef]
114. Kumar, A.A.; Manoj, N.; Onkar, A.K.; Manjuprasad, M. Fluid-Structure Interaction Analysis of a Cropped Delta Wing. *Procedia Eng.* **2016**, *144*, 1205–1212. [CrossRef]
115. Ali, A.; Neely, A.; Young, J.; Blake, B.; Lim, J.Y. Numerical Simulation of Fluidic Modulation of Nozzle Thrust. In Proceedings of the 17th Australasian Fluid Mechanics Conference, Auckland, New Zeeland, 5–9 December 2010; pp. 5–8.
116. Kumar, G.M.; Fernando, D.X.; Kumar, R.M. Design and Optimization of De Lavel Nozzle to Prevent Shock Induced Flow Separation. *Adv. Aerosp. Sci. Appl.* **2013**, *3*, 119–124.
117. Shariatzadeh, O.J.; Abrishamkar, A.; Jafari, A.J. Computational Modeling of a Typical Supersonic Converging-Diverging Nozzle and Validation by Real Measured Data. *J. Clean Energy Technol.* **2015**, *3*, 220–225. [CrossRef]
118. Su, L. CFD Simulation and Shape Optimization of Supersonic Ejectors for Refrigeration and Desalination Applications. Master's Thesis, Washington University, St. Louis, MO, USA, 2015.
119. SA Khan MMohiuddin AS, C.; Fharukh, G.M. Investigation of the Effects of Nozzle Exit Mach number and Nozzle Pressure Ratio on Axisymmetric Flow through Suddenly Expanded Nozzles. *Int. J. Eng. Adv. Technol.* **2019**, *8*, 570–578.
120. Cai, Y.; Liu, Z.; Shi, Z.; Song, Q.; Wan, Y. Residual surface topology modeling and simulation analysis for micro-machined nozzle. *Int. J. Precis. Eng. Manuf.* **2015**, *16*, 157–162. [CrossRef]
121. Belega, B.-A.; Nguyen, T.D. Analysis of Flow in Convergent-Divergent Rocket Engine Nozzle Using Computational Fluid Dynamics. In Proceedings of the International Conference of Science Paper AFASES, Brasov, Romania, 28–30 May 2015; p. 6.
122. Patel, M.S.; Mane, S.D.; Raman, M. Concepts and CFD Analysis of De-Laval Nozzle. *Int. J. Mech. Eng. Technol.* **2016**, *7*, 221–240.
123. Mouli, C.S.; Vardhan, C.N.; Chowdary, G.V.; Jeevan, P. Aerodynamic Design and Performance of Nozzle with Different Mach Numbers Using CFD Analysis. *IRJET* **2017**, *9*, 20–27.
124. Pathan, K.A.; Khan, S.A.; Dabeer, P.S. CFD Analysis of Effect of Flow and Geometry Parameters on Thrust Force Created by Flow from Nozzle. In Proceedings of the 2nd International Conference for Convergence in Technology (I2CT) CFD, Mumbai, India, 7–9 April 2017; pp. 1121–1125.
125. Pathan, K.A.; Khan, S.A.; Dabeer, P.S. CFD analysis of effect of area ratio on suddenly expanded flows. In Proceedings of the 2017 2nd International Conference for Convergence in Technology (I2CT), Mumbai, India, 7–9 April 2017; pp. 1192–1198.
126. Pathan, K.A.; Khan, S.A.; Dabeer, P.S. CFD analysis of effect of Mach number, area ratio and nozzle pressure ratio on velocity for suddenly expanded flows. In Proceedings of the 2017 2nd International Conference for Convergence in Technology (I2CT) CFD, Mumbai, India, 7–9 April 2017; pp. 1104–1110.
127. Ahmed, K.; Dabeer, P.S.; Afghan, S. Optimization of area ratio and thrust in suddenly expanded fl ow at supersonic Mach numbers. *Case Stud. Therm. Eng.* **2018**, *12*, 696–700. [CrossRef]
128. Moorthy, C.; Srinisas, V.; Prasad, V.; Vanaja, T. Computational Analysis of a Cd Nozzle with 'Sed' for a Rocket Air Ejector in Space Applications. *Int. J. Mech. Prod. Eng. Res. Dev.* **2017**, *7*, 53–60. Available online: http://tjprc.org/view-archives.php (accessed on 26 July 2021).
129. Arun, K.; Tiwari, S.; Mani, A. Three-dimensional numerical investigations on rectangular cross-section ejector. *Int. J. Therm. Sci.* **2017**, *122*, 257–265. [CrossRef]
130. Zhang, G.; Kim, H.D. Theoretical and numerical analysis on choked multiphase flows of gas and solid particle through a convergent–divergent nozzle. *J. Comput. Multiph. Flows* **2017**, *10*, 19–32. [CrossRef]
131. Pilon, A.R.; McLaughlin, D.K.; Morris, P.J.; Powers, R.W. Design and Analysis of a Supersonic Jet Noise Reduction Concept. *J. Aircr.* **2014**, *54*, 1705–1717. [CrossRef]
132. Khan, S.A.; Aabid, A.; Ghasi, F.A.M.; Al-Robaian, A.A.; Alsagri, A.S. Analysis of Area Ratio in a CD Nozzle with Suddenly Expanded Duct using CFD Method. *CFD Lett.* **2019**, *11*, 61–71.
133. Fharukh, A.G.M.; Alrobaian, A.A.; Aabid, A.; Khan, S.A. Numerical Analysis of Convergent-Divergent Nozzle Using Finite Element Method. *Int. J. Mech. Prod. Eng. Dev.* **2018**, *8*, 373–382. Available online: www.tjprc.org (accessed on 26 July 2021).
134. Khan, S.A.; Aabid, A.; Baig, M.A.A. CFD analysis of cd nozzle and effect of nozzle pressure ratio on pressure and velocity for suddenly expanded flows. *Int. J. Mech. Prod. Eng. Res. Dev.* **2018**, *8*, 1147–1158. [CrossRef]

135. Khan, S.A.; Aabid, A.; Saleel, C.A. Influence of Micro Jets on the Flow Development in the Enlarged Duct at Supersonic Mach number. *Int. J. Mech. Mechatron. Eng.* **2019**, *19*, 70–82.
136. Aabid, A.; Khan, A.; Mazlan, N.M.; Ismail, M.A.; Akhtar, M.N.; Khan, S.A. Numerical Simulation of Suddenly Expanded Flow at Mach 2.2. *Int. J. Eng. Adv. Technol.* **2019**, *8*, 452–457.
137. Sajali, M.F.M.; Aabid, A.; Khan, S.A.; Fharukh, A.; Sulaeman, E. Numerical Investigation of Flow Field of a Non-Circular Cylinder. *CFD Lett.* **2019**, *11*, 37–49.
138. Khan, S.A.; Aabid, A.; Ahmed, S. CFD Simulation with Analytical and Theoretical Validation of Different Flow Parameters for the Wedge at Supersonic Mach Number. *Int. J. Mech. Mechatron. Eng.* **2019**, *19*, 170–177.
139. Khan, S.A.; Aabid, A.; Mokashi, I.; Al-Robaian, A.A.; Alsagri, A.S. Optimization of Two-dimensional Wedge Flow Field at Supersonic Mach Number. *CFD Lett.* **2019**, *11*, 80–97.
140. Sajali, M.F.M.; Ashfaq, S.; Aabid, A.; Khan, S.A. Simulation of Effect of Various Distances between Front and Rear Body on Drag of a Non-Circular Cylinder. *J. Adv. Res. Fluid Mech. Therm. Sci.* **2019**, *62*, 53–65.
141. Aabid, A.; Afifi, A.; Ali, F.A.G.M.; Akhtar, M.N.; Khan, S.A. CFD Analysis of Splitter Plate on Bluff Body. *CFD Lett.* **2019**, *11*, 5–38.
142. Afifi, A.; Aabid, A.; Khan, S.A. Numerical investigation of splitter plate effect on bluff body using finite volume method. *Mater. Today Proc.* **2021**, *38*, 2181–2190. [CrossRef]
143. Zhang, D.; Tan, J.; Yao, X. Numerical investigation on splitter plate jet assisted mixing in supersonic flow. *Acta Astronaut.* **2021**. [CrossRef]
144. Kharulaman, L.; Aabid, A.; Ahmed, F.; Mehaboobali, G.; Khan, S.A. Research on Flows for NACA 2412 Airfoil using Computational Fluid Dynamics Method. *Int. J. Eng. Adv. Technol.* **2019**, *9*, 5450–5456. [CrossRef]
145. Aabid, A.; Nabilah, L.; Khairulaman, B.; Khan, S.A. Analysis of Flows and Prediction of CH10 Airfoil for Unmanned Arial Vehicle Wing Design. *Adv. Aircr. Spacecr. Sci.* **2021**, *2*, 24. [CrossRef]
146. Khan, S.A.; Fatepurwala, M.A.; Pathan, K.N.; Dabeer, P.S.; Baig, M.A.A. CFD Analysis of Human Powered Submarine to Minimize Drag. *Int. J. Mech. Prod. Eng. Res. Dev.* **2018**, *8*, 1057–1066. [CrossRef]
147. Akhtar, M.N.; Bakar, E.A.; Aabid, A.; Khan, S.A. Numerical Simulations of a CD Nozzle and the Influence of the Duct Length. *Int. J. Innov. Technol. Explor. Eng.* **2019**, *8*, 622–630.
148. Aabid, A.; Chaudhary, Z.I.; Khan, S.A. Modelling and Analysis of Convergent Divergent Nozzle with Sudden Expansion Duct using Finite Element Method. *J. Adv. Res. Fluid Mech. Therm. Sci.* **2019**, *63*, 34–51.
149. Fischer, R.A.; Yates, F. *Statistical Methods for Research Workers*; Oliver and Boyd: Edinburgh, UK, 1935; pp. 241–271.
150. Montgomery, D.C. *Design and Analysis of Experiments*, 8th ed.; John Wiley & Sons, Inc.: Hoboken, NJ, USA, 2013.
151. Quadros, J.D.; Khan, S.A.; Antony, A.J. Investigation of effect of process parameters on suddenly Expanded flows through an axi-symmetric nozzle for different Mach Numbers using Design of Experiments. *IOP Conf. Ser. Mater. Sci. Eng.* **2017**, *184*, 012005. [CrossRef]
152. Quadros, J.D.; Khan, S.A.; Antony, A.J. Base Pressure Behaviour in a Suddenly Expanded Duct at Supersonic Mach Number Regimes using Taguchi Design of Experiments. *Mech. Mech. Eng.* **2018**, *22*, 1077–1098. [CrossRef]
153. Quadros, J.; Khan, S.A.; Antony, A.J. Study of Effect of Flow Parameters on Base Pressure in a Suddenly Expanded Duct at Supersonic Mach Number Regimes using CFD and Design of Experiments. *J. Appl. Fluid Mech.* **2018**, *11*, 483–496. [CrossRef]
154. Quadros, J.D.; Khan, S.A.; Anthony, J.A. Modelling of Suddenly Expanded Flow Process in Supersonic Mach Regime using Design of Experiments and Response Surface Methodology. *J. Comput. Appl. Mech.* **2018**, *49*, 149–160.
155. Quadros, J.D.; Khan, S.; Antony, A. Study of base pressure behavior in a suddenly expanded duct at supersonic mach number regimes using statistical analysis. *J. Appl. Math. Comput. Mech.* **2018**, *17*, 59–72. [CrossRef]
156. Al-khalifah, T.; Aabid, A.; Khan, S.A.; Azami, M.H.B.; Baig, M. Response surface analysis of the nozzle flow parameters at supersonic flow through microjets. *Aust. J. Mech. Eng.* **2021**, *13*, 1–15. [CrossRef]
157. Majil, D.D.; Poojitha, A.; Devi, G. Computational Study on Optimization of Rocket Nozzle. Master's Thesis, University of Southampton, Southampton, UK, 2016; pp. 8–13.
158. Quadros, J.; Khan, S.; Anthony, J.A. Predictive modeling of suddenly expanded flow process in the Supersonic Mach number regime using response surface methodology. *Int. J. Recent Res. Asp.* **2018**, *49*, 149–160.
159. Al-khalifah, T.; Aabid, A.; Khan, S.A. Regression Analysis of Flow Parameters at High Mach Numbers. *Solid State Technol.* **2020**, *63*, 5473–5488.
160. Jagannath, R.; Naresh, N.G.; Pandey, K.M. Studies on Pressure Loss in Sudden Expansion in Flow through Nozzles: A Fuzzy Logic Approach. *ARPN J. Eng. Appl. Sci.* **2007**, *2*, 50–61.
161. Pandey, K.M.; Kumar, A. Studies on Base Pressure in Suddenly Expanded Circular Ducts: A Fuzzy Logic Approach. *Int. J. Eng. Technol.* **2010**, *2*, 379–386. [CrossRef]
162. Pandey, K.M.; Kumar, S.; Kalita, J.P. Wall static pressure variation in sudden expansion in cylindrical ducts with supersonic flow: A fuzzy logic approach. *Int. J. Soft Comput. Eng.* **2012**, *2*, 237–242. [CrossRef]
163. Quadros, J.D.; Khan, S.A.; Sapkota, S.; Vikram, J.; Prashanth, T. On Recirculation Region Length of Suddenly Expanded Supersonic Flows, Using CFD and Fuzzy Logic. *Int. J. Comut. Fluid Dyn.* **2020**, 1–16. [CrossRef]
164. Efe, M.Ö.; Debiasi, M.; Yan, P.; Özbay, H.; Samimy, M. Neural network-based modelling of subsonic cavity flows. *Int. J. Syst. Sci.* **2008**, *39*, 105–117. [CrossRef]

165. De Mattos Secco, B.S.; Ney, R. Artificial neural networks to predict aerodynamic coefficients of transport airplanes. *Aircr. Eng. Aerosp. Technol.* **2017**, *89*, 19–39. [CrossRef]
166. Afzal, A.; Aabid, A.; Khan, A.; Khan, S.A.; Rajak, U.; Verma, T.N.; Kumar, R. Response surface analysis, clustering, and random forest regression of pressure in suddenly expanded high-speed aerodynamic flows. *Aerosp. Sci. Technol.* **2020**, *107*, 106318. [CrossRef]
167. Afzal, A.; Khan, S.A.; Islam, M.T.; Jilte, R.D.; Khan, A.; Soudagar, M.E.M. Investigation and back-propagation modeling of base pressure at sonic and supersonic Mach numbers. *Phys. Fluids* **2020**, *32*. [CrossRef]
168. Uzun, A.; Solomon, J.T.; Foster, C.H.; Oates, W.S.; Hussaini, M.Y.; Alvi, F.S. Flow Physics of a Pulsed Microjet Actuator for High-Speed Flow Control. *AIAA J.* **2013**, *51*, 2894–2918. [CrossRef]
169. Da, X.; Fan, Z.; Fan, J.; Zeng, L.; Rui, W.; Zhou, R. Microjet flow control in an ultra-compact serpentine inlet. *Chin. J. Aeronaut.* **2015**, *28*, 1381–1390. [CrossRef]
170. Kreth, P.; Alvi, F. Microjet based active flow control on a fixed wing UAV. *J. Flow Control. Meas. Vis.* **2010**, *2*, 32–41. [CrossRef]
171. Alvi, F.S.; Elavarasan, R.; Shih, C.; Garg, G.; Krothapalli, A. Control of supersonic impinging jet flows using microjets. *Fluids 2000 Conf. Exhib.* **2000**, 1–8. [CrossRef]
172. Ali, M.Y.; Arora, N.; Topolski, M.; Alvi, F.S.; Solomon, J.T. Properties of Resonance Enhanced Microjets in Supersonic Crossflow. *AIAA J.* **2017**, *55*, 1075–1081. [CrossRef]
173. Tebbiche, H.; Boutoudj, M.S. Active flow control by micro-blowing and effects on aerodynamic performances. Ahmed body and NACA 0015 airfoil. *Int. J. Fluid Mech. Res.* **2021**, *48*, 29–46. [CrossRef]
174. Hosseini, S.S.; Cooperman, A.; van Dam, C.P.; Pandya, S.A. Microjet Configuration Sensitivities for Active Flow Control on Multi-Element High-Lift Systems. *J. Aircr.* **2021**, 1–19. [CrossRef]
175. Hosseini, S.S.; van Dam, C.P.; Pandya, S.A. Surface-Normal Active Flow Control for Lift Enhancement and Separation Mitigation for High-Lift Common Research Model. In Proceedings of the AIAA Scitech 2021 Forum, Vitual Event, 11–15, 19–21 January 2021; Available online: https://arc.aiaa.org/doi/book/10.2514/MSCITECH21 (accessed on 26 July 2021).
176. Verma, S.B.; Manisankar, C. Control of Compression-Ramp-Induced Interaction with Steady Microjets. *AIAA J.* **2019**, *57*, 2892–2904. [CrossRef]
177. Burrows, T.J.; Vukasinovic, B.; Glezer, A. Flow Dynamics Effected by Active Flow Control in an Offset Diffuser. In Proceedings of the AIAA Aviation Forum, Atlanta, GA, USA, 25–29 June 2018; pp. 1–13. [CrossRef]
178. Kreth, P.A. The Effects of High-Frequency, Supersonic Microjet Injection on a High-Speed Cavity Flow. In Proceedings of the 55th AIAA Aerospace Sciences Meeting, Grapevine, TX, USA, 9–13 January 2017. [CrossRef]
179. Uzun, A.; Foster, C.H.; Solomon, J.T.; Oates, W.S.; Hussaini, M.Y.; Alvi, F.S. Simulations of pulsed actuators for high-speed flow control. In Proceedings of the 17th AIAA/CEAS Aeroacoustics Conference (32nd AIAA Aeroacoustics Conference), Portland, OR, USA, 5–8 June 2018. [CrossRef]
180. Afghan, S.; Mohamed, O.; Aabid, A. CFD analysis of compressible flows in a convergent-divergent nozzle. *Mater. Today Proc.* **2021**. [CrossRef]

Article

An MHD Marangoni Boundary Layer Flow and Heat Transfer with Mass Transpiration and Radiation: An Analytical Study

Thippeswamy Anusha [1], Rudraiah Mahesh [1], Ulavathi Shettar Mahabaleshwar [1] and David Laroze [2,*]

[1] Department of Mathematics, Shivagangotri, Davangere University, Davangere 577002, India; anushat.math@gmail.com (T.A.); maheshrudraiah15@gmail.com (R.M.); u.s.m@davangereuniversity.ac.in (U.S.M.)
[2] Instituto de Alta Investigación, Universidad de Tarapacá, Casilla 7D, Arica 1000000, Chile
* Correspondence: dlarozen@academicos.uta.cl

Abstract: This examination is carried out on the two-dimensional magnetohydrodynamic problem for a steady incompressible flow over a porous medium. The $Cu - Al_2O_3$ nanoparticles are added to the water base fluid in order to improve thermal efficiency. The transverse magnetic field with strength B_0 is applied. The governing equations formed for the defined flow form a system of partial differential equations that are then converted to a system of ordinary differential equations upon applying the suitable similarity transformations. On analytically solving the obtained system, the solutions for velocity profile and temperature distribution are obtained in terms of exponential and Gamma functions, respectively. In addition, the physical parameter of interest, the local Nusselt number, is obtained. The results are analyzed through plotting graphs, and the effect of different parameters is analyzed. Furthermore, we observe that the suction/injection parameter enhances the axial velocity. The porous and radiation parameters enhance the temperature distribution, and the suction/injection parameter suppresses the temperature distribution.

Keywords: hybrid nanofluid; porous media; Brinkman ratio; suction/injection; magnetic parameter

1. Introduction

Several studies have focused on the problem of boundary layer flow (BLF) and heat transfer across a stretching/shrinking sheet [1]. Because of its importance in industrial and engineering processes, a significant amount of effort has been devoted to this area in recent years. The application of certain flows in engineering and technological operations includes refrigeration, electrical gadgets with fans, nuclear reactors, polyethylene extraction, steel fabrication, and many more. Crane introduced the idea of flow across a stretching sheet by analytically solving the steady 2D flow through a linearly stretched plat [1]. Wang further generalized this concept to a 3D example [2]. Numerous scholars have since investigated various facets of this form of movement [3–10]. They examined fluid flow and even heat transfer properties of a permeable stretching sheet of convective boundary conditions (BCs), viscous dissipation, and several types of fluid.

Choi was the first to invent the phrase "nanofluid" (NF) in 1995 [11], which is a mixture that improves the physical and chemical properties of a fluid using nanoparticles. Currently, the importance of energy consumption has led scientists to optimize thermal devices. One of the solutions proposed for this purpose is using solid nanoparticles to amend the thermal properties of conventional viscous fluids. Furthermore, a different type of NF, called a hybrid nanofluid (HNF), is being studied to boost the mass transfer coefficient even more. HNF is an enhanced NF composed of two unique nanoparticles, while ordinary NF consists of a special nanoparticle that absorbs the base fluid. The chemical compositions of HNF are then improved, which improves mass transfer efficiency. Most of the studies investigated the BLF and mass transfer. The concept of HNF has been the subject of extensive scientific studies. It was proven that hybrid nanofluids can be an alternative to the single nanofluid, because it can provide more heat transfer enhancement, particularly in the contexts of solar energy, electromechanics and automobile use.

Mahabaleshwar et al. studied the radiation effect on inclined MHD flow and found the exact solution for the flow over the porous media by considering different boundary conditions, such as on the studied MHD flow with CNTs by considering the impact of mass transpiration and radiation, on the flow and heat transfer with chemical reaction in porous media, as well as on the unsteady inclined MHD flow for Casson HNF due to porous media [12,13]. Moreover, Aly and Pop [14] performed a comparison between the significance of HNF and NF on MHD flow and heat transfer by considering the effect of partial slip. The flow of MHD in such a particular case was first explored by Sarpakaya [15] and Mahabaleshwar [16]. Mahabaleshwar et al. [17] investigated the MHD effect on a Newtonian fluid flow due to a super-linear stretching sheet. Fang and Zhang [18] and Hamad [19] examined the MHD flow due to a shrinking sheet and a stretching sheet, respectively. Turkyilmazoglu [20] examined the MHD flow, heat and mass transfer of viscoelastic fluid with slip over the stretching surface, and obtained multiple solutions. Suresh et al. [21,22] investigated the effect of HNF on heat transfer and the formation of HNF out of (Al_2O_3-Cu/H_2O). Vinay Kumar et al. [23] also investigated the MHD flow over a nonlinear stretching/shrinking sheet and the impact of slip on it in a porous medium.

On the other hand, many studies have been conducted on HNF flow, MHD HNF flow due to a quadratic stretching/shrinking sheet, radiative mixed convective flow, and also dusty HNF [24,25]. Furthermore, recent developments and applications of HNF were investigated in Refs. [26,27].

The Marangoni convection is stress due to the transverse gradient of surface tension that is acting along interfaces to produce movements in liquid–liquid or liquid–gas interfaces in some industrial processes. The thermo-Marangoni convection has important applications in the semiconductor and metallurgical industries, as well as in welding and crystal growth [27–29]. Chamkha [30] demonstrated that surface-driven flows, which may be produced not only by Marangoni effects but also by the existence of the buoyancy effects caused by gravity and the external pressure gradient, can produce steady boundary layers along the interface of two immiscible fluids. Motivated by the aforementioned works, the aim of the current study was to examine the 2D MHD steady incompressible flow and heat transfer of HNF over a porous medium. In particular, we included the effect of adding $Cu - Al_2O_3$ nanoparticles to the base fluid water in order to improve thermal efficiency. The thermal conductivity of $Cu - Al_2O_3$ water increases with increasing volume concentration of nanoparticles. The main reason for the increase in thermal conductivity of $Cu - Al_2O_3$ water hybrid nanofluid is the functionalization of Al_2O_3 and Cu nanoparticles, which have a higher thermal conductivity than Al_2O_3 nanoparticles. Thermal radiation was also incorporated in the present study. Because of its impact on processes that operate at high temperatures, thermal radiation has also drawn a lot of interest [31–33]. We performed an analysis to obtain the velocity profile and temperature distribution for this system. The manuscript is arranged as follows: In Section 2, the physical model is presented and, in Section 3, the analytical solutions of the model are obtained. In Section 4, the results are discussed. Finally, the concluding remarks are given in Section 5.

2. Physical Model

The 2D MHD steady incompressible flow and heat transfer of HNF over a porous medium are here considered. As shown in Figure 1, the transverse magnetic field with strength B_0 is applied along the y-axis. In addition, the Cu and Al_2O_3 nanoparticles are added to the water base fluid. This shows that the velocity boundary layer thickness is more than the thermal boundary layer thickness. The ambient temperature of the HNF is kept constant at T_∞. In adopting the standard boundary layer approximation, the leading equations are as follows [27],

$$\frac{\partial u}{\partial x} + \frac{\partial v}{\partial y} = 0 \qquad (1)$$

$$u\frac{\partial u}{\partial x} + v\frac{\partial u}{\partial y} = \frac{\mu_{eff}}{\rho_{hnf}}\frac{\partial^2 u}{\partial y^2} - \frac{\sigma_{hnf}B_0^2}{\rho_{hnf}}u - \frac{\mu_{hnf}}{\rho_{hnf}K}u \qquad (2)$$

$$u\frac{\partial T}{\partial x} + v\frac{\partial T}{\partial y} = \frac{\kappa_{hnf}}{(\rho C_P)_{hnf}}\frac{\partial^2 T}{\partial y^2} - \frac{1}{(\rho C_P)_{hnf}}\frac{\partial q_r}{\partial y} \qquad (3)$$

subject to BCs,

$$\left.\begin{array}{l} v = v_w,\ \mu_{hnf}\frac{\partial u}{\partial y} = \frac{\partial \sigma}{\partial T}\frac{\partial T}{\partial x} \text{ at } y = 0, \\ u \to 0,\ T \to T_\infty \qquad \qquad \text{ as } y \to \infty \end{array}\right\} \qquad (4)$$

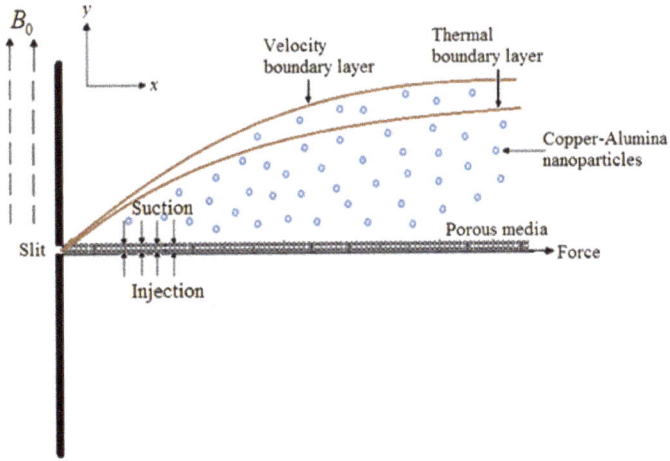

Figure 1. Physical model of the flow.

All mentioned parameters are as described in the nomenclature. The subscript hnf denotes the HNF quantities and are described as below,

$$\frac{\rho_{hnf}}{\rho_f} = (1 - \phi_2)\left(1 - \phi_1 + \phi_1\frac{\rho_{s_1}}{\rho_f}\right) + \phi_2\left(\frac{\rho_{s_2}}{\rho_f}\right)$$

where

$$\frac{\mu_{hnf}}{\mu_f} = \frac{1}{(1-\phi_1)^{2.5}(1-\phi_2)^{2.5}}$$

$$\frac{\sigma_{hnf}}{\sigma_f} = \frac{\sigma_{s_2} + 2\sigma_{bf} + 2\phi_2(\sigma_{s_2} - \sigma_f)}{\sigma_{s_2} + 2\sigma_{bf} - \phi_2(\sigma_{s_2} - \sigma_f)},$$

$$\frac{\sigma_{bf}}{\sigma_f} = \frac{\sigma_{s_1} + 2\sigma_f + 2\phi_1(\sigma_{s_1} - \sigma_f)}{\sigma_{s_1} + 2\sigma_f - \phi_1(\sigma_{s_1} - \sigma_f)}$$

$$\frac{k_{hnf}}{k_f} = \frac{k_{s_2} + 2k_{bf} + 2\phi_2(k_{s_2} - k_f)}{k_{s_2} + 2k_{bf} - \phi_2(k_{s_2} - k_f)},$$

where

$$\frac{k_{bf}}{k_f} = \frac{k_{s_1} + 2k_f + 2\phi_1(k_{s_1} - k_f)}{k_{s_1} + 2k_f - \phi_1(k_{s_1} - k_f)}$$

$$\frac{(\rho C_P)_{hnf}}{(\rho C_P)_f} = (1 - \phi_2)\left(1 - \phi_1 + \phi_1\frac{(\rho C_p)_{s_1}}{(\rho C_p)_f}\right) + \phi_2\frac{(\rho C_p)_{s_2}}{(\rho C_p)_f} \qquad (5)$$

The radiative heat flux is calculated by applying the Rosseland approximation for radiation as follows [12,28],

$$q_r = -\frac{4\sigma^*}{3k^*}\frac{\partial T^4}{\partial y} \qquad (6)$$

It is implicit that the temperature varies within the flow, where the term T^4 is the linear function of the temperature. Therefore, using Taylor series expansion to the term T^4 about T_∞ and ignoring the higher order terms, we acquire

$$T^4 \cong 4T_\infty^3 T - 3T_\infty^4 \qquad (7)$$

Equation (3) reduces to

$$u\frac{\partial T}{\partial x} + v\frac{\partial T}{\partial y} = \frac{\kappa_{hnf}}{(\rho C_P)_{hnf}}\frac{\partial^2 T}{\partial y^2} + \frac{16\sigma^* T_\infty^3}{3k^*(\rho C_P)_{hnf}}\frac{\partial^2 T}{\partial y^2} \qquad (8)$$

Consider the suitable similarity transformations as follows [27]:

$$\begin{array}{l}\psi(\eta) = \xi_2 x f(\eta), \eta = \xi_1 y \\ u = \xi_1\xi_2 x f_\eta(\eta), v = -\xi_2 f(\eta), \theta(\eta) = \frac{T-T_\infty}{ax^2}\end{array} \qquad (9)$$

On using (9), Equations (2) to (4) are converted as

$$\Lambda f_{\eta\eta\eta} + C_1\left(f f_{\eta\eta} - f_\eta^2\right) - \left(C_3 M + C_2 Da^{-1}\right)f_\eta = 0 \qquad (10)$$

$$\frac{1}{C_4}\left(\frac{C_5}{\Pr} + N_R\right)\theta_{\eta\eta} + f\theta_\eta - 2f_\eta\theta = 0 \qquad (11)$$

With the imposed BCs as

$$\left.\begin{array}{l}f(0) = S,\ f_{\eta\eta}(0) = -2(1-\phi_1)^{2.5}(1-\phi_2)^{2.5},\ \theta(0) = 1, \\ f_\eta(\eta) \to 0,\ \theta(\eta) \to 0 \text{ as } \eta \to \infty \end{array}\right\} \qquad (12)$$

where $Da^{-1} = \frac{v_f}{K\xi_1^2\xi_2}$ is the inverse Darcy number; $\Lambda = \frac{\mu_{eff}}{\mu_f}$ is the Brinkman ratio; $M = \frac{\sigma_f B_0^2}{\xi_1^2\xi_2 v_f}$ is the magnetic field; $N_R = \frac{16\sigma^* T_\infty^3}{3k_f k^*}$ is thermal radiation; and C_i, where $i = 1$ to 5, is taken as

$$C_1 = \frac{\rho_{hnf}}{\rho_f},\ C_2 = \frac{\mu_{hnf}}{\mu_f},\ C_3 = \frac{\sigma_{hnf}}{\sigma_f},\ C_4 = \frac{(\rho C_P)_{hnf}}{(\rho C_P)_f} \text{ and } C_5 = \frac{k_{hnf}}{k_f} \qquad (13)$$

The interested physical local Nusselt number Nu_x is given by

$$Nu_x = \frac{x q_w}{k_f(T-T_\infty)} \qquad (14)$$

where q_w is the heat flux given as

$$q_w = -k_{hnf}\left(\frac{\partial T}{\partial y} + q_r\right)_{y=0} \qquad (15)$$

Equations (14) and (15) lead to

$$Nu_x = -\left(\frac{k_{hnf}}{k_f} - k_{hnf}\Pr N_R\right)\xi_1 x \theta_\eta(0) \qquad (16)$$

3. Exact Analytical Solutions

In this section, we compute the analytical solution of the model. We separate the section into two subsections for the velocity and temperature fields.

3.1. Velocity

The exact analytical solution of Equation (10) is in the form

$$f(\eta) = d_1 + d_2 e^{-\alpha \eta} \tag{17}$$

where $\alpha > 0$ is to be determined. On using BCs (12a)

$$d_1 = S - d_2$$
$$d_2 = -\frac{2}{\alpha^2}(1-\phi_1)^{2.5}(1-\phi_2)^{2.5} \tag{18}$$

So, using (17) in Equation (10) gives

$$\Lambda \alpha^3 - SC_1 \alpha^2 - \left(C_3 M + C_2 Da^{-1}\right)\alpha - 2C_1(1-\phi_1)^{2.5}(1-\phi_2)^{2.5} = 0 \tag{19}$$

3.2. Temperature Distribution $\theta(\eta)$

Using Equation (17) and applying a new variable $\varepsilon = -e^{-\alpha \eta}$ in Equation (11),

$$\varepsilon \theta_{\varepsilon\varepsilon}(\varepsilon) + (p - q\varepsilon)\theta_\varepsilon(\varepsilon) + 2q\theta(\varepsilon) = 0 \tag{20}$$

with BCs as

$$\theta(0) = 0, \; \theta(-1) = 1, \tag{21}$$

where $p = 1 - \frac{n}{\alpha}\left[S + \frac{2}{\alpha^2}(1-\phi_1)^{2.5}(1-\phi_2)^{2.5}\right]$ and $q = \frac{2n}{\alpha^3}(1-\phi_1)^{2.5}(1-\phi_2)^{2.5}$, where $n = \frac{C_4}{\left(\frac{C_5}{P_r} + N_R\right)}$.

To solve Equation (20), we deploy the Laplace transformation to obtain

$$S(q - S)\Theta_S(S) + [3q + S(p-2)]\Theta(S) = 0 \tag{22}$$

Here, $\Theta(S) = L[\theta(\varepsilon)]$. Integrating Equation (22) gives

$$\Theta(S) = \frac{C(S-q)^{(p+1)}}{S^3} \tag{23}$$

In order to obtain the solution of Equation (20), we apply the inverse Laplace transformation and use the convolution property to acquire

$$\theta(t) = \frac{C}{2\Gamma[-1-p]} \int_0^t \frac{(t-w)^2}{w^{p+2}} exp(qw)dw, \text{ here } p < -1 \tag{24}$$

where C is integrating constant can be determined by using the BCs $\theta(-1) = 1$ in Equation (24) to obtain,

$$C = \frac{2\Gamma[-1-p]}{\int_0^{-1} \frac{(1+w)^2}{w^{p+2}} exp(qw)dw} \tag{25}$$

Therefore, Equation (24) becomes

$$\theta(\varepsilon) = -\frac{\int_0^\varepsilon \frac{(\varepsilon-w)^2}{w^{p+2}} exp(qw) dw}{\int_{-1}^0 \frac{(1+w)^2}{w^{p+2}} exp(qw) dw} \qquad (26)$$

Operating the integration of Equation (26) gives the final expression for $\theta(t)$

$$\theta(\varepsilon) = \frac{q^2\varepsilon^2 \Gamma[-p-1,0,-q\varepsilon] + 2q\varepsilon\, \Gamma[-p,0,-q\varepsilon] + \Gamma[-p+1,0,-q\varepsilon]}{q^2 \Gamma[-p-1,0,q] - 2q\Gamma[-p,0,q] + \Gamma[-p+1,0,q]} \qquad (27)$$

In terms of the similarity variable η, Equation (27) becomes:

$$\theta(\eta) = \frac{q^2 e^{-2\alpha\eta}\, \Gamma[-p-1,0,qe^{-\alpha\eta}] - 2qe^{-\alpha\eta}\, \Gamma[-p,0,qe^{-\alpha\eta}] + \Gamma[-p+1,0,qe^{-\alpha\eta}]}{q^2 \Gamma[-p-1,0,q] - 2q\Gamma[-p,0,q] + \Gamma[-p+1,0,q]} \qquad (28)$$

Differentiating Equation (28), we obtain

$$\theta_\eta(0) = \frac{2\alpha q\{\Gamma[-p,0,q] - q\Gamma[-p-1,0,q]\}}{q^2 \Gamma[-p-1,0,q] - 2q\Gamma[-p,0,q] + \Gamma[-p+1,0,q]} \qquad (29)$$

Therefore, from Equation (16,) Nusselt number becomes

$$Nu_x = -\left(\frac{k_{hnf}}{k_f} - k_{hnf} \mathrm{Pr} N_R\right) \xi_1 x \frac{2\alpha q\{\Gamma[-p,0,q] - q\Gamma[-p-1,0,q]\}}{q^2 \Gamma[-p-1,0,q] - 2q\Gamma[-p,0,q] + \Gamma[-p+1,0,q]} \qquad (30)$$

In the next section, we analyze these results.

4. Results and Discussion

We examined the 2D MHD steady incompressible flow and heat transfer due to a porous medium containing $Cu - Al_2O_3$ nanoparticles in the base fluid by applying a magnetic field of strength B_0 to the fluid flow. The addition of nanoparticles enhances the thermal efficiency of the flow system. The leading equations form the system of PDEs and are then converted into the system of ODEs by adopting suitable similarity transformations. The system is analytically solved to obtain the solutions for the velocity profile and temperature distribution in terms of exponential and Gamma functions, respectively. In all plots, the dotted lines refer to the behavior of the base fluid, while the solid lines refer to the behavior of HNF for $Cu - Al_2O_3$.

Figure 2 demonstrates the axial velocity $f(\eta)$ for various Da^{-1}. We found that the velocity declines as Da^{-1} increases. Panels (a)–(c), where the velocity is examined for $S = 0$ show that there is no permeability. For suction $S = 1$ and injection $S = -1$, we observe that as S increases from injection to suction, the axial velocity for HNF very quickly coincides with the base fluid as Da^{-1} increases. In all cases, the profile of $f_\eta(\eta)$ has a decreasing nature, and it becomes constant to zero at a certain point of η.

Figure 3 depicts the velocity $f_\eta(\eta)$ as a function of η for various values of Λ. The velocity increases with an increase in Λ. Panels (a)–(c), where the velocity is examined for $S = 0$, show that there is no permeability. For suction $S = 1$ and injection $S = -1$, we observe that the difference between the axial velocity of the base fluid and HNF is larger in the case of suction and smaller in the case of injection. In all cases, the axial velocity is less for an HNF than the base fluid.

Figure 4 shows the profile of $f_\eta(\eta)$ for various M. It can be seen that $f_\eta(\eta)$ is smaller for larger values of M. Panels (a)–(c), where the velocity is examined for $S = 0$, show that there is no permeability. For suction $S = 1$ and injection $S = -1$, we observe that the difference between the axial velocity of the base fluid and the HNF is more in the case of suction and less in the case of injection.

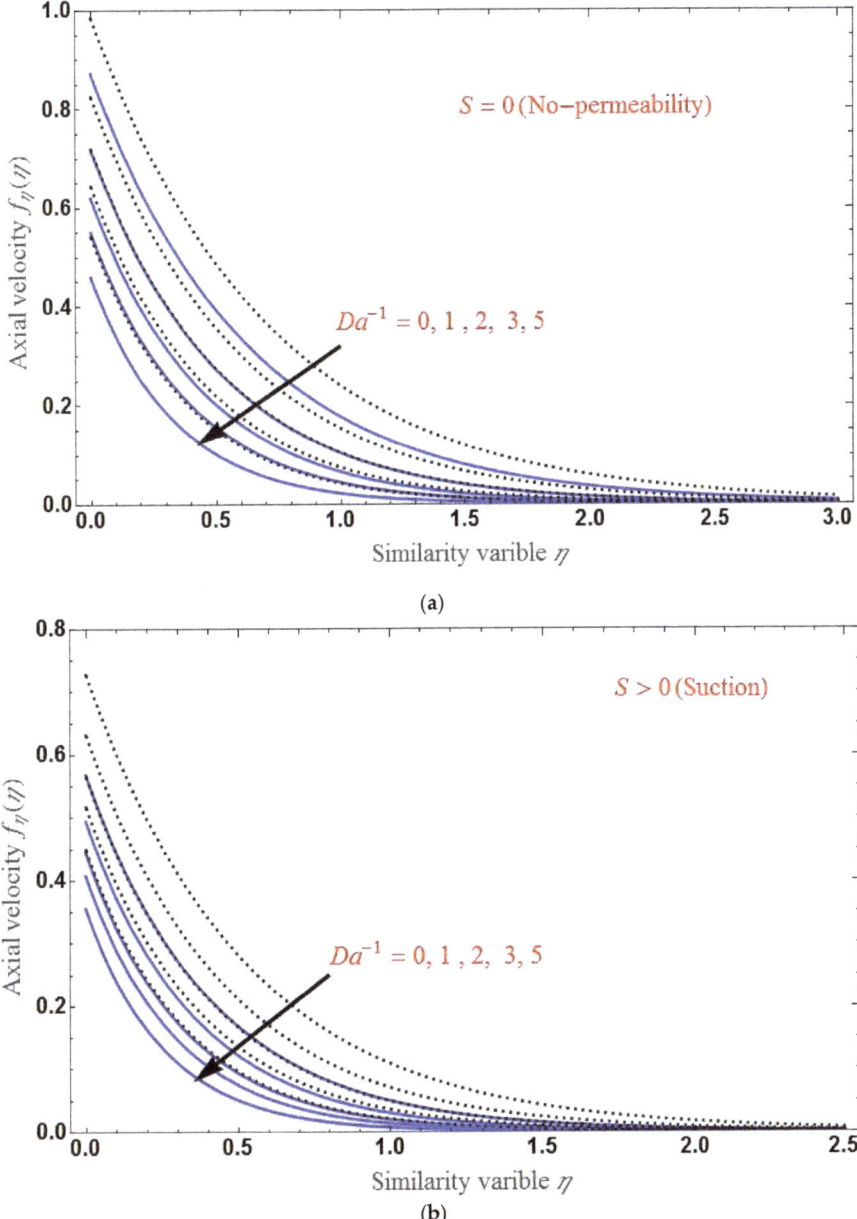

Figure 2. *Cont.*

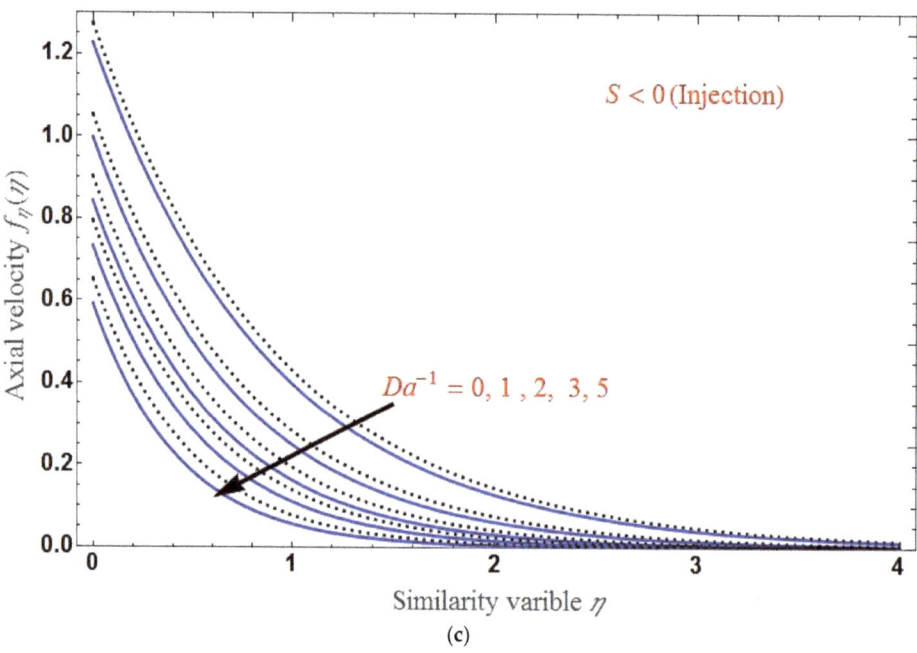

(c)

Figure 2. The axial velocity as a function of η for various values of Da^{-1} and for three different regimes of S: (**a**) $S = 0$, (**b**) $S = 1$, and (**c**) $S = -1$. The other fixed parameters are: $\Lambda = M = 1$, $\phi_1 = 0.1$, $\phi_2 = 0.04$.

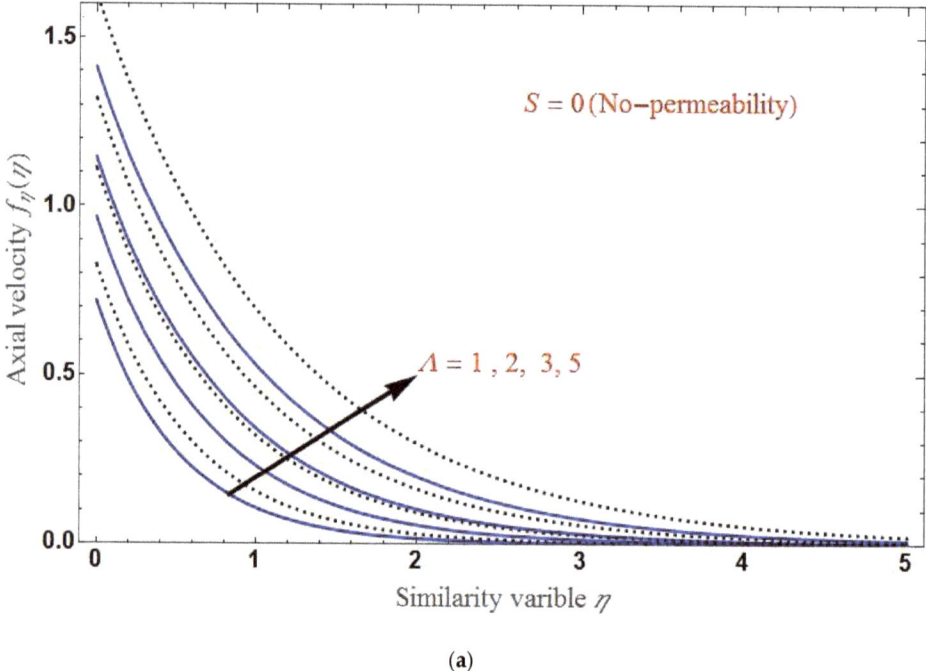

(a)

Figure 3. *Cont.*

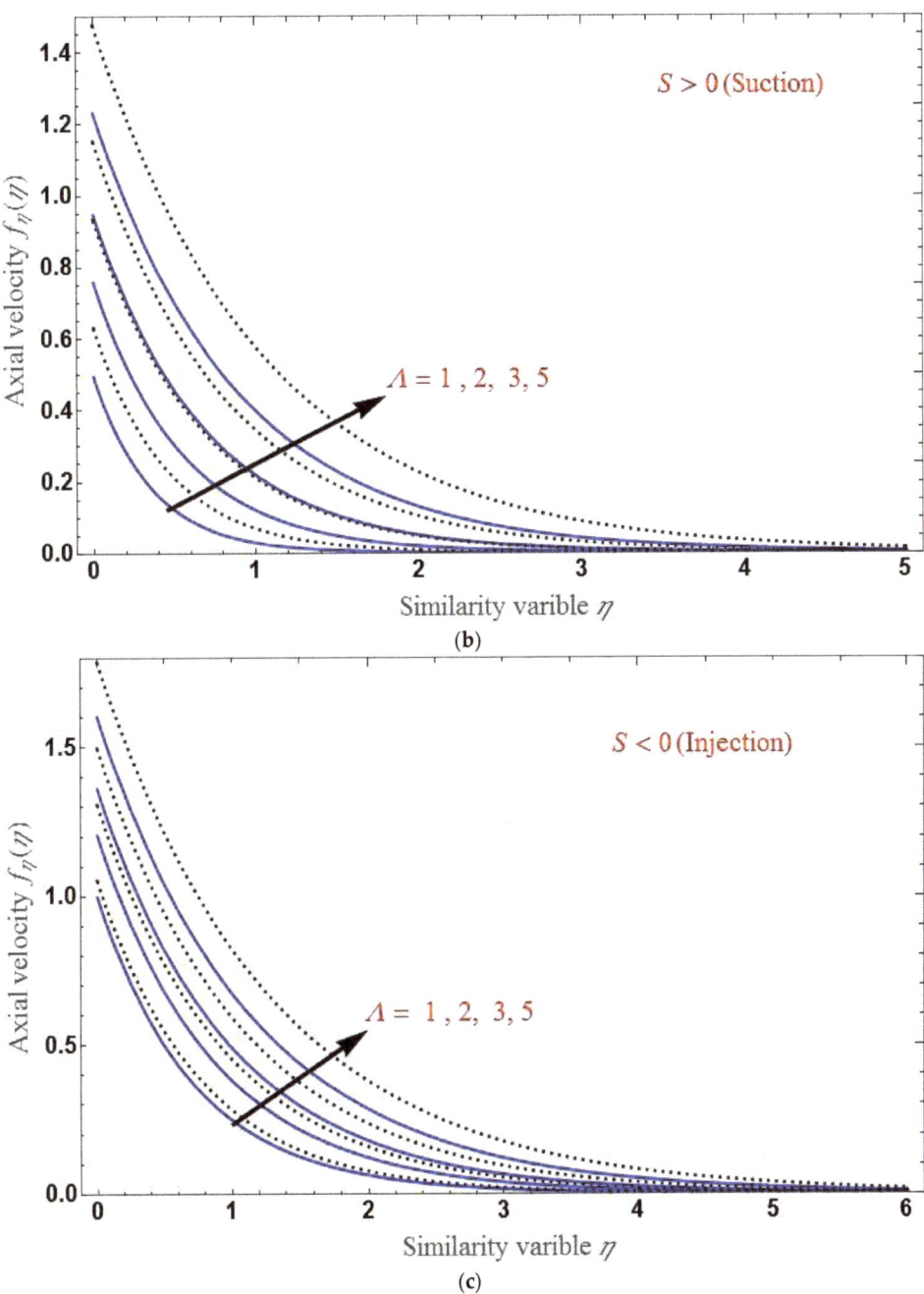

Figure 3. The axial velocity as a function of η for various values of Λ and for three different regimes of S: (a) $S = 0$, (b) $S = 1$ and (c) $S = -1$. The other fixed parameters are: $Da^{-1} = M = 1$, $\phi_1 = 0.1$, $\phi_2 = 0.04$.

The effect of S is shown in Figure 5 for both the suction and injection cases. Clearly, the velocity profile $f_\eta(\eta)$ decreases for a larger S. Panel (a) shows that the difference between the axial velocity of the base fluid and the HNF is larger in the case of suction and smaller in the case of injection as in panel (b). At $\eta = 0$, the axial velocity is different for various values of each parameter.

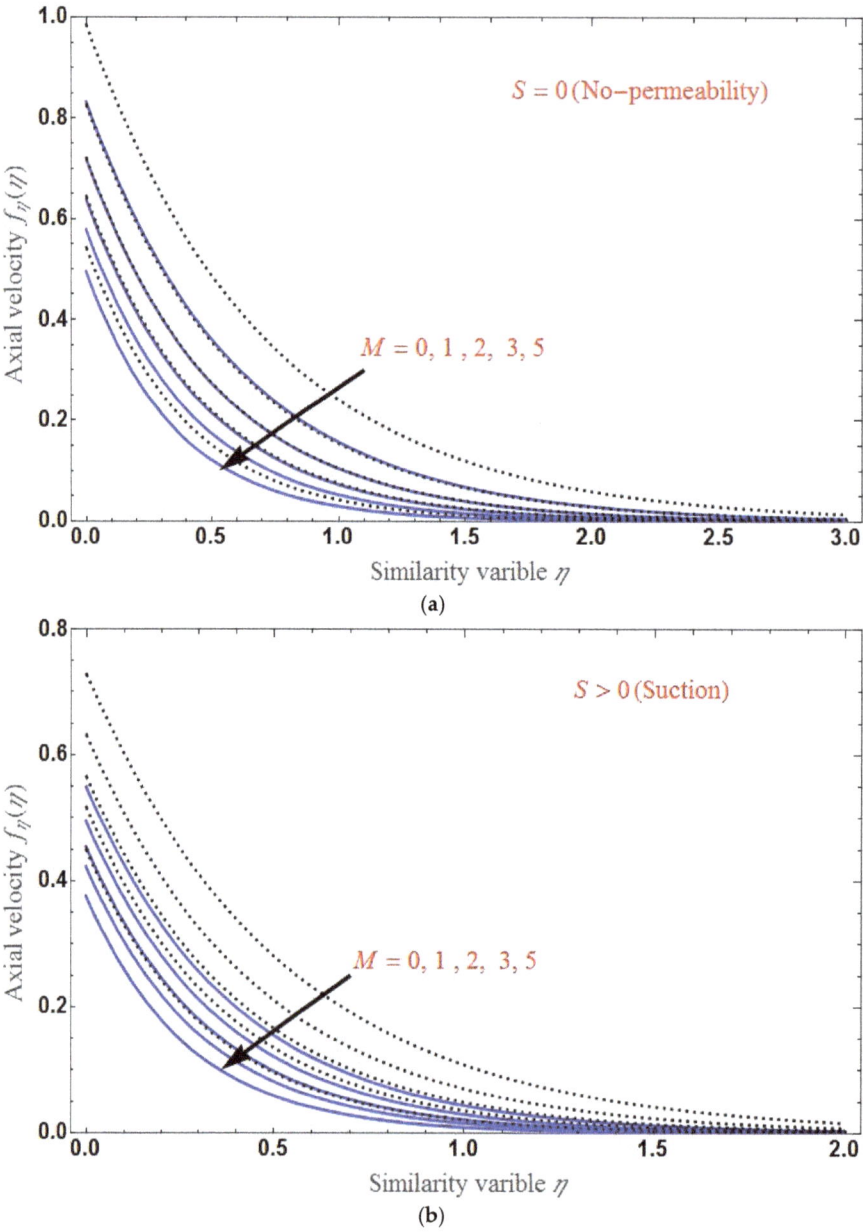

Figure 4. Cont.

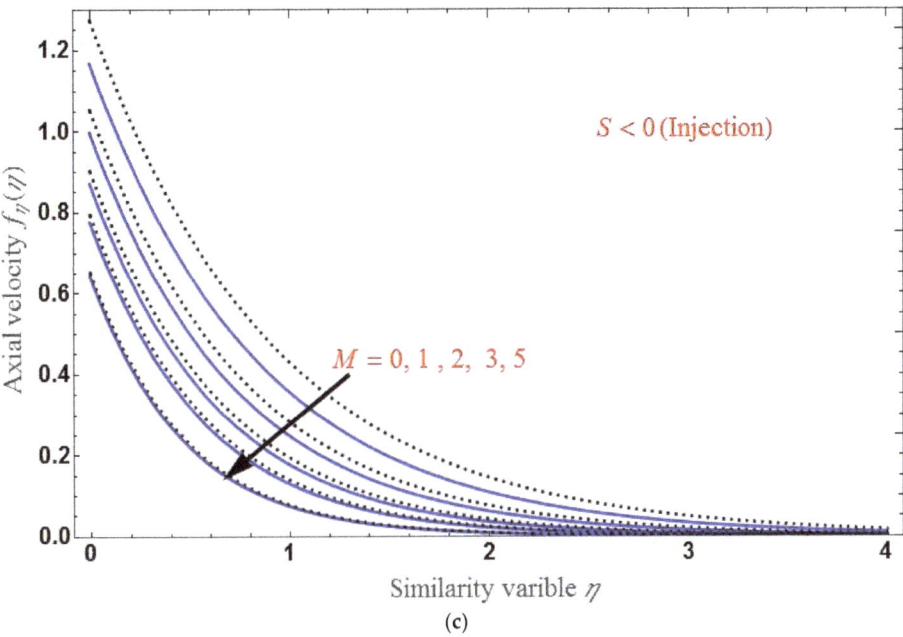

(c)

Figure 4. The axial velocity as a function of η for various values of M and for three different regimes of S: (**a**) $S = 0$, (**b**) $S = 1$ and (**c**) $S = -1$. The other fixed parameters are: $\Lambda = Da^{-1} = 1$, $\phi_1 = 0.1$, $\phi_2 = 0.04$.

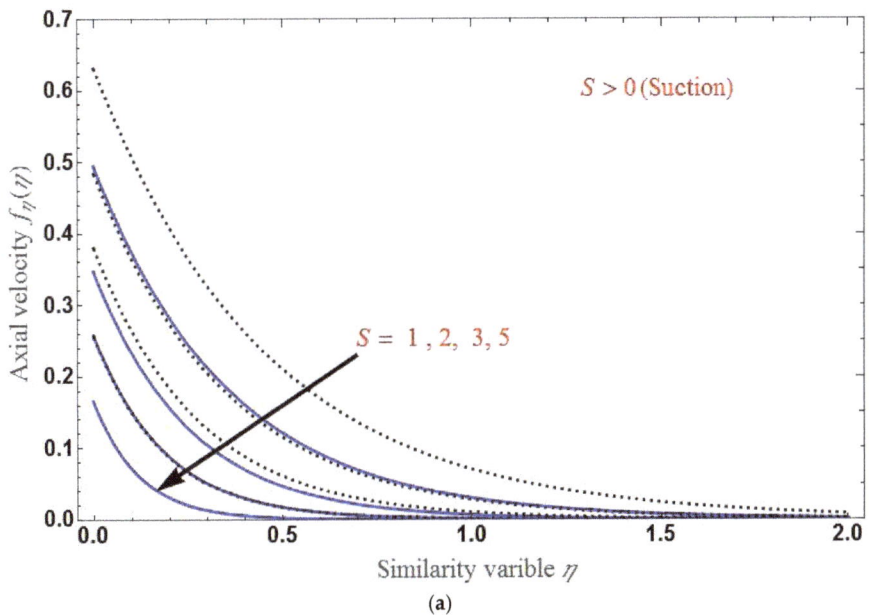

(a)

Figure 5. *Cont.*

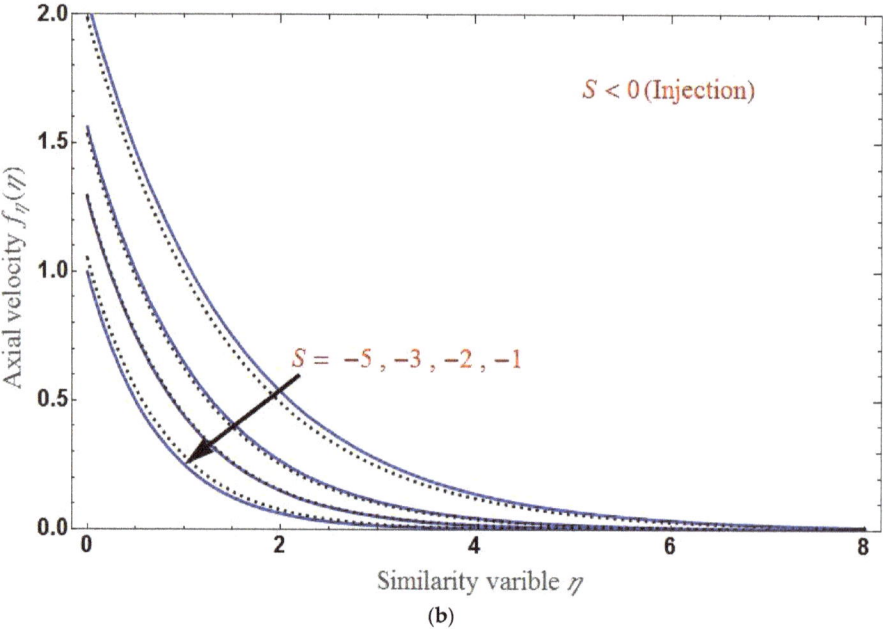

Figure 5. The axial velocity as a function of η for various values of S, (**a**) for suction case and (**b**) for injection case. The other fixed parameters are: $\Lambda = Da^{-1} = M = 1$, $\phi_1 = 0.1$, $\phi_2 = 0.04$.

Figure 6 displays the temperature profile $\theta(\eta)$ for various values of Da^{-1}. We observe that $\theta(\eta)$ increases as Da^{-1} increases. Panels (a)–(c), where the temperature is shown for $S = 0$, show that there is no permeability. For suction $S = 1$ and injection $S = -1$, we observe that $\theta(\eta)$ for HNF coincides with the base fluid as Da^{-1} increases. In all cases, we found that $\theta(\eta)$ decreases and that becomes constant to zero at a certain point of η. The temperature profile has the same value at $\eta = 0$ irrespective of the parameters' values. Moreover, as the thermal rate increases upon adding nanoparticles to the base fluid, we can see from the figures that $\theta(\eta)$ becomes more of an HNF than a base fluid.

Figure 7 shows the temperature profile $\theta(\eta)$ for different values Λ. We observe that $\theta(\eta)$ decreases with the enhancement in Λ. Panels (a)–(c) show the regimens of S. In panel (a) for $S = 0$, there is no permeability, whereas in panel (b) suction ($S = 1$) and finally in panel (c) for injection case ($S = -1$). We found that there is an achievement of enhancement of heat transfer upon using HNF and that Casson fluid will suppress the temperature distribution.

Figure 8 shows the temperature profile as a function of η for different values of M. We see that $\theta(\eta)$ increases as M increases. Panels (a)–(c) show θ for $S = 0$, $S = 1$, and $S = -1$, respectively. We observe that HNF has much more thermal conductivity than the base fluid. Furthermore, we observe that as the value of S increases, the domain of the temperature distribution decreases.

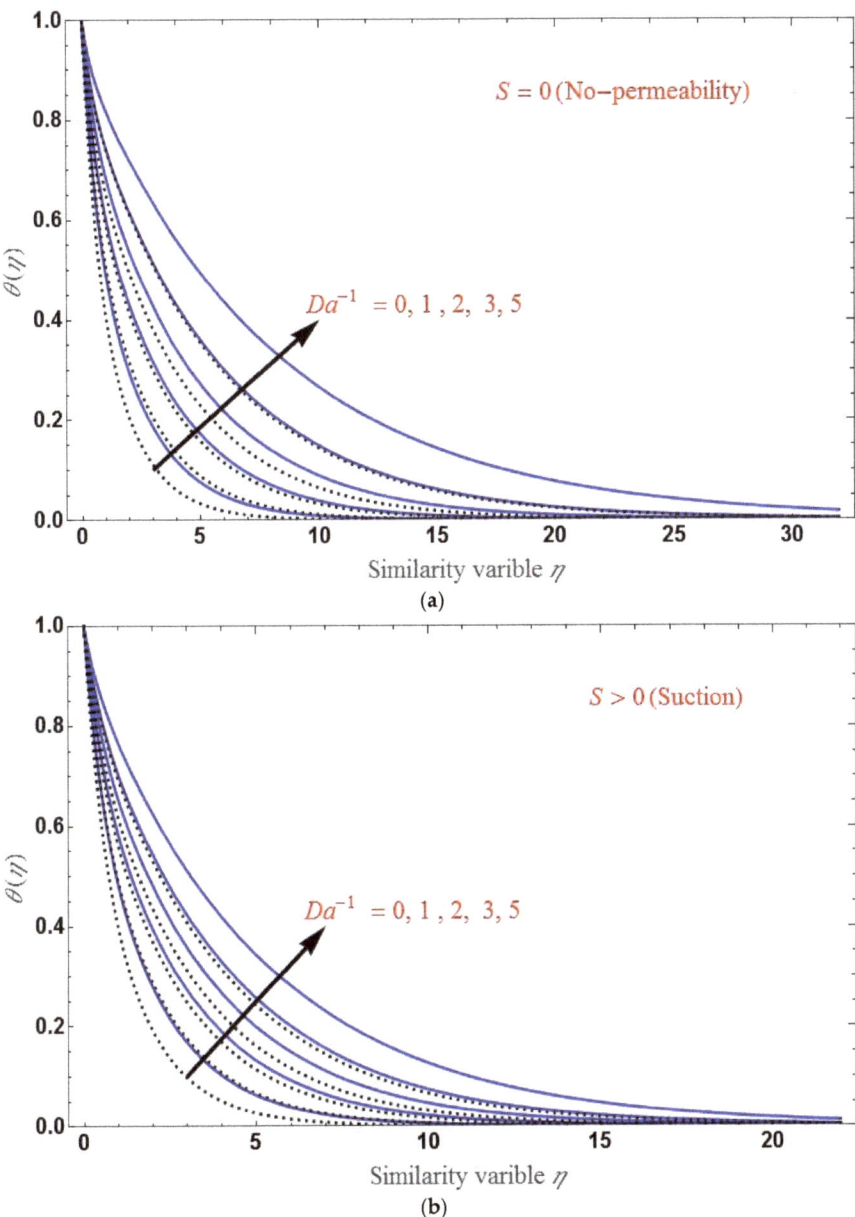

Figure 6. *Cont.*

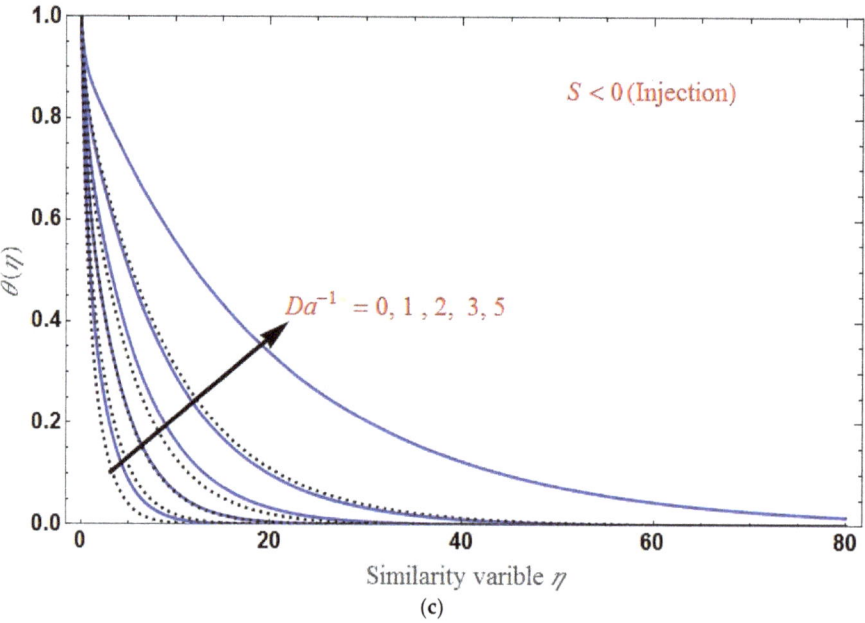

Figure 6. The temperature distribution $\theta(\eta)$ as a function of η for various values of Da^{-1} and for three different regimes of S: (**a**) $S = 0$, (**b**) $S = 1$, and (**c**) $S = -1$. The other fixed parameters are $M = \Lambda = N_R = 1$, $\phi_1 = 0.1$, $\phi_2 = 0.04$.

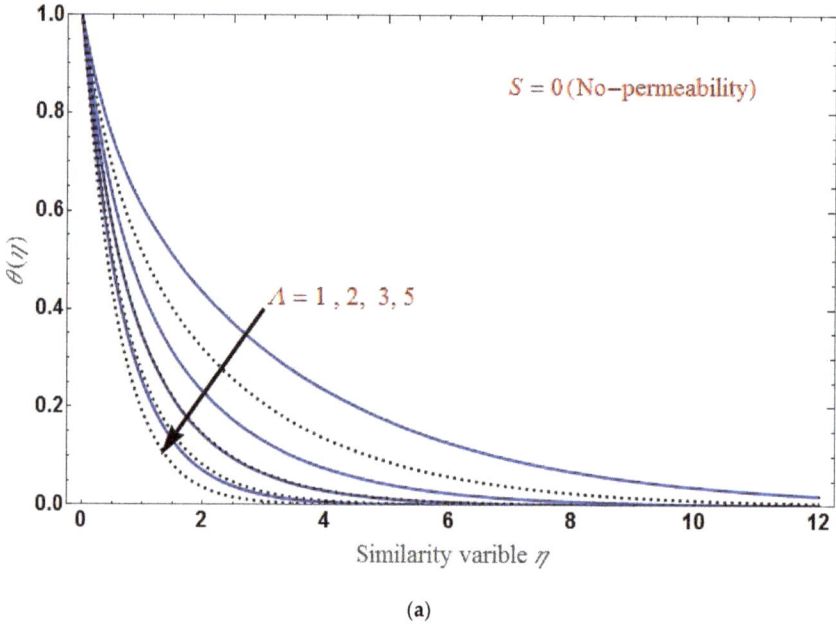

(**a**)

Figure 7. *Cont.*

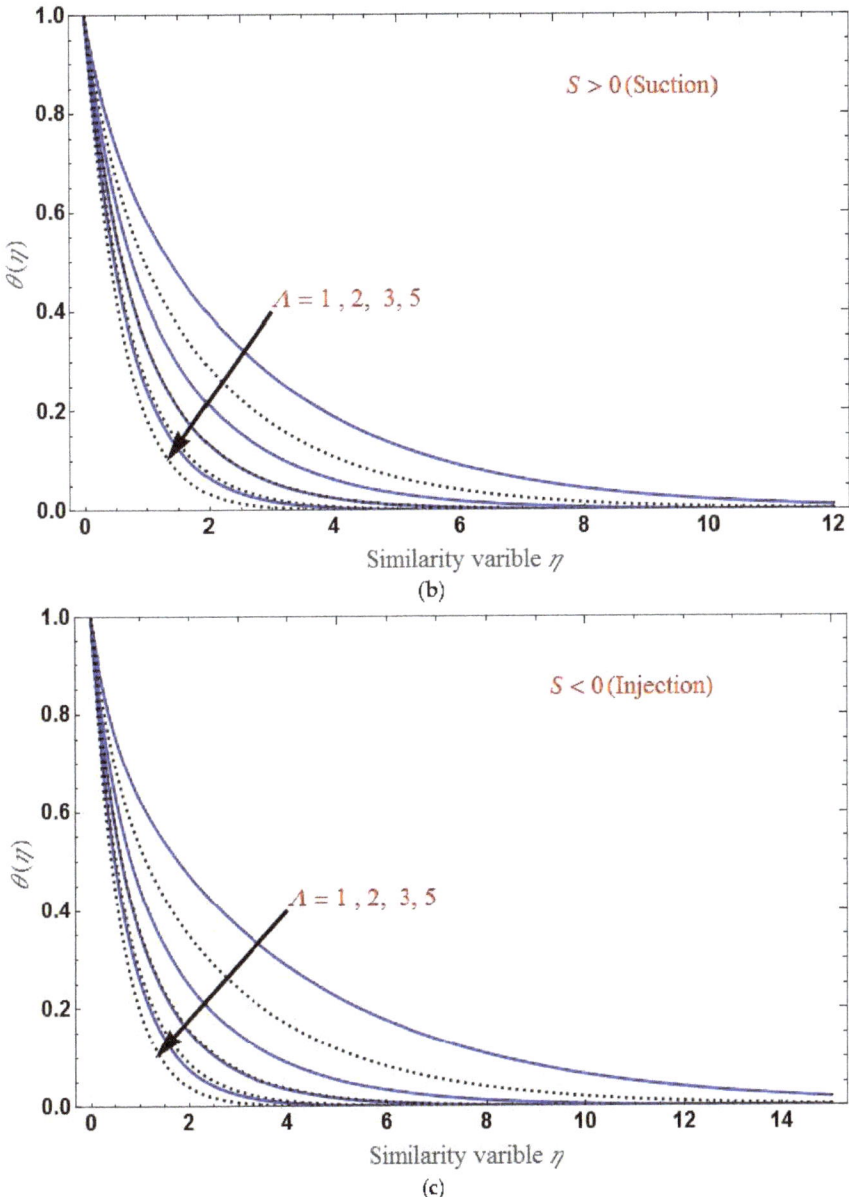

Figure 7. The temperature distribution $\theta(\eta)$ as a function of η for various values of Λ and for three different regimes of S: (**a**) $S = 0$, (**b**) $S = 1$ and (**c**) $S = -1$. The other fixed parameters are: $M = Da^{-1} = N_R = 1$, $\phi_1 = 0.1$, $\phi_2 = 0.04$.

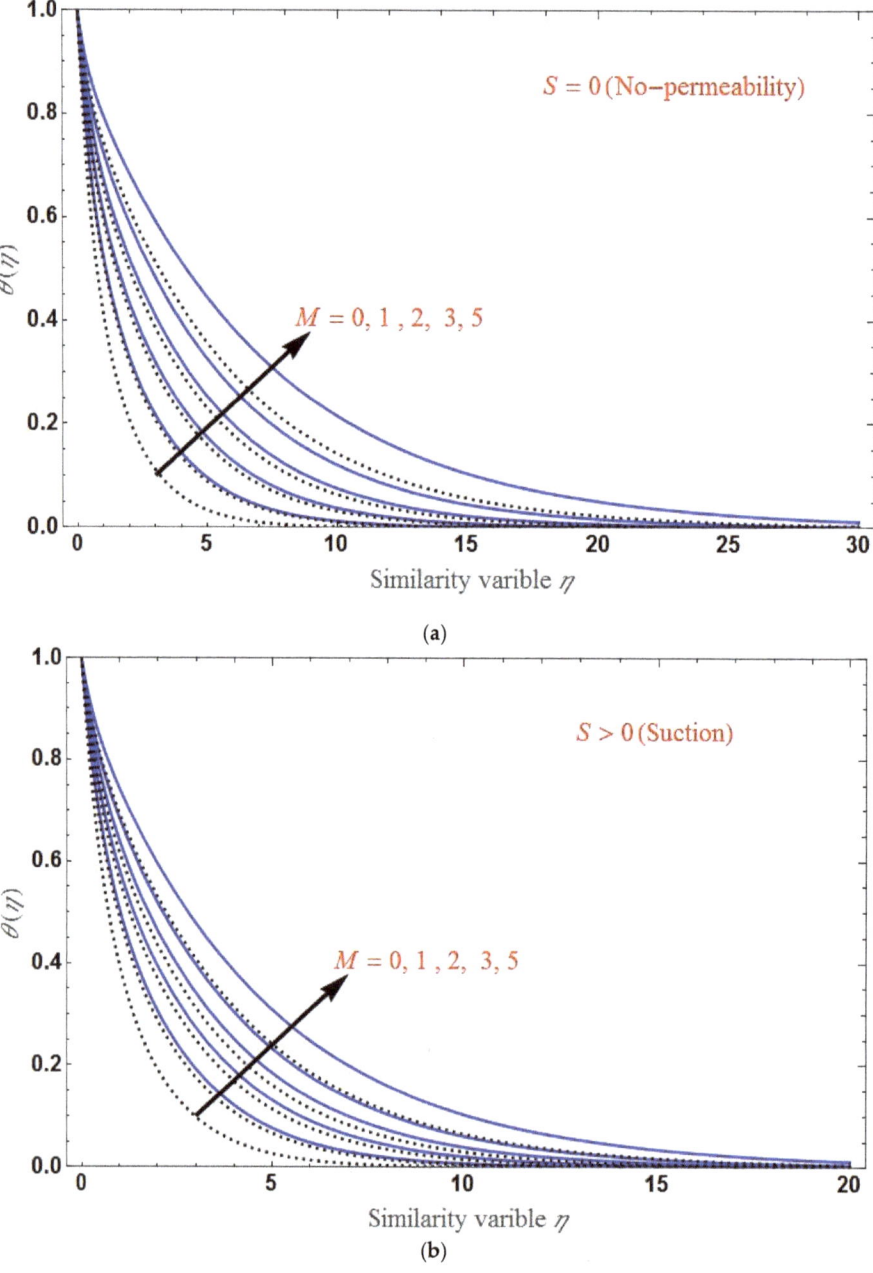

Figure 8. *Cont.*

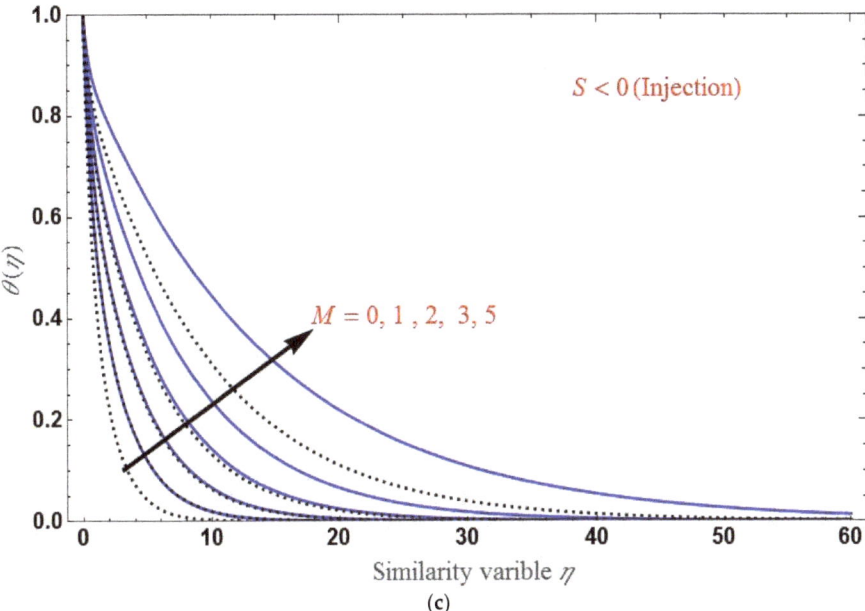

Figure 8. The temperature profile $\theta(\eta)$ as a function of η for several values of M and for three different regimes of S: (**a**) $S = 0$, (**b**) $S = 1$, and (**c**) $S = -1$. The other fixed parameters are: $\Lambda = Da^{-1} = N_R = 1$, $\phi_1 = 0.1$, $\phi_2 = 0.04$.

Figure 9 shows the temperature distribution $\theta(\eta)$ for various values of the radiation parameter N_R. We can observe that $\theta(\eta)$ increases as the effect of radiation increases. Panels (a)–(c) show the profile for cases with no permeability, with suction and injection. We note that the domain of the temperature is larger in the case of injection and smaller in the case of no-permeability. Furthermore, by observing each plot of temperature distribution, the effect of radiation on the change of heat transfer rate is less than those of the magnetic field and Casson fluid, i.e., the difference in temperature distribution between base fluid and HNF is not that much more significant.

Finally, the effect of various values of S on the temperature profile is shown in Figure 10 for suction and injection cases. $\theta(\eta)$ decreases when as S increases, as shown in the figure. The domain of $\theta(\eta)$ is larger in the case of injection than in suction.

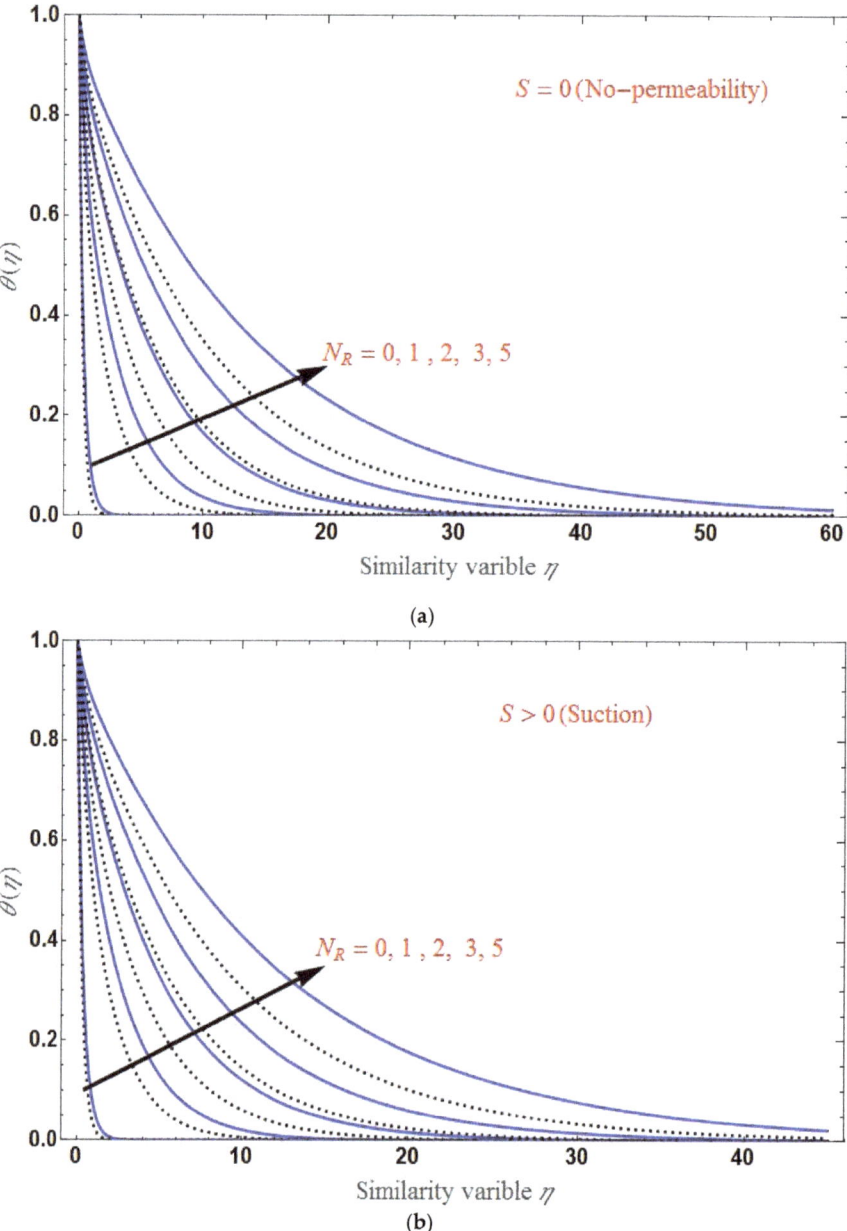

Figure 9. *Cont.*

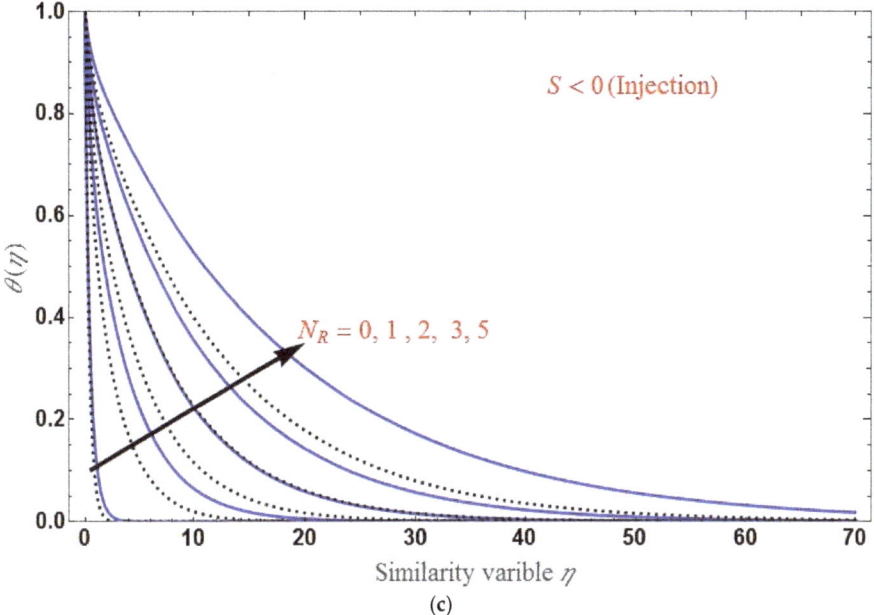

(c)

Figure 9. The temperature distribution $\theta(\eta)$ as a function of η for different values of N_R and for three different regimes of S: (**a**) $S = 0$, (**b**) $S = 1$, and (**c**) $S = -1$. The other fixed parameters are: $M = \Lambda = Da^{-1} = N_R = 1$, $\phi_1 = 0.1$, $\phi_2 = 0.04$.

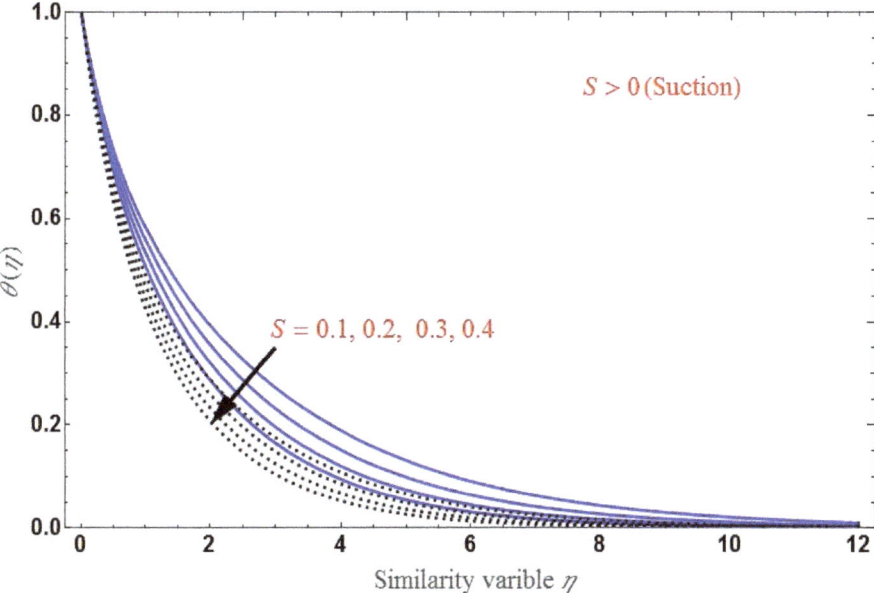

Figure 10. *Cont.*

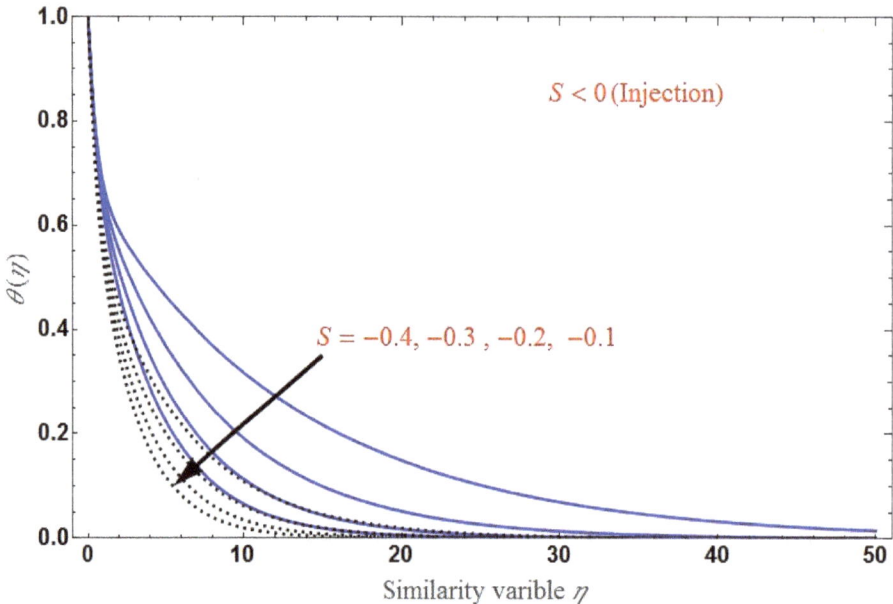

Figure 10. The temperature profile $\theta(\eta)$ as a function of η for several values of S. The other fixed parameters are: $M = \Lambda = Da^{-1} = N_R = 1$, $\phi_1 = 0.1$, $\phi_2 = 0.04$.

5. Concluding Remarks

In the present study, we examined 2D MHD steady incompressible flow and heat transfer over a porous medium containing $Cu - Al_2O_3$ nanoparticles in a base fluid. The addition of nanoparticles enhances the thermal efficiency of the flow system. The system was analytically solved to obtain the solutions for the velocity profile and temperature distribution in terms of exponential and Gamma functions, respectively. In addition, the effect of different physical parameters was examined by using graphical representations. The following observations were made:

- The axial velocity declines with increasing porous parameter or magnetic field, and the suction/injection parameter increases with increasing Brinkman ratio.
- The temperature distribution increases for higher values of the porous parameter, magnetic field, or radiation; it decreases with an increase in the Brinkman ratio or suction/injection parameter.
- At $\eta = 0$, the axial velocity is different for various values of each parameter.
- The axial velocity is smaller for hybrid nanofluid than for base fluid.
- $\theta(\eta)$ is the same for every value of varying parameters at $\eta = 0$.
- As the thermal rate increases upon adding nanoparticles to the base fluid, the figures showed that $\theta(\eta)$ will be larger for hybrid nanofluid than for base fluid.

In the future, we plan to conduct a similar investigation on a non-Newtonian fluid with mass transfer problems. We postulate that adding the effect of viscous dissipation and various physical parameters can uncover another interesting phenomenon.

Author Contributions: Conceptualization: U.S.M.; methodology: U.S.M. and D.L.; software: T.A. and R.M.; formal analysis: T.A., R.M. and U.S.M.; investigation: T.A., R.M., U.S.M. and D.L.; writing—original draft preparation: U.S.M.; writing—review and editing: D.L. All authors have read and agreed to the published version of the manuscript.

Funding: D.L. acknowledges partial financial support from Centers of Excellence with BASAL/ANID financing, grant nos. AFB180001, CEDENNA.

Data Availability Statement: Data sharing is not applicable to this article.

Conflicts of Interest: The authors declare that they have no known competing financial interests or personal relationships that could have appeared to influence the work reported in this paper.

Nomenclature

Symbol	Explanation	SI unit
Latin symbols		
B_0	applied magnetic field	$[\text{wm}^{-2}]$
C_P	specific heat at constant pressure	$[\text{JKg}^{-1}\text{K}^{-1}]$
Da^{-1}	inverse Darcy number	$[-]$
f	similarity variable	$[-]$
k^*	mean absorption coefficient	$[\text{m}^{-2}]$
K	permeability of porous medium	$[\text{m}^{-2}]$
Pr	Prandtl number	$[-]$
q_r	radiative heat flux	$[\text{Wm}^{-2}]$
q_w	local heat flux at the wall	$[-]$
M	magnetic parameter	$[-]$
N_R	radiation parameter	$[-]$
Nu_x	local Nusselt number	$[-]$
T	temperature	$[\text{K}]$
$S > 0 / < 0$	suction/injection velocity	$[-]$
(x, y)	coordinate axes	$[\text{m}]$
(u, v)	velocities along x- and y-directions	$[\text{ms}^{-1}]$
Greek symbols		
α	thermal diffusivity	$[\text{m}^2\text{s}^{-1}]$
Γ	gamma function	$[-]$
κ	thermal conductivity of fluid	$[\text{WKg}^{-1}\text{K}^{-1}]$
η	similarity variable	$[-]$
μ_f	dynamic viscosity of fluid	$[\text{kgm}^{-1}\text{S}^{-1}]$
μ_{eff}	effective viscosity	$[\text{kgm}^{-1}\text{S}^{-1}]$
ν	kinematic viscosity	$[\text{m}^2\text{s}^{-1}]$
ρ	density	$[\text{Kgm}^{-3}]$
σ	electrical conductivity	$[\text{Sm}^{-1}]$
σ^*	Stefan–Boltzmann constant	$[-]$
ϕ	nanoparticle volume fraction	$[-]$
ψ	stream function	$[-]$
Λ	Brinkman ratio	$[-]$
Subscripts		
f	base fluid	$[-]$
hnf	nanofluid	$[-]$
Abbreviations		
BCs	boundary conditions	$[-]$
BLF	boundary layer flow	$[-]$
MHD	magnetohydrodynamics	$[-]$
HNF	hybrid nanofluid	$[-]$
Cu	copper	$[-]$
Al_2O_3	aluminum oxide	$[-]$

References

1. Crane, L.J. Flow past a stretching plate. *Z. Angew. Math. Phys.* **1970**, *21*, 645–647. [CrossRef]
2. Wang, C.W. The three-dimensional flow due to a stretching flat surface. *Phys. Fluids* **1984**, *27*, 1915–1917. [CrossRef]
3. Rosca, A.V.; Pop, I. Flow and heat transfer over a vertical permeablestretching/shrinking sheet with a second order slip. *Int. J. Heat Mass Transf.* **2013**, *60*, 355–364. [CrossRef]
4. Nandy, S.K.; Mahapatra, T.R. Effects of slip and heat generation/absorptionon MHD stagnation flow of nanofluid past a stretching/shrinking surface withconvective boundary conditions. *Int. J. Heat Mass Transf.* **2013**, *64*, 1091–1100. [CrossRef]

5. Kumaran, V.; Banerjee, A.K.; Vanav Kumar, A.; Pop, I. Unsteady MHD flow and heat transfer with viscous dissipation past a stretching sheet. *Int. Commun. Heat Mass Transf.* **2011**, *38*, 335–339. [CrossRef]
6. Turkyilmazoglu, M. Multiple solutions of heat and mass transfer of MHD slip flow for the viscoelastic fluid over a stretching sheet. *Int. J. Therm. Sci.* **2011**, *50*, 2264–2276. [CrossRef]
7. Ishak, A.; Nazar, R.; Pop, I. Unsteady mixed convection boundary layer flowdue to a stretching vertical surface. *Arab. J. Sci. Eng.* **2006**, *31*, 165–182.
8. Ishak, A.; Nazar, R.; Pop, I. MHD boundary-layer flow due to a movingextensible surface. *J. Eng. Math.* **2008**, *62*, 23–33. [CrossRef]
9. Ishak, A.; Nazar, R.; Pop, I. Magnetohydrodynamic (MHD) flow and heattransfer due to a stretching cylinder. *Energy Convers. Manag.* **2008**, *49*, 3265–3269. [CrossRef]
10. Yacob, N.A.; Ishak, A.; Pop, I.; Vajravelu, K. Boundary layer flow past astretching/shrinking surface beneath an external uniform shear flow with aconvective surface boundary condition in a nanofluid. *Nanoscale Res. Lett.* **2011**, *6*, 314. [CrossRef]
11. Choi, S.U.S. Enhancing Thermal Conductivity of Fluids with Nanoparticles. In *Developments and Applications of Non-Newtonian Flows*; Siginer, D.A., Wang, H.P., Eds.; ASME: New York, NY, USA, 1995; pp. 99–105.
12. Mahabaleshwar, U.S.; Anusha, T.; Sakanaka, P.H. Suvanjan Bhattacharyya, Impact of inclined Lorentz force and Schmidt number on chemically reactive Newtonian fluid flow on a stretchable surface when Stefan blowing and thermal radiation are significant. *Arab. J. Sci. Eng.* **2021**, *46*, 12427–12443. [CrossRef]
13. Mahabaleshwar, U.S.; Lorenzini, G. Combined effect of heat source/sink and stress work on MHD Newtonian fluid flow over a stretching porous sheet. *Int. J. Heat Technol.* **2017**, *35*, S330–S335. [CrossRef]
14. Aly, E.H.; Pop, I. MHD flow and heat transfer near stagnation point over a stretching/shrinking surface with partial slip and viscous dissipation: Hybrid nanofluid versus nanofluid. *Powder Technol.* **2020**, *367*, 192–205. [CrossRef]
15. Sarpkaya, T. Flow of non-Newtonian fluids in a magnetic field. *AIChEJ* **1961**, *7*, 324–328. [CrossRef]
16. Mahabaleshwar, U.S. Combined effect of temperature and gravity modulations on the onset of magneto-convection in weak electrically conducting micropolar liquids. *Int. J. Eng. Sci.* **2007**, *45*, 525–540. [CrossRef]
17. Mahabaleshwar, U.S.; Vinay kumar, P.N.; Sakanaka, P.H.; Lorenzini, G. An MHD effect on a Newtonian fluid flow due to a superlinear stretching sheet. *J. Eng. Thermophys.* **2018**, *27*, 501–506.
18. Fang, T.; Zhang, J. Closed-form exact solutions of MHD viscous flow over a shrinking sheet. *Commun. Nonlinear Sci. Numer. Simul.* **2009**, *14*, 2853–2857. [CrossRef]
19. Hamad, M.A.A. Analytical solution of natural convection flow of a nanofluid over a linearly stretching sheet in the presence of magnetic field. *Int. Commun. Heat Mass Transf.* **2011**, *38*, 487–492. [CrossRef]
20. Turkyilmazoglu, M. Multiple analytic solutions of heat and mass transfer of magnetohydrodynamic slip flow for two types of viscoelastic fluids over a stretching surface. *J. Heat Transf.* **2012**, *134*, 071701. [CrossRef]
21. Suresh, S.; Venkitaraj, K.; Selvakumar, P.; Chandrasekar, M. Effect of Al2O3–Cu/water hybrid nanofluid in heat transfer. *Exp. Therm. Fluid Sci.* **2012**, *38*, 54–60. [CrossRef]
22. Suresh, S.; Venkitaraj, K.; Selvakumar, P.; Chandrasekar, M. Synthesis of Al2O3–Cu/water hybrid nanofluids using two step method and its thermo physical properties. *Colloids Surf. A* **2011**, *388*, 41–48. [CrossRef]
23. Vinay Kumar, P.N.; Mahabaleshwar, U.S.; Nagaraju, K.R.; MousaviNezhad, M.; Daneshkhah, A. Mass transpiration in magneto-hydrodynamic boundary layer flow over a superlinear stretching sheet embedded in porous medium with slip. *J. Porous Media* **2019**, *22*, 1015–1025. [CrossRef]
24. AmiraZainal, N.; RoslindaNazar, R.; Naganthran, K.; Pop, I. Stability analysis of MHD hybrid nanofluid flow over a stretching/shrinking sheet with quadratic velocity. *Alex. Eng. J.* **2020**, *60*, 915–926.
25. Khan, U.; Shafiq, A.; Zaib, A.; Baleanu, D. Hybrid nanofluid on mixed convective radiative flow from an irregular variably thick moving surface with convex and concave effects. *Case Stud. Therm. Eng.* **2020**, *21*, 100660. [CrossRef]
26. Sarkar, J.; Ghosh, P.; Adil, A. A review on hybrid nanofluids: Recent research, development and applications. *Renew. Sustain. Energy Rev.* **2015**, *43*, 164–177. [CrossRef]
27. Aly, E.H.; Ebaid, A. Exact analysis for the effect of heat transfer on MHD and radiation Marangoni boundary layer nanofluid flow past a surface embedded in a porous medium. *J. Mol. Liq.* **2016**, *215*, 625–639. [CrossRef]
28. Mahabaleshwar, U.S.; Nagaraju, K.R.; Vinay Kumar, P.N.; Azese, M.N. Effect of radiation on thermosolutal Marangoni convection in a porous medium with chemical reaction and heat source/sink. *Phys. Fluids* **2020**, *32*, 1136902. [CrossRef]
29. Napolitano, L.G. Marangoni boundary layers. In Proceedings of the 3rd European Symposium on Material Science in Space, Grenoble, France, 24–27 April 1979.
30. Chamkha, A.J.; Pop, I.; Takhar, H.S. Marangoni mixed convection boundary layer flow. *Meccanica* **2006**, *41*, 219–232. [CrossRef]
31. Cortell, R. Radiation effects for the Blasius and Sakiadis flows with a convective surface boundary condition. *Appl. Math. Comput.* **2008**, *206*, 832–840.
32. Mahabaleshwar, U.S.; Sarris, I.E.; Hill, A.A.; Lorenzini, G.; Pop, I. An MHD couple stress fluid due to a perforated sheet undergoing linear stretching with heat transfer. *IJHM* **2017**, *105*, 157–167. [CrossRef]
33. Mahabaleshwar, U.S.; Nagaraju, K.R.; Vinay Kumar, P.N.; Nadagoud, M.N.; Bennacer, R.; Sheremet, M.A. Effects of Dufour and Soret mechanisms on MHD mixed convective-radiation non-Newtonian liquid flow and heat transfer over a porous sheet. *Therm. Sci. Eng. Prog.* **2019**, *16*, 100459. [CrossRef]

MDPI
St. Alban-Anlage 66
4052 Basel
Switzerland
www.mdpi.com

Applied Sciences Editorial Office
E-mail: applsci@mdpi.com
www.mdpi.com/journal/applsci

Disclaimer/Publisher's Note: The statements, opinions and data contained in all publications are solely those of the individual author(s) and contributor(s) and not of MDPI and/or the editor(s). MDPI and/or the editor(s) disclaim responsibility for any injury to people or property resulting from any ideas, methods, instructions or products referred to in the content.

www.ingramcontent.com/pod-product-compliance
Lightning Source LLC
LaVergne TN
LVHW070425100526
838202LV00014B/1531